THE NEW NATURALIST LIBRARY

A SURVEY OF BRITISH NATURAL HISTORY

TREES

THE NEW NATURALIST LIBRARY

TREES

PETER A. THOMAS

WILLIAM
COLLINS

This edition published in 2022 by William Collins,
an imprint of HarperCollins*Publishers*

HarperCollins*Publishers*
1 London Bridge Street
London SE1 9GF

WilliamCollinsBooks.com

HarperCollins*Publishers*
1st Floor, Watermarque Building, Ringsend Road
Dublin 4, Ireland

First published 2022

A CIP catalogue record for this book is available
from the British Library.

Set in FF Nexus

Edited and designed by
D & N Publishing
Baydon, Wiltshire

Printed in Bosnia-Herzegovina by GPS Group

Hardback
ISBN 978-0-00-830451-5

Paperback
ISBN 978-0-00-830453-9

Contents

Editors' Preface

You may think that you know trees, but until you have read this book you don't. Dr Peter Thomas ranges wider and deeper here into the natural history of trees than any other book of its kind, without departing for a moment from our steadfast rule that a New Naturalist is written for everyone. Did you know that trees grow mostly at night? Or, that a large broadleaf tree contains within its canopy a light-capturing area of more than 350 square kilometres? Read Chapter 3 to discover how this is possible and why it matters. Do you know why the spruce wood used by Stradivarius to make his unmatched violins was so special? Which European city is built on foundation posts of ancient alder wood? How does water get to the top of a 100 m tall tree? How do tree roots behave, and which ones give the insurance industry the jitters?

Anyone professionally involved with trees will find much of practical interest. For example, why staking a newly planted sapling too high up the stem will weaken it when the stake is removed. And why the strongest oak timber is fast-grown in Britain, but the strongest spruce is slow-grown in Scandinavia. Conifers and broadleaves also react in different ways to mechanical stresses caused by supporting their branches against gravity and wind. The reaction wood that forms on trunks and at branch junctions comes in two kinds: generally, conifers push while broadleaves pull. The function of reaction wood formation in the trunk is to bend a leaning tree back to the upright, and astonishingly the genes responsible for this behaviour are activated within half an hour of leaning.

The ecological relationships of trees – with each other, with fungi and insects and with soil and climate – are all covered in fascinating detail. If you have read certain other books on trees lately, it may surprise you to learn that trees compete fiercely with their neighbours, especially when young. A young white spruce growing up through a stand of aspen suffers damaging corporal punishment from aspen's wind-whipped boughs. Mature trees may share water and sugar with neighbours, but in nature thousands of saplings die of light starvation and water deprivation to make room for a single survivor that grows to maturity. If we must anthropomorphise the relationship among forest trees, it is more like a stand-off between battle-worn warriors than any kind of comradeship. Understanding trees for what they really are does not diminish the respect we have for them. As this book amply illustrates, how could it?

Trees are a big subject and Peter Thomas accepts no artificial boundaries in pursuing it. Large shrubs are also small trees, so they are included. The native trees of Britain are comparatively few, mainly through geographic accident rather than climatic limitation, as demonstrated by the very many introduced species that will grow and even naturalise in these islands. So, you will meet a global supporting cast of tree species in this book, appearing around our own familiar natives. This cosmopolitan approach was pioneered by the late Oliver Rackham in his New Naturalist, *Woodlands*, published almost 50 volumes ago. As he proved, it's natural to cherish these islands, but of no merit to be insular. Peter Thomas' *Trees* is otherwise different from *Woodlands*, except in this respect: it is also a revelation and a landmark of a book.

Author's Foreword and Acknowledgements

As many other authors of New Naturalist volumes, I owe a great debt to the series. I was never much of a naturalist as a youth but decided to do a degree in biology because I was interested in insects, particularly in how they were built and moved. Curiosity about how insects carried pollen led me to read *The Pollination of Plants* (1973), by Michael Proctor and Peter Yeo. This book was such an eye-opener that I jumped at the chance of doing a module on plant ecology taught by Michael in my final year as an undergraduate at Exeter University. I had a good dose of 'plant blindness' before this – not even really aware that larches lost their leaves in winter – but I was captivated by Michael's knowledge and insight into plant ecology. It all made wonderful sense and was fascinating! My interest in trees was stimulated by a different person: my wife. We met at high school in north Kent and, being a country girl, she taught me the difference between a beech and a hornbeam and introduced me to the beauty of trees on the North Downs. Plant ecology and tree appreciation came together in an MSc in ecology at Aberdeen University. Thanks to the inspiration of Charles Gimingham, this was followed by a PhD in eastern Canada on the regeneration of trees after fire. The rest is history!

In Michael Proctor's last New Naturalist, *Vegetation of Britain and Ireland* (2013), he states in his foreword that looking at nature prompts 'that classic New Naturalist question, "why?"' And that's where this book starts. Trees are the world's oldest, biggest and tallest living organisms and they cannot run away from things that want to eat them or from harsh aspects of the environment. Developing a large and old woody skeleton is clearly a very successful way of living, and yet that same woody skeleton tends to hide what goes on inside. As Terry Pratchett was fond of saying, we can often see trees as 'stiff weeds' or as part of the backdrop to the more exciting bits of natural history. Watching a tree for an hour may give the impression that it isn't doing very much, and yet it is constantly monitoring what is going on around it, even in winter, and reacting in quite astonishing ways and sometimes just as quickly as an animal would. But trees mostly have little need to react at lightning speed and this can make them strange things to understand.

Peering inside a tree leads to all sorts of questions, such as how a tree's immune system works, or whether trees 'talk' to each other. There are also many questions about how the tree grows – why do roots stop growing in summer, or how deep do the roots of a large tree need to go to hold it up and provide it with enough water?

Many studies are showing that green spaces and trees in particular are very important to our wellbeing by beneficially affecting us psychologically and physiologically – they are good things to have around us. But our interaction goes beyond this to include other questions such as: if trees were wiped out, would we run out of oxygen, and perhaps most importantly, can we plant enough trees to save us from climate change? Other questions range from the mundane (why are leaves on a railway line such a nuisance?) to the more topical (how can trees be useful against the effects of terrorist bombs?).

There is currently a huge amount of new research being published on these and other questions, and we understand so much more about how trees work than we did even just a few years ago. The aim of this book is to incorporate this new knowledge to help give a deeper understanding of just how a tree works. Much of this new knowledge has yet to reach textbooks and there are few good summaries elsewhere. Moreover, the information is contained in a vast array of journals on everything from alternative medicine and psychology to geophysics and applied acoustics to peasant and First World War studies. This can make it difficult to track down information. To help, I have included quite a few references in the text, since the original publications are often the only place to go for more detail. In some cases, these are older but classic publications, and in others, they are the most recent, which will allow the reader to track back through the subject should they wish. I apologise to those readers who find that these references get in the way, and I hope you can skip by the names and dates. Tree biology like many areas of science has accrued a wealth of technical terms. I have included a number of these in italics (with explanations!) in order to help the reader should they want to read further afield.

As with most New Naturalists, the emphasis is on Britain and Ireland, but to put our wonderful trees into perspective, I have used examples from around the world, from the Arctic to the tropics. Given the range of trees covered, the scientific names of trees are included where the common name is first used, and occasionally elsewhere, on the grounds that both are useful, especially for more exotic trees. Again, I hope these do not interfere with the flow of the text for those not needing or wanting both names. Common names are also linked to scientific names in the index to help if the latter are needed.

Although primarily about trees, I have ventured a little into the ecology of woodlands, where it is relevant to understanding why trees act the way they do and to help put the value of and threats to trees into perspective. I have used the terms *woodland* and *forest* almost interchangeably. Both terms, of course, refer to areas of trees, with the main difference being that woodlands are smaller and more open, while forests are larger and often with a closed canopy, except where openings have been created by trees that have fallen or died.

Trees are all around us and are very much a part of the lives of most of us. Hopefully this book will help you understand them better, increasing your enjoyment of a set of woody organisms that really are indispensable to us and the world. The first two chapters deal with the origins and value of trees, followed by nine chapters on what goes on inside a tree over a typical year, and a final five chapters on the longevity of trees and the issues that this brings.

All photographs were taken by me in the environs around Keele in Staffordshire unless otherwise stated. Many others result from research trips and holidays. Strangely, not many people ask to see my holiday snaps. Photographs and diagrams were also kindly supplied by Will Blozan, Dave Emley BEM, Richard Hobbs, Shinsuke Koibe, Nick Mott, Ross Newham of East Malling Research, David Orwig, Chris Sanders VMH and Matthew Tosdevin.

So many people have helped me over the years in appreciating the beauty and inner workings of trees and their interactions with people. In particular I thank Andy Hirons for his expert knowledge, Mark Smith for information on train wheels and leaves, and Walt Koenig for discussions on the acorn woodpeckers of California. I am also very grateful to Alan Crowden, who helped me with my first book for Cambridge University Press and who has been a positive influence ever since. I am really grateful to Jonathan Silvertown, who read and commented on the whole book, greatly improving it, and to all the people who read draft sections and chapters of this book, including (in alphabetical order) Margaret Buxton, Dave Emley BEM, Michael Hawkes, Andy Hirons, Graham Lees, Linda and Pete Norbury, Rosemary Payne, Lynn Pickering, Chris Sanders VMH and Sally Vaughan – and especially Kath and Tony Thompson. It has been a pleasure and a privilege to work with Hazel Eriksson, David Price-Goodfellow and Jennifer Dixon in the production of this book. The staff of Harvard Forest are thanked for always being there to answer difficult tree questions and Harvard University for access to their libraries and for allowing me to spend quality time thinking and writing about trees at Harvard Forest. Errors, of course, are my own.

This book would not have been possible without the support of my wife, Judy, who endured long silences during the Covid lockdowns as I worked on the manuscript, and who read every word of the book with refreshing honesty and good humour.

CHAPTER 1

Setting the Tree Scene

Trees are amazing things that provide us with many products while giving us shade and adding beauty to our surroundings. Towering above us, trees, not surprisingly, are the largest and longest-living things in the world. Their very size and age lead to problems that we animals do not readily appreciate, such as how they can possibly arrange thousands of leaves to receive the best light, and how they can take up huge amounts of water from the soil. They also have to detect and respond to attacks when they can't run away and yet survive for sometimes thousands of years. Fortunately, we are increasingly learning their inner secrets to how they live and survive. Yet they can still puzzle us, even to the extent of defining just what a tree is!

DEFINITION OF A TREE

What is a tree? Surely this is a stupid question, since we all recognise a tree when we see one! The problem here is that most plant groups occur within a single plant family, sharing a common ancestor and so sharing characteristics found in the whole group. Thus orchids are all in the Orchidaceae family, grasses in the Poaceae, and you can tell by the structure of the flowers. However, the woody tree habit has evolved independently in a wide range of families. Trees are found in at least 20 families in temperate areas and so probably hundreds worldwide. Even the 220 native woody plants of the Canary Islands have at least 38 independent origins (Lens *et al.* 2103). Many unrelated plants have evolved the woody habit, allowing them to grow taller than neighbours in the competition for light, which makes it harder to pin down a good biological definition of a tree. Moreover, the woody habit is just as easily lost through evolution by some plants as it is gained by others so it is even harder to pin down any genetic differences between woody and non-woody plants – there are no unique tree genes (Groover 2005). The best we can say is that a tree is a perennial woody plant with a self-supporting main trunk and branches, which grows fatter over time and forms a distinct crown (Hirons & Thomas 2018). The 'wood' itself is defined as a mixture of cellulose fibres forming tubes that are strengthened with lignin.

Others would argue that trees also tend to be tall, and so the IUCN's Global Tree Specialist Group defines a tree as a woody plant with usually a single stem growing to a height of at least 2 m, or, if multi-stemmed, then with at least one vertical stem that is 5 cm in diameter at chest height (Beech *et al.* 2017). These definitions are helpful but, except for the perennial bit, there are many exceptions. A number of trees like Hazel *Corylus avellana* normally have multiple thin trunks rather than one main stem; strangler figs that germinate in the crown of another tree and grow 'strangling' roots down to the ground are not self-supporting for the first part of their life; and many trees such as those growing near the polar treelines would never attain more than a few centimetres in height. When it comes down to it, a tree is just recognisable as being a tree. Indeed, Lord Denning, who had to define a tree as part of a judgement under the UK Town and Country Planning Act 1990, stated that 'anything that one would ordinarily call a tree is a "tree" within this group of sections in the Act'.

As far as the remit of this book goes, while trees can be towering giants over 100 m tall, they can also be small arctic willows just a few centimetres high – they both share a woody skeleton that they reuse each year and so work in roughly the same way. Equally important, using the above definitions, some plants are not considered to be trees. Lianas and vines are not self-supporting and so are not really trees (although some will be discussed as we go along), and plants with woody stems which die down to the ground each year, such as Asparagus *Asparagus officinalis*, do not have a 'perennial woody stem'. Similarly, bananas are not trees because they have no wood – the trunk is made from non-woody leaf stalks squeezed together. Nor are bamboos, since they are just very hard grasses that do not grow fatter over time, even though they can be up to 25 m tall and 25 cm thick.

What then is the difference between a tree and a shrub? This is really a horticultural distinction that separates growth forms such that a 'tree' has a single stem more than 6 m tall that branches at some distance above ground. By contrast, a shrub has multiple stems from or below ground level and is less than 6 m tall. From a biological viewpoint there is no real difference in how they work, so both are included in this book.

Structure of a tree

Trees are primarily made up of three main parts: leaves, shoots and roots. These will be considered in more detail later in this book, but for now a simple overview may help. The leaves produce the food of the tree – sugars – by photosynthesis. The 'shoots' are made up of the trunk (biologically, the stem) and branches which hold the leaves up to the light, forming the *crown* of the tree, and also hold the flowers and fruits. The crowns of several trees combine to make up the woodland *canopy*. The trunk and branches also act as a conduit for water to be brought from the roots to the leaves via the wood (*xylem*), and for sugars to be transported around the tree in the inner bark (*phloem*). The water flowing through the wood contains nutrients such as nitrogen dissolved from the soil and also carries plant hormones which transmit messages from the roots to the crown.

The roots have the dual role of holding the tree up and absorbing water from the soil. The above-ground part of the tree is perhaps intuitively much bigger and heavier than the roots below ground, but, as shown in Table 1, which uses Scots Pine *Pinus sylvestris* as an example, as a general rule of thumb the weight (biomass) of the fine roots that take up water and nutrients is roughly equal to the weight of the leaves (or, in this example, needles). As the tree matures, the trunk makes up a progressively larger proportion of the biomass while the coarse or big woody roots that hold the tree up stay around the same percentage of the biomass through the life of the tree. The weight of the fine roots, which absorb most of the water, and the weight of the needles make up a decreasingly small proportion of the tree's biomass. Reproduction, in the guise of the weight of cones, is a very small part of the tree's total biomass.

TABLE 1. Amount of biomass (as a percentage of the whole tree) in Scots Pine *Pinus sylvestris* at three different ages.

Biomass	Sapling (15 years)	Pole (35 years)	Mature (100 years)
Needles	9	9	4
Cones	<1	<1	<1
Stem (trunk)	32	54	72
Branches	33	14	10
Coarse roots	11	14	11
Fine roots	15	7	2

Data from Helmisaari *et al.* (2002).

Types of trees

Trees come in two main types (Fig. 1): gymnosperms and angiosperms. Both of these are seed plants but only the second produces flowers. The oldest group are the gymnosperms (from the Greek *gymnos*, meaning 'naked', and *sperma*, meaning 'seed'; *gymnos* is also the root of 'gymnasium', since Greeks exercised naked). These are the *conifers and their relatives*, all of which have seeds borne in cones (or fruits in junipers and yews) that are unenclosed such that the seeds can be seen without breaking anything open. Although they are seed plants they do not have flowers as such.

The other main group are the angiosperms (from the Greek *angeion*, meaning 'case', and *sperma*, meaning 'seed'). These are the flowering plants: the name comes from the fact that they also have fruits (or cases) enclosing the seeds. This includes many of the common *broadleaf* trees, such as oaks, beeches and ashes that mostly have flat leaves, in contrast to the conifers that mostly have needle leaves. There are many names that one could use for the angiosperms, the most common being hardwoods (the wood is generally harder than the conifers but there are exceptions) and broadleaf trees (the leaves are mostly but not always flat and broad). Neither term is completely accurate, but we will use broadleaf trees as shorthand for angiosperm trees. These broadleaf trees can be divided into the dicotyledonous (dicot) trees and the monocotyledonous (monocot) trees. The monocots are defined as having a single seed leaf (cotyledon) and include a number of trees in a few families, notably:

- Palms (the family Arecaceae, previously Palmae or Palmaceae)
- Dragon trees *Dracaena* species, cordyline palms *Cordyline* spp., European butcher's broom *Ruscus* spp., Yuccas such as the Joshua Tree *Yucca brevifolia* (Aspagagaceae)
- Screw pines *Pandanus* spp. (Pandanaceae)
- Grass trees *Xanthorrhoea* spp. of Australia and aloes *Aloe* spp. from South Africa (Xanthorrhoeaceae, but this is likely to be absorbed into the family of Asphodelaceae in the future)
- Traveller's Palm *Ravenala madagascariensis* (Strelitziaceae).

The dicots with two seed leaves form the rest of the familiar broadleaf trees.

FIG 1. The two main types of trees. Conifers and their relatives make up the gymnosperms, including (a) the pines (Pinaceae) but also including the flat-leaved Maidenhair Tree *Ginkgo biloba* (Ginkgoaceae), (b) monkey puzzles (Araucariaceae), junipers, redwoods, cypresses and many more in the Cupressaceae, and the yews (Taxaceae). The angiosperms, or broadleaf trees for convenience, mostly have flat, wide leaves like (c) the oak but include some with needle-like leaves such as (d) gorse *Ulex* spp. and broom (*Cytisus* and *Spartium* species). Photographs of (a) Hartweg's Pine *Pinus hartwegii* in Honduras, (b) Monkey Puzzle *Araucaria araucana* in Staffordshire but native to Chile and Argentina, (c) Georgian Oak *Quercus iberica* in the Caucasus of Azerbaijan and (d) Gorse *Ulex europaeus* on the coast of Devon.

GEOLOGICAL HISTORY: WHERE DO OUR TREES COME FROM?

Plants colonised land around 465 Ma (*mega-annum*, or millions of years ago) in the Middle Ordovician, and the first vascular plants (those with internal plumbing) appeared by 400 Ma in the basal Devonian (formerly the Silurian). These vascular plants had simple stems just a few millimetres wide and a few centimetres tall with no leaves, typified by *Cooksonia*. It was another 40–50 Ma before leaves were common. These early land plants were small because the vascular tissue was just for conducting water up the plant without giving any great structural strength. Taller plants – the first trees – became possible with the evolution of stiffer vascular tissue and a more developed root system to give support. These first trees, reproducing by spores like their primitive ancestors, evolved in the mid-Devonian around 393–383 Ma and were capable of living for several decades and reaching up to 30 m tall and 1 m wide. They spread to produce widespread forests dominated by the first tree-like plant, *Archaeopteris*, not to be confused with *Archaeopteryx*, the bird-like dinosaur. This was a turning point that helped change the world's climate.

A wider diversity of trees, referred to as the progymnosperms, evolved in the mid-Devonian. Within 100 million years, the coal-producing swamps of the Carboniferous (359–299 Ma) were dominated by lush forests. We would have recognised the tree ferns from today's forests but the others – fast-growing giant horsetails and clubmosses – have long since disappeared, leaving us with just a few small relatives. The horsetails such as *Calamites* were up to 15 m tall and 30 cm in diameter but would have been dwarfed by the clubmosses (notably *Lepidodendron*) which reached 45 m in height and 2 m in diameter (Fig. 2). These can be considered the dinosaurs of the plant world: an extinct group that achieved a huge size (Thomas & Cleal 2018). The forests were accompanied by giant insects (such as dragonflies with wingspans up to 63 cm), all made possible by oxygen levels in the atmosphere rising up to 35 per cent (compared to the 21 per cent today). This rise in oxygen may have been due to the evolution of lignin which provides structural support to wood, but which microbes had not yet evolved the ability to digest, leading to carbon being locked up and oxygen released in photosynthesis left in the atmosphere (see David Beerling's book *The Emerald Planet* [2007] for a very readable account of these changes). In these forests the first seed plants appeared, called pteridosperms or seed ferns. They resembled tree ferns but, importantly, produced seeds rather than spores. From these, the first primitive conifers appeared around 300 Ma at the end of the Carboniferous. The earliest-known conifer was named *Swillingtonia denticulata*

FIG 2. An artist's model in the Singapore Botanic Gardens illustrating how the giant clubmoss *Lepidodendron* would have looked in the Carboniferous, though they are known to have grown much larger than these 3 m tall models.

after Swillington Quarry near Leeds in Yorkshire where it was found, dating back to 301 Ma (from Scott & Chaloner 1983).

Comparatively soon after this in the mid-Permian (290–248 Ma), the global climate became much drier, and by 248 Ma the world reached the third great mass extinction event, arguably the biggest of the 'big five' extinctions. Up to 90 per cent of all marine invertebrates with a skeleton (and thus ones that would be most easily fossilised) and 54 per cent of all marine families became extinct. Plants are more versatile – they can regrow from surviving bits, including stumps and seeds, and so were not affected anywhere near as drastically by any of these extinctions. However, the associated change in the climate at the end of the Permian led to big changes in dominance: the lush swamp forests of the Carboniferous declined, and gymnosperms took over the world's forests, making up over 60 per cent of the global flora. At the same time, trees such as cycads, ginkgos and monkey puzzles appeared, many of which are now fossilised in the petrified forest of Arizona from the late Triassic, 200 Ma (Fig. 3).

FIG 3. Sections of fossilised tree in the Petrified Forest National Park, Arizona, USA. These were growing in the late Triassic around 200 Ma and became buried in river sediments that preserved the trunks. Water flowing through the sediments deposited silica along with colourful minerals such as manganese and copper, which gradually replaced the woody structure, turning it to stone but preserving the original structure of the wood.

The next notable appearance was the pines around 180–135 Ma (Jurassic), with all the other major conifer families appearing by the end of the late Cretaceous around 65 Ma. This means that many of the conifers evolved to share the world with the dinosaurs. Conifers were hugely successful, dominating the world's forests from around 245–67 Ma. The seeds of their decline began during the early Cretaceous around 140–135 Ma, with the rapid evolution of the early broadleafed plants (angiosperms) – so rapid that Darwin called their origin an 'abominable mystery' (Buggs 2021). Many of these would have been herbaceous (i.e. non-woody). The woody angiosperms probably arose from a now extinct conifer group that had insect-pollinated cones (van der Kooi & Ollerton 2020). The magnolias are probably some of the earliest types of broadleaf trees that are still alive. By 95 Ma at the start of the late Cretaceous, a number of trees we would recognise today were around: elms, birches, laurels, magnolias, planes, maples, oaks, willows and, within another 20 million years, the palms, and around 60 Ma, the eucalypts appeared. It is likely that in this period of the late Cretaceous all current families of trees evolved. When the dinosaurs were disappearing (by 65 Ma) and through to the mid-Cenozoic (25 Ma), broadleaf trees had spread to dominate the world, helped by the overall warm global climate, the disruption of the huge asteroid impact at Chicxulub in Mexico and lower atmospheric oxygen levels approaching the 21 per cent of today – before then, closed forests would have been unlikely due to frequent fires fuelled by the high oxygen levels (Belcher *et al.* 2021, Carvalho *et al.* 2021).

Fortunately for the conifers, the end of the late Eocene around 35 Ma saw the development of the polar ice caps. A cooler polar climate allowed conifers to dominate and diversify through the northern regions of the globe. There were some losses, since this northern-cooling spelled the end of a subtropical climate in the far north, and trees such as the Dawn Redwood (Fig. 4), which had been very extensive around the world *c.*45 Ma (mid-Eocene), were now left as relicts in small areas of eastern Asia.

The distribution of trees around the world has been greatly affected by plate tectonics creating continental drift, the movement of the continents

FIG 4. A piece of 45-million-year-old subfossil wood from a Dawn Redwood *Metasequoia glytostroboides* from Axel Heiberg Island in the Arctic. The island is at 79 °N, above the current northern limit of trees at 69–72 °N. The island was in almost the same place in the Eocene as it is now, but the climate near the poles was much warmer, allowing the trees to thrive. The wood was preserved in fine sediments and then by permafrost in what is now a polar desert. Although a little compressed, it is still largely organic and can be split and burnt and has been described as mummified wood.

around the world over geological history. At the end of the mid-Permian period, around 250 Ma, most of the Earth's landmasses were joined together into the supercontinent of Pangaea. By the time the broadleaf trees had evolved, Pangaea had broken into two parts: Laurasia, which contained what would become the northern hemisphere continents, and Gondwanaland, which contained what is now Australia, Africa, South America, India and Antarctica. This breakage left the pines primarily in the northern hemisphere and helps explain why the broadleaves of the northern and southern hemispheres are so different from each other. As Laurasia and Gondwanaland themselves broke apart, the ancestors of many broadleaf families were taken around each hemisphere and with time they evolved into different species in different continents. This explains why a number of genera are found throughout the northern hemisphere, as their ancestors had moved across Laurasia, but there has been time for different species to evolve on each continent. For example, the genus of tulip trees is found across the northern hemisphere, but the Chinese Tulip Tree *Liriodendron chinense* in Asia is different from the Tulip Tree *L. tulipifera* in North America. As another example, the Chinese Witchhazel *Hamamelis mollis* is distinct from the American *H. virginiana*.

The spreading of trees across Laurasia has resulted in some interesting similarities in different parts of the world (Yih 2012). The most notable is the *E. Asia–E. North America floristic disjunction* where the flora of these two areas is remarkably similar and distinct (or disjunct) from anything in between. This results in 67 per cent of plant genera in Maine, USA also occurring on Honshu island, Japan. Several genera, such as *Ginkgo* and *Metasequoia*, are now endemic to East Asia but occur as fossils in North America, and the reverse is true for *Sequoia* and *Taxodium*. The reason for this similarity appears to be that the flora in common spread across Laurasia via the Bering land bridge. After this, the climate in the middle of the continent was changed by mountain formation in western North America, leaving the two ends both similar to and different from anywhere else.

This also gives rise to biomes, areas of the world that have similar-looking vegetation because they are growing in similar climates. The interesting result is that if you were flown around the world and dropped in, say, the oak woodlands of eastern North America, England or Japan, you might not know which continent you were on unless you could identify the plants down to species (Fig. 5), since the forests have a similar appearance.

FIG 5. (a) Mixed woodland with Red Oak
Quercus rubra of New England, USA,
(b) Mongolian Oak *Q. mongolica* in Hokkaido,
Japan and (c) Pedunculate Oak *Q. robur* in
Epping Forest on the London-Essex border.
Very different parts of the world but all in the
same biome type and visually very much alike
unless you know your species.

GLACIAL HISTORY – WHY DOES BRITAIN HAVE SO FEW NATIVE TREES?

As described above, changes in climate and movement of continents have resulted in movement of trees around the world. We might like to think of our forests as being fairly static now but, of course, this movement is still happening as modern climate change is causing trees to move once again. A warming climate is rather ironic, since geologically we are still in the Quaternary Ice Age that started 2.6 Ma and was made up of some eight waves of glacial advance and retreat which marked the Pleistocene epoch from 2.6 Ma to 11,700 years ago. Ice ages tend to occur at regular intervals of 100,000 years with warm interglacial periods lasting 15,000–20,000 years. Since the last glacial maximum at the end of the Pleistocene, we have been in the current warmer interglacial period (the Holocene).

Various trees have been present in Britain in previous interglacial periods. Norway Spruce *Picea abies* was found in Britain in the Chelford Interstadial around 45,000 years ago (Holyoak 1983), named after Chelford Quarry in Cheshire (Fig. 6) where I once spent a happy afternoon pulling out stumps, logs and cones

FIG 6. Norway Spruce *Picea abies* around 45,000 years old buried under the sands of Chelford Quarry, Cheshire. Although old by human standards, the wood is not fossilised and is still composed of cellulose and lignin, and so is burnable.

that had been exposed by the removal of the overlying sand. However, after the last glacial period ended around 15,000–11,700 years ago, Norway Spruce was pushed out of Britain and never returned until humans introduced it back as a forestry tree.

At its maximum around 18,000 years ago, the last glacial period produced an ice sheet up to 5 km thick extending from the Arctic down to below the Great Lakes, reaching New York, London and Berlin. With so much water bound up in ice, sea level dropped by around 130 m. In the southern hemisphere ice extended up from Antarctica to cover Chile and much of Argentina. All this ice pushed plants and animals towards the equator into refuge areas. Much of Britain would have been devoid of plants as our trees found refuge mainly in western France and northern Spain, and also in Italy and even the Balkan Peninsula in southeast Europe. Scots Pine *Pinus sylvestris* has been identified as having taken refuge in Spain (Sinclair *et al.* 1999) before moving back into Britain.

A number of molecular techniques have been used to reconstruct the migration route of species (Newton *et al.* 1999), but chloroplast DNA has proved particularly useful. Chloroplasts, responsible for photosynthesis in plant cells, have their own DNA. They are inherited only from the mother tree, and there is no mixing of genes (recombination). Moreover, chloroplasts have a slow mutation rate, so they are ideal for tracking postglacial movement by looking for similarities across Europe. Using this evidence points to Black Poplar *Populus nigra* migrating to Britain after the last ice age from Italy, Austria and Hungary, following the Danube river, with some input from trees finding refuge in Spain (Cottrell *et al.* 2005). As well as looking at chloroplast DNA from living trees, it is also possible to isolate it from subfossil wood – wood that has been stored in peat bogs, lake bottoms or permafrost for more than 10,000 years (Lendvay *et al.* 2018) – so it is possible to track genetic changes through time as well as spatially after the last glaciation.

As the climate warmed around 10,000 years ago, animals, trees and other plants migrated polewards again. In North America the mountains tend to run north to south, so migration northwards after the retreat of the ice was unchecked, producing eastern forests dominated by at least 10 major tree species. In eastern Asia at similar latitudes, where the tree fauna is richer and the mountains run in a similar direction, forests can easily contain 20 major species. In Europe, however, the main mountain ranges, such as the Pyrenees, Alps and Carpathians, tend to run east to west and acted as distinct barriers to migration, which resulted in Europe having almost half the number of trees and large shrubs seen in eastern North America. Britain and Ireland are even more impoverished, because the land bridge joining us to Europe was submerged

TABLE 2. Trees that arrived unassisted to Britain and Ireland after the last ice age and thus considered to be native. These are given in approximate order of arrival, with Juniper arriving first. While 33 species are listed, some people would, for example, recognise more species of elms and whitebeams, and others would include some of the larger shrubs listed in Table 3.

Common name	Scientific name
Juniper	*Juniperus communis*
Downy Birch	*Betula pubescens*
Silver Birch	*B. pendula*
Aspen	*Populus tremula*
Scots Pine	*Pinus sylvestris*
Bay Willow	*Salix pentandra*
Alder	*Alnus glutinosa*
Hazel	*Corylus avellana*
Small-leaved Lime	*Tilia cordata*
Bird Cherry	*Prunus padus*
Goat Willow	*Salix caprea*
Wych Elm	*Ulmus glabra*
Grey Willow	*Salix cinerea*
Rowan	*Sorbus aucuparia*
Sessile Oak	*Quercus petraea*
Ash	*Fraxinus excelsior*
Holly	*Ilex aquifolium*
Pedunculate Oak	*Quercus robur*
Hawthorn	*Crataegus monogyna*
Black Poplar	*Populus nigra* ssp. *betulifolia*
Yew	*Taxus baccata*
Whitebeam	*Sorbus aria*
Midland Hawthorn	*Crataegus laevigata*
Crab Apple	*Malus sylvestris*
Wild Cherry	*Prunus avium*
Strawberry-tree	*Arbutus unedo*
Field Maple	*Acer campestre*
Wild Service-tree	*Sorbus torminalis*
Large-leaved Lime	*Tilia platyphyllos*
Small-leaved/Smooth-leaved Elm	*Ulmus minor*
Beech	*Fagus sylvatica*
Hornbeam	*Carpinus betulus*
Box	*Buxus sempervirens*

8,300 years ago (now the English Channel), giving very little time for trees to reinvade.

This leaves Britain with around five dominant tree species and a total of 32–35 native species of trees (Table 2), depending on what you count as a tree or a shrub (Table 3), whether hybrids are included and how many species of elms and whitebeams you recognise. For example, the genus *Sorbus* contains three distinct species –Whitebeam *S. aria*, Rowan *S. aucuparia* and Wild Service-tree *S. torminalis* (listed in Table 2). However, some species of *Sorbus* are *apomictic*; that is, they can produce seeds without fertilisation from pollen, so all the offspring have the same genetic makeup as their mother. The result can be a group, or *clone*, of similar-looking plants that look a little different from another clone. Some classify these apomicts with the nearest of the three distinct species, others call them *microspecies* (literally 'little species') akin to varieties within the three species (Jauhar & Joshi 1970), while others call them all proper species. The number of *Sorbus* species can therefore range from three to, currently, 52 (Rich *et al.* 2010). A similar situation is found in brambles *Rubus* species. A slightly different problem

TABLE 3. Large shrubs native to Britain and Ireland that could easily be counted as trees, arranged in increasing order of their normal maximum height (figures in brackets are the maximum recorded, taken from Stace 2019).

Common name	Scientific name	Height (m)
Purple Willow	*Salix purpurea*	1.5–5
Wild Privet	*Ligustrum vulgare*	3 (5)
Sea-buckthorn	*Hippophae rhamnoides*	3 (14)
Dogwood	*Cornus sanguinea*	4
Guelder-rose	*Viburnum opulus*	4
Tea-leaved Willow	*Salix phylicifolia*	4 (5)
Blackthorn	*Prunus spinosa*	4 (6)
Spindle	*Euonymus europaeus*	5 (10)
Wayfaring-tree	*Viburnum lantana*	6
Laurel-leaved Willow	*Salix* × *laurina*	6 (13)
Elder	*Sambucus nigra*	6 (16)
Buckthorn	*Rhamnus catharticus*	7
Alder Buckthorn	*Frangula alnus*	7
Broad-leaved Osier	*Salix* × *smithiana*	9 (15)

exists in roses *Rosa* and willows *Salix* where it is difficult to separate out hybrids from actual species. If all woody plants are included – so shrubs and even small shrubby plants like heathers (*Calluna* and *Erica* species) – our woody flora goes up to 160 native woody species. However, hybrids and another 309 microspecies could be added to this list!

NATIVE, NATURALISED AND INTRODUCED TREES

Native species (sometimes called indigenous or autochthonous species) are defined as those that arrived after the last glaciation by natural processes – the spread of seeds or vegetative parts by wind, water or animals; in other words, without human help. The evidence for this is usually taken from the presence of pollen or macrofossils, such as leaves or seeds, in postglacial deposits in peat bogs or at the bottom of lakes. This covers most of the commonest trees in Britain that normally come to mind.

A tree can be introduced to an area outside of its natural range by humans, either intentionally or accidentally, where it becomes an *introduced,* non-native or alien species. This can be taken further to distinguish between *archaeophytes* (introduced in ancient times) and *neophytes* (introduced in recent history). The date usually used to separate these is 1492, often rounded to 1500, representative of the time when Christopher Columbus landed in the New World and the start of extensive movement of plants around the world. For archaeophytes it is often hard to distinguish whether they are native or introduced, since this information is lost in the mists of time. For some, such as the Sweet Chestnut *Castanea sativa*, it is more exact; it is almost certain that it was introduced by the Romans to feed both soldiers and their horses.

Some of these introduced plants are sufficiently at home that they can reproduce and spread into native vegetation and persist indefinitely – these are referred to as *naturalised* species (Table 4) and include the neophytes white poplar, sycamore and Turkey Oak *Quercus cerris*. It could be argued that to spread into new habitats requires a naturalised species to be able to produce seeds, but some, such as the English elm, which spreads primarily by suckering from the roots, are included because they are so common. These naturalised species are now honorary members of our tree flora and capable of spreading (Fig. 7), and we can add another 220 naturalised species of woody plants to the 160 species that are native.

If a species is introduced but not naturalised, it is classified as an *exotic* and will not spread beyond where it has been planted. This includes most of the

FIG 7. One of the southern beeches, Rauli *Nothofagus procera*, native to Chile and western Argentina but very much at home and naturalised in Britain, here forming a dense shrub layer on acid soils of Keele University, Staffordshire.

trees planted in ornamental gardens. But again, it is not always clear-cut: Horse-chestnut *Aesculus hippocastanum* is native to the Balkan Peninsula of southeast Europe, and while it is widely naturalised through Europe, it is only partially so in Britain. It can be self-sown in open scrub and waste ground but is rarely naturalised in woodlands (Thomas *et al.* 2019) and so can be classified as exotic or naturalised depending upon where one looks. Just how many exotic species we have in Britain is discussed at the end of this chapter.

Exotic trees can be welcome for a variety of reasons. For example, evergreen trees such as the many cypresses can provide shelter to insects and birds in winter, and mostly they are aesthetically pleasing. Naturalised trees, however, have the annoying habit of not staying where they are planted, and some can be downright invasive. In past decades this has led to a knee-jerk reaction that only native trees have a place in our woodlands, and naturalised trees should certainly be removed from conservation areas. It's easy to see why. Take Sycamore *Acer pseudoplatanus* – it is invasive (good at producing seedlings, as you'll know if you have one nearby – Fig. 8), its leaf litter swamps woodland flora, and the tree casts deep shade. It is often seen as being capable of invading and taking over, particularly in Ash woodlands, since Ash *Fraxinus excelsior* and Sycamore have similar ecological needs, although Ash prefers moister soils. However, the

TABLE 4. Trees and shrubs from different parts of the world that are naturalised in the British Isles. These are the commoner or most obvious of the current 220 naturalised species, and so the list is by no means complete and will alter over time as climate changes and new trees start reproducing and spreading without help. Those listed are mostly neophytes (introduced after 1500 – shown in blue), with some archaeophytes (in green) that were introduced earlier. There is still discussion over whether some such as Common Lime and Service-tree may be native. Notes based on Stace (2019).

Common name	Scientific name	Notes on naturalisation
Europe		
European Silver-fir	*Abies alba*	Often self-sown
Cappadocian Maple	*Acer cappadocicum*	Seedlings and suckers in SE England
Norway Maple	*A. platanoides*	Common in open and wooded habitats
Sycamore	*A. pseudoplatanus*	One of the most abundant trees in Britain
Horse-chestnut	*Aesculus hippocastanum*	Open and wooded habitats in the lowlands
Red Horse-chestnut	*A. carnea*	SE England
Italian Alder	*Alnus cordata*	Frequently self-sown
Laburnum	*Laburnum anagyroides*	Open ground
European Larch	*Larix decidua*	Commonly self-sown
Norway Spruce	*Picea abies*	Occasionally self-sown
Corsican Pine	*Pinus nigra* ssp. *laricio*	Escape from plantations
Maritime Pine	*P. pinaster*	Often self-sown
White Poplar	*Populus alba*	Common; suckers well
Grey Poplar	*P.* × *canescens*	Rarely naturalised
Cherry Plum	*Prunus cerasifera*	Widespread
Cherry Laurel	*P. laurocerasus*	Sometimes in woodland
Portugal Laurel	*P. lusitanica*	Scattered
St Lucie Cherry	*P. mahaleb*	N England
Caucasian Wingnut	*Pterocarya fraxinifolia*	Occasional
Turkey Oak	*Quercus cerris*	Acidic sand in the south
Lucombe Oak	*Q.* × *crenata* (previously *Q.* × *hispanica*)	Occasional south of the Midlands
Holm Oak	*Q. ilex*	Occasional south of the Midlands
Rhododendron	*Rhododendron ponticum*	Widely naturalised
European White-elm	*Ulmus laevis*	Occasional in W England, Wales and Guernsey
Sweet Chestnut	*Castanea sativa*	In S England and the Channel Islands
Walnut	*Juglans regia*	In warm areas
Medlar	*Mespilus germanica*	Hedges
Dwarf Cherry	*Prunus cerasus*	Hedges and woodland

Common name	Scientific name	Notes on naturalisation
Wild Plum	*P. domestica*	Scrubby areas
Plymouth Pear	*Pyrus cordata*	Possibly native
Wild Pear	*P. pyraster*	Hedges and woodlands; Midlands and south
Almond Willow	*Salix triandra*	Central and S England; planted for basketry
Osier	*S. viminalis*	Common in lowlands; planted for biomass
Service-tree	*Sorbus domestica*	Possibly native
Common Lime	*Tilia × europaea*	Possibly native
English Elm	*Ulmus procera*	Widely introduced, spreads by suckering

North America

Grey Alder	*Alnus incana*	Occasionally self-sown especially in the N
Juneberry	*Amelanchier lamarckii*	On sandy soils
Lawson's Cypress	*Cupressus lawsoniana* (previously *Chamaecyparis lawsoniana*)	Commonly self-sown
Black Walnut	*Juglans nigra*	Self-sown mainly along River Thames in Surrey
Western Balsam Poplar	*Populus trichocarpa*	Suckering in Lanarkshire, lowland Scotland
Sitka Spruce	*Picea sitchensis*	Often self-sown
Lodgepole pine	*Pinus contorta*	Occasionally self-sown
Pin Cherry	*Prunus pensylvanica*	Surrey
Rum Cherry	*P. serotina*	Scattered
Red Oak	*Quercus rubra*	Often self-sown
False-acacia	*Robinia pseudoacacia*	Only in the south
Western Red-cedar	*Thuja plicata*	Frequently self-sown
Western Hemlock	*Tsuga heterophylla*	Often self-sown

Asia

Butterfly-bush	*Buddleja davidii*	Wasteland, scrubs and walls
Japanese Larch	*Larix kaempferi*	Sometimes self-sown
Foxglove-tree	*Paulownia tomentosa*	SE London
Fuji Cherry	*Prunus incisa*	Rare in oak woodland on clay
Domestic Apple	*Malus domestica*	Hedges, scrub and waste ground

South America

Rauli	*Nothofagus procera* (some now call it *N. alpina* or *Lophozonia alpina*)	Widely self-sown
Roblé Beech	*N. obliqua* (or *Lophozonia obliqua*)	Often self-sown

FIG 8. Abundant seedlings of the naturalised Sycamore *Acer pseudoplatanus* forming a seedling bank at the edge of a wooded area, waiting for years with little growth until they either die or a gap opens in the canopy above. They would seem to have the competitive edge and be likely to take over the woodland, but this is not always the case.

evidence points to Sycamore and Ash going through a cycle, taking it in turns to dominate (Thomas 2016) and so not being as domineering as once thought. As described in Chapter 2, Sycamore is criticised for holding only 43 species of insects compared to 423 on our two native species of oak, underlining the value of native-only trees. Yet the biomass of insects held by a Sycamore in spring is probably very similar to that of an oak, making them just as valuable to birds looking to feed hungry chicks in spring.

Other naturalised trees are more troublesome. Rhododendron *Rhododendron ponticum* was introduced into Britain from the Mediterranean around 1763 as it formed valuable shelter for game birds on large estates, as well as being ornamental. It has, however, proved to be very invasive of woodland. A survey in 2011 showed that it then covered 98,700 hectares, roughly 3.3 per cent of Britain's total woodland (NFI 2016). It is also a pernicious invader of uplands such as in Snowdonia National Park where it is being very troublesome (Snowdonia Rhododendron Partnership 2015). It casts a very dense shade, and the foliage and leaf litter are toxic to many insects and large animals, while a large bush can produce upwards of a million seeds per year and so it can spread rapidly. Also, as described in Chapter 15, it is a carrier of the virulent pathogen *Phytophthora*. In the right place, and controlled, it is a useful shrub, but as an invasive naturalised species it is very problematic.

Some trees are unhelpful for more indirect reasons. Turkey Oak is native to southern Europe and Southwest Asia and is naturalised over much of Europe. It was introduced into Britain in 1735 and was naturalised by 1905 if not before, particularly in southern England. It can be fairly invasive in oak woodlands in Britain, but the biggest concern is the role it plays in the spread of knopper galls caused by the gall wasp *Andricus quercuscalicis* (Fig. 9). The wasp was accidentally introduced and has been spreading in Britain since the 1960s. In early spring the wasp produces a sexual generation on the male flowers of Turkey Oak and then an asexual generation on the developing acorns of British native oaks in mid- to late May, and the distinctive galls are detectable by mid-July. These replace the normal acorn, so reducing the seed crop. In southern England, this causes a loss of between 30–50 per cent of the acorn crop (Collins *et al.* 1983) with likely consequences for the long-term production of new oak trees, all due to the naturalised Turkey Oak.

The idea of a naturalised species becomes muddier when native trees are moved outside of their native ranges. For example, the Large-leaved Lime *Tilia platyphyllos* is native mostly in old woodlands on calcareous soils in a few places in England and the Welsh borders, but it has been widely planted along avenues and on large estates since at least the sixteenth century, from where it has spread

FIG 9. Brown knopper galls caused by the gall wasp *Andricus quercuscalicis* that replace, or almost so, the acorns, seen below two healthy acorns of Pedunculate Oak *Quercus robur*.

into woodland outside of its natural range. Similarly, Beech *Fagus sylvatica* is found over much of Britain and Ireland are but is doubtfully native outside of the southeast of both England and Wales. From a conservation point of view, are these to be removed from conservation areas as naturalised introductions?

This also raises the interesting question of the introduction of genetic material. For example, in Britain we have a long history of planting oaks sourced from mainland Europe and, in particular, Hungary. This could be viewed as diluting the genetic distinctiveness of British oaks that has developed over the last 10,000 years. Indeed, there is an argument for conserving the genetic distinctiveness of different regions or provenances within countries; for example, not moving oaks from southern England to the north. On the other hand, moving trees around can be viewed as increasing the genetic diversity of British oaks, making them more resilient to future changes. If trees are to persist under climate change, we should almost certainly be bringing in trees that are adapted to warmer, drier conditions further south in Europe. This is discussed further in Chapter 16.

NUMBERS OF TREE SPECIES

Estimates of the total number of tree species in the world have been made by various people and it is usually set somewhere between 45,000 and 100,000. A careful piece of work by Emily Beech of Botanic Gardens Conservation International and colleagues published in 2017 brought the total of known tree species to 60,065, representing 20 per cent of all seed plant species (gymnosperms and angiosperms). This uses the IUCN definition of a tree mentioned at the start of this chapter and so includes just trees and not shrubby things below 2 m tall. Within this estimate, nearly half of all tree species (45 per cent) are found in just 10 families, with the 3 most tree-rich families being Fabaceae (5,405 species), Rubiaceae (4,827) and Myrtaceae (4,330, including 747 species of eucalypts). But bear in mind that new species are still being found. Some of these were just hidden in difficult geography, such as the Wollemi Pine *Wollemia nobilis* discovered by David Noble in 1994 in the Wollemi National Park, just 200 km west of Sydney, Australia. Others have just been overlooked, such as *Incadendron esseri*, first identified in 2017 in the South American Andes, an abundant tree that just merged into the background and was not previously named (Wurdack & Farfan-Rios 2017). This was not just a new species but a whole new genus. There are likely to be others out there!

Given the role of geology and plate tectonics in the evolution of tree species, it is perhaps not surprising that almost 58 per cent of all tree species are endemic to just one country (and so found nowhere else). The countries with the most endemics tend to be those with the most diverse vegetation, such as Brazil, Australia and China, or are isolated islands such as Madagascar, Papua New Guinea and Indonesia.

Overall, the tropics have the most trees species: Brazil has 8,715 tree species, followed by Colombia with 5,776 species and Indonesia with 5,142 species. Indeed, a forest reserve on the equator in Sarawak, Malaysian Borneo has been found to contain 1,008 tree species in a single 50-hectare plot. To put this into perspective, the entire landmass of North America contains just 700 species of trees. But that is still somewhat more than the rather paltry but much-loved 32–35 species of Britain! At least this makes tree identification easier.

Having said that, the impoverishment of tree species in Britain is an accident of history rather than unsuitable conditions, and many introduced exotic trees will thrive in our climate. Alan Mitchell, an expert in tree identification who died in 1995, and with whom I had the privilege of teaching, wrote in the preface of his iconic tree identification book (Mitchell 1978) that there are 500 introduced species and 200 varieties of various species that 'can easily be encountered by anyone looking for trees in parks and gardens'. Moreover, if specialist tree collections in arboreta are included, this total rises to around 1,700 species of trees. So maybe we can be forgiven if tree identification in Britain is not always as straightforward as it would seem to be.

CHAPTER 2

The Value of Trees

CULTURAL IMPACT

If you were to weigh all the animals of the world in one hand, and all the plants in the other, how would their weights compare? Fortunately, this calculation has been done for us (Bar-On *et al.* 2018, Elhacham *et al.* 2020)! It has been estimated that the biomass of all life on Earth weighs in at around 1,100 Gt (gigatons, or 10^9 tons, or thousands of millions). Of this, plants make up a staggering 82 per cent of the weight of biomass (around 900 Gt), while animals (including humans) form less than 0.5 per cent of the world's biomass (just 4 Gt). The other 17.5 per cent of biomass is formed of bacteria, fungi, viruses and protists (single-celled organisms). Plants are obviously very important to the world just by their sheer abundance. Better still if you are interested in trees, it is estimated that trees form around three-quarters of the world's plant biomass, or just over 60 per cent of the total biomass on Earth.

Within this huge amount of biomass of trees, it has been calculated by Tom Crowther and colleagues (2015) that there are in the order of 3.04 trillion trees around the world (1.3 trillion in the tropics and subtropics, 0.74 trillion in the northern conifer forests and 0.66 trillion in temperate forests). This figure was based on sampling forests around the world and modelling global numbers and is likely to be an underestimate, since there are also many individual trees in drier parts of the world that were likely under-represented. For example, data from high-resolution satellites that can spot individual trees has shown that 1.3 million square kilometres of the arid western Sahara contains at least 1.8 billion trees (Brandt *et al.* 2020). We also need to add in the many trees in urban areas, which in Britain and Ireland are is likely to be more than half a billion.

Whatever the true weight and number of trees around us, they are obviously hugely important to us and to the wellbeing of our planet. Our whole life is pervaded with trees and wood, from the words we use to the things we eat to the possessions we have around us. So despite what the author Terry Pratchett has suggested, trees are more than just 'stiff weeds'. Expressions such as being 'out of the woods' (out of danger or difficulty) go back to a primeval fear of becoming lost in the forest. 'Knocking on wood' or 'touching wood' is still widely heard and probably derives from the pagan belief that vengeful spirits inhabited wood and that, by touching it, you could appease them or prevent them from hearing. Many people around the world have worshipped, and still do worship, sacred trees and imbue them with spiritual significance. This includes the dryads, or tree spirits, of Greek mythology and the *kodama* spirits of Japan that similarly inhabit trees.

Many common words also stem from trees. The Romans gave successful athletes wreaths of laurel leaves (*Laurus nobilis* – the noble laurel: Fig. 10), and the honour was extended to poets and scholars in the Middle Ages, resulting in 'poet laureate'. Romans studying at the feet of their tutors were named after the laurel berry, bacclaureus, resulting in a baccalaureate high-school qualification and a bachelor's university degree, and since Roman students were not allowed to marry, we get unmarried bachelor males. A wooden 'beam' and the German word for 'tree', *baum*, are derived from the same root in Old High German. Written records in Europe were once made on sheets of Beech wood *Fagus sylvatica*. 'Beech' is the Anglo-Saxon word for 'book', and the Beech is still called *bok* in Swedish and *beuk* in Danish. There are other examples that could be included, but that's probably enough!

Looking around most homes, we are usually never far away from wood, from toilet seats to picture frames, furniture to kitchen utensils. The last of these are

FIG 10. Bay Laurel
Laurus nobilis.

FIG 11. The value of wood and trees in culture. (a) A sculpture made of charred wood by David Nash at the Yorkshire Sculpture Park, and (b) the remains of Betty Kenny's yew tree in Shining Cliff Woods, Derbyshire, believed to have inspired the nursery rhyme 'Rock-a-bye Baby on the Tree Top'. Betty, who was really Kate Kenyon, lived in the woods with her husband, a charcoal burner, during the late 1700s. They raised eight children under the shelter of this tree, and legend has it that the babies were rocked to sleep in a hollowed-out bough, hence the nursery rhyme.

FIG 12. We are happy to manipulate the shapes of trees (a) to fit within the built landscape and (b) to create aesthetically appealing shapes. (a) A willow *Salix* species in Bruges, Belgium and (b) Japanese Yew *Taxus cuspidata*, Tomakomai Citizens Park, Hokkaido, Japan.

made of woods that do not have a strong flavour, such as sycamore, maple or poplar. We also, of course, value wood and trees culturally and historically (Fig. 11), and for their aesthetic appeal (Figs. 12 and 13).

FIG 13. Trees are used to mould our landscape by gardening to (a) create a green relaxing space using Yew *Taxus baccata* and Box *Buxus sempervirens*, and (b) a more formal view using pleached Caucasian Limes *Tilia × euchlora*. (a) Ifield Manor, Somerset and (b) Erddig, Wrexham, a National Trust property.

TRADITIONAL USES: FORESTRY, FIBRE AND FOOD

Modern technology has provided us with many new materials, but we are still heavily reliant on wood. Forests cover 3,950 million hectares, making up a third of our world's land area. Around three-quarters of this is natural forest and just a quarter is classified as human-created plantations. However, there is growing evidence that even 12,000 years ago more than 90 per cent of temperate and tropical woodlands were being used and shaped by humans (Ellis *et al.* 2021). Currently, just over half of the world's forests contribute to the production of forest products, including both wood and non-wood (e.g. food) products (Ramage *et al.* 2017). In 2017, the global production of wood was almost 3,800 million cubic metres (m³) according to the Food and Agriculture Organization. Of this, around 80 per cent was from conifers. To help put this large amount of wood into perspective, a single conifer in a European plantation at maturity contains around 1–2 m³ of wood in the trunk.

Wood is in high demand in Britain. In 2013, we produced 3.6 million cubic metres of sawn wood but also imported 5.5 million cubic metres, making Britain the world's third-largest net importer of forest products, after China and Japan. Most of the imported wood still comes from clear-cut harvesting where a whole section of forest is cleared of all trees – not the most natural of operations but very efficient and with the extra advantage that the trunks are fairly similar in size and easy to market. Most of the felling is done with a harvesting machine

FIG 14. A tree harvester at Cannock Chase, Staffordshire. The working head on the extendable arm can hold a tree while a saw (near the forester's foot in the picture) cuts the tree. The head can then remove all branches and, with the aid of an onboard computer, cut the tree into sections of various lengths to get the most value from the tree.

operated by one person, which can harvest up to 60,000 tonnes of trees per year, the equivalent of 24 chainsaw operators (Fig. 14).

Wood was the building material most readily available to past generations (Fig. 15a), being used for huts built of sticks and mud (wattle and daub) through to sturdy wooden Tudor buildings and their interiors. Wood is still the most-used construction material in many parts of the world (Fig. 15b). It has a high tensile strength, and being lightweight compared to steel, it can support its own weight better, allowing for bigger spaces with fewer supports. Stronger composite boards

FIG 15. (a) Water pipes made from elm logs bored from end to end and fitted together, probably dating from the eighteenth century and unearthed in London in 1969. The scale is in inches. (b) An old beam being repurposed in the renovation of a wooden beamed building in Transylvania, Romania.

and beams can be made by gluing laminates together for greater stability and size. This includes 'glulam', made from several laminations glued together with the grain running the same way, which is excellent for making curved or long beams, as well as plywood with the grain running at right angles in different laminates.

Wood also has high heat resistance, so a wooden building requires less insulation. In a building fire, wooden beams remain strong at high temperatures for longer periods than steel, even when charring on the outside, improving safety.

Current trends: Flooring to firewood to commercial biofuels

Green wood straight from a tree has a moisture content of around 100 per cent where there is the same weight of water as there is of woody material. Water in wood consists of 'free water' inside the tubes of the wood and 'bound water' that is chemically attached to the cell walls of the wood. Left outside and given long enough, the free water will evaporate, leaving wood at around 25–35 per cent moisture content. But bring the wood into a heated, dry house and the moisture content will drop to 8–12 per cent. This is good news for our homes, since most fungi cannot operate below 20 per cent moisture, and wood-boring insects such as woodworm (usually the Common Furniture Beetle *Anobium punctatum*) prefer timber over 18 per cent moisture and do not prosper below 12 per cent.

The biggest problem in using wood for structures or for an aesthetically pleasing wood finish is that it will shrink and expand depending upon moisture and temperature. Wood dried from green to 12 per cent moisture tends to shrink longitudinally along the grain by 0.1–0.2 per cent, but across the grain (as in a cross-section of a tree) it can be 3–10 per cent, depending on species and environmental conditions. If you have a solid wood floorboard 15 cm wide, this could make the difference of 4.5–15 mm width between a board cut green to one used in house conditions. Moreover, the small changes in humidity and temperature normal in any house, between day and night and across seasons, can cause significant movement.

This change in dimensions is often largely unnoticeable in rooms with fairly constant humidity such as bedrooms or living rooms, and in furniture where changes of a few fractions of a millimetre go unnoticed. Nevertheless, floorboards and parquet flooring are laid with small gaps to allow for changes in size. However, the rise in wooden flooring in bathrooms and kitchens where the boards are designed to fit tightly together creates a problem, especially where water is liable to be spilt and humidity varies greatly. The normal contraction and expansion are avoided by using either laminate wood or engineered wood rather than solid wood. Laminate wood is made from small particles of wood glued together, creating high-density fibreboard (HDF) or chipboard, depending on particle size. These are structurally very stable and usually topped with a film of wood-effect plastic and a transparent surface layer. This outer coating helps resist the absorption of water and humidity and the release of formaldehyde from the glue bonding the wood particles together. Engineered flooring, also called wood veneer flooring, uses plywood as the base with a thin veneer of solid wood glued to the upper surface. Plywood gains dimensional stability from the longitudinal grain going in two directions and is very much more resistant to expansion and contraction than solid wood. But both HDF and plywood will move a little. This is usually accommodated by laying the flooring not quite to the walls, giving room for the whole floor to expand and contract.

Wood is also in increasing demand for firewood. In 2018, Britain had somewhere between 1–1.5 million wood-burning stoves (and perhaps a million more open fires), representing around 10 per cent of all households, as well as hundreds of biomass boilers run on wood chips or pellets, all needing wood or solid fuel. It is difficult to be precise, but it seems that up to 200,000 more stoves are currently being sold in the UK each year. In March 2016, almost 600,000 of the wood-fuel users were in the densely populated southeast of England. While most people used logs that were presumably bought commercially, 18 per cent acknowledged that they gathered wood from the countryside, which may have been burnt while still wet. Modern stoves are very efficient but will still release particulate matter, especially if wet wood is burnt. According to Defra, in 2017, domestic wood-burning was responsible for 36 per cent of the UK's PM2.5 (particulate matter less than 2.5 μm, or 0.0025 mm, in diameter). The European Environment Agency estimated that in 2012, PM2.5 caused 37,800 premature deaths in the UK. Wood does not have to be burnt directly to get energy, since it can also be used to produce bioethanol. At present, 250 litres of petrol can be produced from one tonne of wood, enough to propel a family car the length of England five times.

Food and drink

Most of the fruits we eat come from woody plants, with the exception of strawberries, pineapples and a few others (Fig. 16). Nuts also primarily come from trees (except the peanut), including pine nuts, which are really the seeds from about 20 different pine species that produce seeds big enough to be worth collecting. The Stone Pine *Pinus pinea* is collected in Europe and the Pinyon Pine *P. monophylla* in North America, named by Spanish explorers after the *P. pinea* seeds they ate at home. Commercially bought seeds are now most likely to be

FIG 16. Date seller in a market in Tripoli, Libya, offering a huge number of varieties.

from the Korean Pine *P. koraiensis* grown in China. A decade or so ago, I awoke to find a strong, unpleasant metallic taste in my mouth that got much worse when eating or drinking and stayed with me for two weeks. This is often referred to as 'pine mouth', since the cause is eating pine nuts from the Chinese White and Red Pines (*P. armandii* and *P. massoniana*). Seeds from these pines were not previously exported but as demand grew in the west, some slipped in. The Chinese authorities have now made the exportation of these species much less common so you may never experience such delights!

Many of our spices come from woody plants, including the bark of the Cinnamon tree *Cinnamomum verum*, flower buds of the Clove tree *Eugenia caryophyllus*, and seeds of the Nutmeg *Myristica fragrans*. Many of these spices are quite toxic in high doses, which is perhaps unsurprising given that the compounds involved have evolved to dissuade insects and other pests from eating the various plant parts. Nutmeg, native to the Spice Islands (now called the Maluka Islands) of Indonesia, is a good example. The nutmeg grows inside a fleshy fruit about the size of an egg, which is eaten by people in Southeast Asia (Fig. 17). The seed inside is surrounded by a bright red aril which is used as the

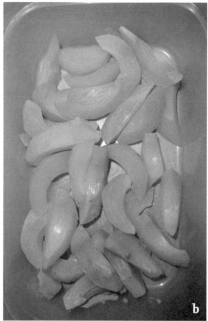

FIG 17. The nutmeg that we use as a spice is the seed of the Nutmeg tree *Myristica fragrans* found widely across the tropics. (a) The seed is contained inside a bright red aril. In Malaysia the outer fruit is eaten as a snack. (b) The fruit here has been crystallised and is fairly hard and crunchy, while on the right (c) it has been dried and further sweetened and is like eating desiccated coconut. Both have a strong nutmeg flavour but too much can cause respiratory problems, particularly in asthmatics. Photograph (a) by W.A. Djatmiko from West Java, Indonesia, reproduced under the GNU Free Documentation License.

spice mace, a milder form of nutmeg. The seed itself, hidden inside a black seed coat, is the nutmeg we use. In addition to a high oil content, the seed contains an abundance of myristicin, a compound also found in smaller amounts in the herbs fennel and dill. This is a psychoactive compound able to change the brain's function. One teaspoon of freshly grated nutmeg (c.2.4 g – half a nutmeg) causes hallucinations but unfortunately also nausea, vomiting and circulatory problems. Five teaspoons (c.12 g – 2.5 nutmegs) is lethal. Fortunately, the amount we would typically have in, say, a rice pudding is just pleasant and not lethal. Just to add some extra spice to Sunday afternoon after the rice pudding, it has been shown that nutmeg acts as an aphrodisiac (Tajuddin *et al.* 2003) – in mice at least! The potential toxicity of nutmeg underlines the effectiveness of these defensive compounds in keeping animals from eating the seed before it can germinate. Similar observations can be made about the abundant caffeine in the fruits of Coffee (*Coffea arabica* and *C. canephora,* sometimes called *C. robusta*) and leaves of Tea *Camellia sinensis*: the more you imbibe, the greater the effects. For us humans, tea and coffee have many benefits, but too much caffeine can affect our heart and digestion and also cause anxiety and a number of other problems.

Many compounds with a medicinal value have also been derived from trees, including quinine *Cinchona* species, or 'Jesuit's bark', from Peru and Ecuador, used to prevent and treat malaria. Quinine is very bitter; hence the development of tonic water which goes so well with gin – a palatable way of taking your medicine. Extracts from the Maidenhair Tree *Ginkgo biloba* are thought to improve blood flow to the brain, countering memory loss, particularly for those suffering from Alzheimer's disease. Paclitaxel, formerly called taxol, is used very successfully in the chemotherapy of various cancers. It was originally produced from the bark of the Pacific Yew *Taxus brevifolia,* causing great concern for its future in the wilds of western North America, but fortunately it can now be artificially synthesised.

We obviously use fruits and even the sugars from palms (Chapter 3) to either make or flavour various beverages – would life be the same without sloe gin, flavoured with Blackthorn fruit *Prunus spinosa*? And then there is retsina, the Greek wine flavoured with pine resin that would originally have been widely used as a preservative and is perhaps the ancient taste of wine. We also use wood for flavouring via its use in barrels for wine, sherry, port and assorted spirits. Most barrels are made of various oak species usually classified as:

- American, mostly using American White Oak *Quercus alba* and relatives, but not Red Oak *Q. rubra,* since the wood has no tyloses, the balloon-like outgrowths that block the water-conducting tubes in the wood – see Chapter 6 – so the barrels would leak.

- European, using Sessile and English Oaks (Q. *petraea* and Q. *robur*).
- Asian, including Daimyo Q. *dentata*, Mizunara Q. *crispula* and Mongolian Q. *mongolica* Oaks.

It is telling that other tree species can be used to make barrels, but often these are lined to keep the drink away from the wood. What then is so special about oak? The main benefit is the wood chemistry and the 'extractables' that leach into the drink to give it colour and flavour. Whisky coming from the still is colourless, and the wonderful hue and complex flavour that develop over time in the barrel are due to the extractables (Fig. 18). For example, lactones imbue coconut flavours and phenols a vanilla flavour; see Mosedale (1995) for a wider discussion on extractables of different oaks and flavours. Toasting – the heating process used to help bend the oak staves – and charring – a much more serious burning of the inside of barrels used for bourbon, which leaves the oak like black alligator skin – also affect the flavours. This burning partially breaks down the components of the wood, producing varying amounts of the compounds furfural, maltol and ethoxylactone, giving malty, caramel notes to the spirit, all adding to the developing flavour. The same process also happens with wines, helping

FIG 18. Barrels of whisky at the Glenfarclas distillery, Speyside, Scotland. To get the best flavours from the wood takes time, and whisky can spend more than two decades in the barrel. Photograph by Richard Hobbs.

develop and change the flavour, and these barrels can then be used for whisky to produce spirits with wine overtones. The permutations are almost equally endless and enjoyable!

Oil from the fruits of the Olive *Olea europaea* has a long European history although it is currently facing an uncertain future due to the bacterium *Xylella fastidiosa* (Chapter 15). Yet in the news we probably hear more about palm oil from the seeds of the Oil Palm *Elaeis guineensis*, native to tropical West Africa and now used in more than half of the products available in a typical supermarket, including margarines, candles and cosmetics. The race to produce this lucrative oil is causing the destruction of large areas of rainforest, but in some places such as Malaysia, Oil Palms are being planted on old rubber plantations (Fig. 19) and provide a good income for locals without further deforestation.

FIG 19. Oil Palm *Elaeis guineensis* plantations in northern Malaysia, replacing a former rubber plantation.

ECOSYSTEM SERVICES

We can look at trees and forests and consider their value in terms of *ecosystem services*, defined as what trees and forests provide that we humans find useful (UK National Ecosystem Assessment 2011). The Food and Agriculture Organization (FAO) note that nearly 1.6 billion people – more than 25 per cent of the world's population – directly rely on trees and forests for their livelihoods. These people and others in cities gain services that include the provision of resources such as wood and other forest products, including food, medicines and fresh water. Cultural services include the aesthetic, recreational and spiritual values of trees and woodlands. They also include regulating services such as the improvement of air quality by removing pollutants and irritants, especially important in urban areas. In the USA, trees and forests removed 17.4 million tonnes of air pollution in 2010, a service valued at $6.8 billion (£5.6 billion) (Nowak *et al.* 2014). Up to 70 per cent of gaseous pollutants such as ozone, sulphur dioxide and nitrogen dioxide can be filtered out by street trees. Moreover, trees can also trap particulate matter such that 1 hectare of woodland in urban areas can filter out up to 15 tonnes of particles each year. This includes PM2.5 (particulate matter less than 2.5 μm in diameter) produced by wood-burning stoves, diesel engines and forest fires, which, as discussed above, is a source of human ill health and death. Once caught, particles are subsequently washed to the ground by rain. A study of the West Midlands conurbation showed that planting trees on a quarter of available urban area, and thus increasing the percent cover of trees from 3.7 to 16.5, reduced particulate matter concentrations in the air by 10 per cent, effectively removing 110 tonnes per year from the atmosphere (McDonald *et al.* 2007).

Pollution capture is greatest when planting conifers (which have a high leaf surface area) and broadleaf trees with rough or hairy leaves, such as bean trees *Catalpa* and some elms and oaks (Fig. 20). It is also important that tree shape and size are matched to the planting area. Large trees in the canyons created by narrow streets with tall buildings can trap pollutants, actually making pollution worse. This is especially important as some trees, especially willows, poplars and oaks, produce large amounts of volatile organic compounds (VOCs) that react with nitrogen oxides from vehicles to produce harmful ozone, which can be trapped at street level. On the positive side, trees shade us from harmful ultraviolet light, especially UV-B, which is involved in sunburn. It has been found in Canada that trees such as a Norway Maple variety *Acer platanoides* 'Crimson King', Copper Beech *Fagus sylvatica* 'Purpurea', Hackberry *Celtis occidentalis* and Swamp White Oak *Quercus bicolor* reduced UV-B by two-thirds, while others such as the Maidenhair Tree and Red Maple *Acer rubrum* reduced it by half. It may be

FIG 20. Downy Oak *Quercus pubescens*, native to Spain, southern Europe and western Asia, has naturally downy leaves which give them a pale, matt look. These are better at capturing small particulate pollutants from the air than the smooth-leaved oaks native to northern Europe.

tempting to think that this is because the first two trees listed have red or copper foliage which reduces UV but it appears to be more a case of density and spread of the crown (Sivarajah *et al.* 2020).

In addition to the physical benefits that trees give us, they are also important in screening us to a degree from noise and reducing our stress. Noise reduction is better provided by trees with rounder, thicker, larger leaves that are grown in opaque hedges with a high enough leaf density that you can't see through them, and that are tall enough to be above your ears and preferably at least 2 m wide (Van Renterghem *et al.* 2012). Taller trees, separating us from traffic noise by mainly stems, are obviously much less effective (Fig. 21). Individual trunks scatter and absorb less than 10 per cent of noise below 1 kHz, although trees with rougher barks – generally conifers – are slightly better than broadleaf trees (Li *et al.* 2020). Even so, dense belts of trees are needed to make a difference, and even then the deep bass rumble of traffic is difficult to filter out with barriers. Interestingly, not being able to see the traffic gives a perception of quieter conditions, and the more solid the barrier the quieter we feel it is. Studies have shown that natural-looking and aesthetically pleasing barriers such as hedges are perceived as doing less to reduce noise levels than solid metal and concrete barriers (Joynt & Kang 2010).

Trees give significant financial benefits. Urban areas with more trees have higher property values and reduced energy costs, since the trees hold in heat in winter and give cooling shade in summer. Trees also reduce the need for storm water infrastructure, since they act like sponges, slowing the speed at which water falls to the ground through the canopy. This helps more water to soak into the soil rather than running directly to drains, and since the soil releases

FIG 21. (a) The trees in this square are valuable for giving shade and a sense of calm but do very little to reduce traffic noise, since the foliage is above ear height. The trunks themselves do little to reduce noise pollution. (b) Conversely, in this city park, the abundant foliage that comes down towards ground level in this thick bank of trees will do much to reduce noise pollution from traffic in the city. (a) Reims, France and (b) Sydney, Australia.

it slowly, the trees help in reducing peak flooding and soil erosion (Roy *et al.* 2012, Mullaney *et al.* 2015). Trees also reduce the amount of water reaching drains simply because they use a significant amount of it to grow (Chapter 6).

More broadly, forests offer regional services such as their role in creating weather patterns and in increasing rainfall. Forests evaporate enormous amounts of water, producing clouds that release rain downwind. In the Amazon rainforest, a third of the rainfall comes from water evaporating from within the forest area (Staal *et al.* 2018). The amount of water evaporated from global land surfaces, much of it from wooded areas, is in the order of 190 km^3 of water per day. This is just 16 per cent of that evaporated from the oceans but is still 80 km^3 more water than flows down the world's rivers each day (Rodell *et al.* 2015). Moreover, chemicals released by trees, particularly monoterpenes, help to 'seed' clouds by acting as nuclei for water to condense around, increasing rainfall. Not surprisingly then, global precipitation over land amounts to 300 km^3 of water per day.

The ecosystem services described above, and the many others that trees provide, are obviously valuable but can be very difficult to quantify in monetary terms beyond saying that 'trees are good for us'. Yet if a developer wants to remove a park to build a shopping complex, they can clearly state the financial gain for the city. This disparity of argument was behind the development of tools such as i-Tree (see https://www.itreetools.org) and the citizen science project Treezilla (https://treezilla.org). These tools attempt to put a financial value on the ecosystem services provided by the trees of an area. The value can be arrived at by looking at how much these services would cost if they were done another way; for example, calculating the cost of removing carbon dioxide from the air of a city by a mechanical or chemical means. Another way is to work out the value of a reduction in the risk of, say, flooding through the savings in insurance premiums resulting from lower flood risks (Catney & Henneberry 2022). Using these algorithms, i-Tree has been used to establish that the estimated 8.4 million trees of London are each year worth £126.1 million in pollution removal, £146.9 million in carbon storage, and all the services together are worth £132.7 million each and every year. Within London there are 8.9 million people, so roughly a tree for every person. The trees cover 20 per cent of land area, and 47 per cent of London is green space worth £91 billion in gross assets (Greater London Authority 2018). This gives something concrete to use in arguments for keeping and valuing the trees around us.

It can, however, be much harder to put a value on the beauty and enjoyment that trees provide. But it can be done. Woodlands give opportunities for recreation such as walking or bird-watching or mountain biking. Putting a financial value on these can involve calculating the cost savings in health care resulting from people being fitter and less stressed. In this way it has been estimated that the social and environmental benefits of woodlands in Great Britain are worth approximately £30 billion (Willis *et al.* 2003).

The whole concept of ecosystem services is useful but not without its problems. It is very anthropocentric, being concerned with what we humans can get from an ecosystem rather than what is good for the future of the planet. It also brings everything down to a consideration of money, potentially shifting decision making from citizens and governments to corporations, and it can be simplistic in assigning a monetary value to things such as aesthetic enjoyment that surely have a value beyond price. Jonathan Silvertown (2015) and Richard Gunton and colleagues (2017) give good reviews of these issues.

Problematic services

It is worth pointing out, however, that trees are not always our friends. Honeydew from lime *Tilia* species can be slippery and is well known to set to a sticky mass on cars and park benches, and there have been instances of falls, broken limbs and consequent litigation. Fruits and other debris such as petals can cause issues when tramped into buildings. Trees have also been felled, for example, when a threat is perceived, such as when fruit trees in a school driveway might drop fruit on a waiting parent and result in litigation. Shade cast by large trees on small gardens has also been the cause of many neighbourly feuds and further litigation.

Trees roots can cause the upheaval of roads and paths, cause subsidence of buildings, especially on shrinkable clay, and a myriad of other issues. Fortunately, we are getting better at solving these problems without wholesale slaughter of trees. For example, in a World Heritage Site in Israel, a historic burial cave around 1,800 years old was under threat from tree roots penetrating the cave ceiling (Fig. 22), leading to a risk of cave collapse (Jakoby *et al.* 2020). Rather than indiscriminately fell the woodland above the cave, DNA barcoding was used to identify that the roots were primarily from two species of *Pistacia*. This allowed a selective felling of these trees above the cave while maintaining the rest of the woodland.

Explosives and polemobotany

We can gain services from trees in perhaps unexpected ways. Paul Warnstedt and colleagues at the University of the Bundeswehr Munich, Germany have investigated the use of trees to help protect urban areas against terrorist threats, and specifically the problems of bombs (Gebbeken *et al.* 2018). Strategic placing of clumps or lines of trees, planted in the ground or in containers, can help prevent

FIG 22. Penetration of tree roots through the limestone ceilings of the Beit She'arim necropolis, Galilee, Israel. (a) Entrance to one of the burial caves, showing the Mediterranean woodland around 5 m above the cave. (b) Roots can be seen penetrating from the ceiling while (c) roots of *Pistacia palaestina* can be seen growing through the limestone on top of the entrance of another cave. (d) and (e) Cave ceilings affected by roots, showing several penetration points. From Jakoby *et al.* (2020).

vehicles from getting close to an intended target but can also act to dissipate the pressure wave from an explosion, reducing the damage caused. Evergreens are best at this as they provide year-round protection, and conifers with needle leaves are better than broadleafed trees. If protection of people is the main aim, these trees need be no taller than a hedge and may indeed give more protection than mature trees with more open space at the bottom of the canopy.

The German team demonstrated this by detonating the equivalent of 5 kg of TNT 5 m away from hedges of Thuja that were just over 1 m tall and 2 m long by 0.55 m thick and hedges of Cherry Laurel of the same sort of size (Fig. 23). They found that the small leaves of the Thuja suffered little visual damage even after four repeat explosions! Cherry Laurel by contrast lost most of their leaves after the first. Most importantly, Thuja reduced the pressure of the explosion 0.5 m behind the hedge by 62 per cent and at 5 m by 36 per cent, whereas Cherry

FIG 23. Hedges of (a) Thuja *Thuja occidentalis* and (b) Cherry Laurel *Prunus laurocerasus* before (upper) and after (lower) being exposed to the equivalent of 5 kg of TNT exploded 5 m from the hedges. The Thuja looks remarkably unaffected while the broader leaves of the Cherry Laurel have been extensively shredded. From Gebbeken *et al.* (2018).

Laurel reduced the pressure by 31 per cent and 14 per cent, respectively, with less consistency between repeated explosions. So, while large leaves are better at sound reduction caused by traffic and other city noises, smaller leaves of conifers are better at blast reduction.

This opens up a much older field of *polemobotany* – the study of plants utilised for or impacted upon by military activity. A number of tests were made after the Second World War on damage caused to trees within a few kilometres of nuclear bomb detonations, including 23 nuclear bombs detonated at Bikini Atoll in the Pacific Ocean (Brown *et al.* 1953, Fons & Storey 1955). Such large explosions tended to cause the canopy to break from the trunk in conifers and broadleaf trees, but the broadleaf trees regrew from surviving stumps and roots. By contrast, Coconut Palms *Cocos nucifera* just bent under the pressure wave when more than 3 km away from the epicentre. The survival of trees in such extreme situations was seen after the nuclear bomb was dropped on Hiroshima on 6 August 1945, when it was found that at least 34 species of woody plants survived the blast. The 170 or so surviving trees within 2 km of the epicentre are called the 'Atomic Bombed Trees', or *hibakujumoku*. Nuclear detonations have, however, left their mark in other ways. Trees around the world contain a radiocarbon (^{14}C) 'bomb peak' that ended after the Partial Nuclear Test Ban Treaty of 1963 which prohibited all nuclear-bomb testing above ground. This 1945–63 peak has proved useful, for example, in dating tropical trees that do not produce annual rings.

In the trench warfare of the First World War, trees were soon shredded and broken by the relentless shelling. Wearn and colleagues (2017) writing about the Somme battlefield noted that the 'trees added a layer of unease', as the shattered stems created a melancholy sight with a negative psychological impact. Many of these trees regrew from the base, and a century later this survivorship reflects the original composition of the woods. At some sites, additional trees were planted as national memorials. For example, Delville Wood on the Somme was replanted after the war with oaks from Cape Colony, South Africa to form the South African National Memorial site. Others were planted to promote 'economic healing'; for example, from 1921–22, Germany provided 153,000 small conifer trees, 1.4 tonnes of acorns and 180 kg of other tree seeds for replanting the forests around Verdun in northeast France, part of the western front (Wearn 2016).

In other sites with great historical significance, the original landscape and woods are maintained as a form of remembrance. This is especially true of trees, since they are often the oldest living things on the landscape and are evocative symbols of endurance, a constant link between those who fought in the war and those of the current generation. This preservation of the landscape is clearly seen at Auschwitz in what is now Poland, where the trees at this former concentration camp are a direct link with its terrible history. Birches and Lombardy Poplars *Populus nigra* 'Italica' with some oaks and sweet chestnuts made up the majority of the trees. Birches and poplars are fairly short-lived, and, as they begin to die off, this is seen as a poignant loss to the site.

WELLBEING AND TREES

Trees are valuable items in our environment that are not just useful for the products and physical services they provide but are also essential for our mental and physical wellbeing (Fig. 24). This is enshrined in the Town and Country Planning Act of 1990 where trees are only considered for their amenity value. Subsequent regulations go further and specify that Tree Preservation Orders (TPOs) can include their importance for nature conservation or a response to climate change but 'These factors alone would not warrant making [a

FIG 24. Trees provide beauty, which is good for our wellbeing, whether this is (a) a single species planted for effect, such as Sargent's Cherry *Prunus sargentii* along a road, or (b) more varied plantings of a former stately home, providing a range of shapes, colours, textures and sizes in early autumn. (a) Aomori Prefecture, Honshu, Japan and (b) Killerton, Devon, now a National Trust property.

TPO]' (Ministry of Housing, Communities & Local Government 2014). The amenity value of trees around our homes and places of work is important to our wellbeing, particularly in urban areas where larger green spaces may be less available than in rural areas. To put this into perspective, Song *et al.* (2016) pointed out that humans evolved in a wooded landscape over 6–7 million years, and so 'if we define the beginning of urbanization as the rise of the industrial revolution … humans have spent over 99.99 per cent of their time living in the natural environment'.

No wonder it is being increasingly confirmed that trees and green spaces contribute to our wellbeing and sociability. Part of this is due to the psychological benefits of looking at trees. One of the first studies on this effect was published by Roger Ulrich in the journal *Science* in 1984. He looked at the recovery of patients in a Pennsylvania hospital after having had their gall bladders removed. Patients on one side of each ward could see deciduous trees out of the window while those on the other side of the room could see just a brick wall. He found that those patients looking at the trees had a less negative attitude, quicker recovery, fewer complications, and needed fewer painkillers. The same effect can be had by looking at photographs of trees and green spaces. In a study in Taiwan, Ying-Hsuan Lin and colleagues (2014) showed people pictures of cityscapes with no trees and the same pictures to which trees had been added digitally (Fig. 25). The individuals had to recall sequences of numbers backwards; those seeing trees before the test and especially those who had the trees pointed out to them did better at the test. Moreover, the pictures with trees gave a perceived sense of calm and enhanced mood. Other studies have found that green spaces can reduce our pulse rate, blood pressure and levels of the stress hormone cortisol (Ochiai *et al.* 2015).

FIG 25. Views of streets in Taiwan that have no trees (upper) and the same pictures where trees have been added digitally. From Lin *et al.* (2014).

Tree-lined streets also encourage us to walk for recreation and to walk to work, with consequent benefits for our health. A study in Sacramento, California showed that having trees within 250 m of home led to better health with lower levels of obesity and better neighbourhood 'social cohesion' (Ulmer *et al.* 2016), simply because people were out walking. Most interestingly, lowest income groups benefit the most from green space near their home, where street trees within 100 m around the home significantly reduced the probability of being prescribed antidepressants (Marselle *et al.* 2020). Yet in many urban areas it is the wealthiest areas that have the most trees while the poorest areas have bricks and concrete. Part of the problem is that trees planted in low-income areas are likely to be vandalised and not last long. To develop a sense of ownership and pride, various initiatives have been started whereby the local residents get to plant and look after their own trees. An excellent example is in the London Borough of Hackney where 1,000 street trees were planted between 2006 and 2010, involving 1,800 residents and including community tree champions for each street. The loss of planted trees was staggeringly low at just 1 per cent (TDAG 2012). The community decided what to plant, and over the period of the scheme almost 250 different tree species were planted. They also took over the care of the trees with hip-level watering tubes and free watering cans to ensure that watering was easy. This sort of initiative gives the community a sense that 'you can actually do something instead of ignore the stuff around us' (Chawla 2020). Generally, even heavily built-up urban areas (Fig 26a) are improved by the presence of trees even when the architecture is pleasing to the eye (Fig 26b).

Trees and green spaces can also benefit us physiologically. Chemicals released by trees, mainly monoterpenes, are breathed in and interact with our brains, leading to feelings of reduced tension, mental stress, aggression and depression, and an increased feeling of wellbeing. For example, the oil from Hinoki Cypress *Chamaecyparis obtusa*, widely used to scent soap, toothpaste and cosmetics in Japan, has been found to have positive effects on our brains and induce a feeling of 'comfortableness' (Ikei *et al.* 2016).

These combined effects mean that being out amongst trees is good for us psychologically and physiologically. This is enshrined in the Japanese concept of *shinrin-yoku*, forest-air breathing or forest bathing, which is becoming commoner in the west (Fig. 27). This advocates that standing amongst trees and slowly breathing is very relaxing. A short lunchtime walk of 1.8 km through green areas can improve sleep (7 hours sleep compared to 6.5 hours in those that did not walk in woodland), and a 3-day forest visit can have beneficial effects on the immune system for up to 30 days (Hansen *et al.* 2017).

FIG 26. (a) A tree-lined street in Barcelona, Spain, which even above the canopy of leaves adds to the sense of peace in a busy city. (b) Matsumae Castle in Hokkaido, Japan, dating from 1639, which even in early spring is beautifully framed by the surrounding trees of Matsumae Park, containing some 10,000 cherry trees of around 250 different varieties.

THE VALUE OF TREES AS HABITATS

Which trees hold most wildlife?

Professor Richard Southwood of Oxford University published an influential paper in 1961 in which he showed that the number of insect species that feed on the leaves of a tree species is linked to the length of time a tree has been found in Britain. This relationship has been seen to hold true in a number of other European countries and as far afield as Hawaii. Kennedy & Southwood (1984) took this a little further and looked at the commonness of a tree in the British Isles (excluding Ireland) matched to the number of insect species, in this case including mites. This was a rough-and-ready-type calculation, since estimating tree abundance from records is difficult and it also relies on there being good records of insect species found on different trees. Despite these problems they found that host tree abundance is a good indicator of total insect species richness – the more of a tree there is in Britain, the more insects and mites call

FIG 27. A sign encouraging people to breathe in the fragrant air of the woodland in northern Honshu, Japan. This is called *shinrin-yoku*, or forest-air breathing, a popular form of relaxation in Japan.

it home. Closely related trees also tend to have higher numbers of invertebrates shared between them than trees with no other species in the same genus or family (Table 5). Plus, within a region, the larger an area dominated by a species, the more of the insects associated with them will be found in that area.

One of the ways in which these data have been used is to support the idea that native trees are better than introduced trees, since they tend to hold more insect species than the introduced species, as can be seen by the clumping of introduced species towards the bottom of Table 5. But this is, of course, only one aspect of the conservational interest of a tree. The introduced Sycamore may hold just 43 species of insects and mites compared to 423 species on our two native oak species but it does hold a lot of them, particularly aphids. In fact, the weight of insects on a Sycamore is probably similar to that on an oak and both can thus provide an abundant food supply. A Blue Tit *Cyanistes caeruleus* looking for food to feed hungry young chicks would find abundant food on a Sycamore and an oak, and the introduced tree is thus of value for wildlife. Another argument in favour of introduced species is that many insects are not specific to only one species of tree, and an introduced tree such as Sycamore is capable of hosting many of the 51 species of insects and mites associated with, for example, the native Field Maple.

TABLE 5. Number of insect and mite species found on different trees in Britain. Blue indicates native species (listed in Table 2) and green introduced species.

Common name	Scientific name	Leaf miners	Gall formers	Sap feeders	Chewers	Total
Willow	Salix (5 spp.)	23	39	81	307	450
Oak	Quercus (2 spp.)	35	62	88	238	423
Birch	Betula (2 spp.)	43	8	47	236	334
Hawthorn	Crataegus monogya	23	5	43	134	205
Poplar	Populus (4 spp.)	19	24	31	115	189
Scots Pine	Pinus sylvestris	7	3	30	132	172
Blackthorn	Prunus spinosa	12	6	34	99	151
Alder	Alnus glutinosa	18	8	35	80	141
Elm	Ulmus (2 spp.)	13	16	32	63	124
Wild Apple	Malus sylvestris	21	0	32	63	116
Hazel	Corylus avellana	14	11	21	60	106
Beech	Fagus sylvatica	5	10	11	72	98
Norway Spruce	Picea abies	2	3	24	41	70
Ash	Fraxinus excelsior	3	5	23	37	68
Rowan	Sorbus aucuparia	10	5	6	37	58
Lime	Tilia (2 spp.)	1	9	18	29	57
Hornbeam	Carpinus betulus	12	4	12	23	51
Field Maple	Acer campestre	6	10	13	22	51
Sycamore	Acer pseudoplatanus	5	8	11	19	43
European Larch	Larix decidua	1	2	9	26	38
Juniper	Juniperus communis	4	6	8	14	32
Sweet Chestnut	Castanea sativa	3	0	1	7	11
Holly	Ilex aquifolium	1	0	2	7	10
Horse-chestnut	Aesculus hippocastanum	0	2	5	2	9
Walnut	Juglans regia	0	3	2	2	7
Yew	Taxus baccata	0	2	1	3	6
Holm Oak	Quercus ilex	4	0	1	0	5
False-acacia	Robinia pseudoacacia	0	0	1	1	2

Data from Kennedy & Southwood (1984).

It is also interesting to note that the native Holly *Ilex aquifolium* has just 10 species of insects and mites associated with it, poorer than many introduced species – but it provides a huge amount of shelter in winter for woodland birds. Moreover, trees support much more wildlife than just insects, including epiphytes hitching a ride on the outside of the tree, such as lichens, mosses and algae, through to wood decay invertebrates, and offer food for animals and leaf litter for soil organisms (Table 6). Here as well it is clear that introduced species have much to offer in terms of wildlife value. The authors of the data in Table 6 make the point that the heartwood of Sweet Chestnut and False-acacia 'decays in a very similar way of oak and therefore supports some of the invertebrates more associated with decaying oak' (Alexander *et al.* 2006). All this underlines the fact that every tree species in an area will have some wildlife value, whether native or not.

TABLE 6. The value of trees and shrubs widespread in Britain as habitats and food sources for different groups of organisms. A greater number of asterisks indicates greater value for wildlife. The trees are in the order of decreasing number of insect feeders used in Table 5. Blue indicates native species and green introduced species.

Common name	Scientific name	Myco-rrhizal fungi	Wood-decay fungi	Wood-decay inverts	Biomass of foliage inverts	Leaf litter	Pollen and nectar	Fruit and seeds	Epiphytes
Willow	*Salix* (5 spp.)	****	***	***	***	***	*****	*	****
Oak	*Quercus* (2 spp.)	*****	*****	*****	*****	***	*	*****	*****
Birch	*Betula* (2 spp.)	*****	****	****	****	***	*	****	****
Hawthorn	*Crataegus monogya*	***	**	***	***	*****	*****	****	*
Poplar	*Populus* (4 spp.)	***	***	***	***	***	*	*	*
Scots Pine	*Pinus sylvestris*	*****	***	****	****	*	*	****	*
Blackthorn	*Prunus spinosa*	***	**	***	***	****	****	****	*
Alder	*Alnus glutinosa*	***	***	**	****	***	*	****	**
Elm	*Ulmus* (2 spp.)	***	****	***	***	****	*	*	*****
Wild Apple	*Malus sylvestris*	***	**	***	***	****	****	****	***
Hazel	*Corylus avellana*	**	***	***	***	****	*	***	****
Beech	*Fagus sylvatica*	*****	*****	*****	*	*	*	*****	*****
Norway Spruce	*Picea abies*	*****	**	***	***	*	*	****	*
Ash	*Fraxinus excelsior*	***	***	*****	*	*****	*	*	*****

continued overleaf

TABLE 6. *continued*

Common name	Scientific name	Myco-rrhizal fungi	Wood-decay fungi	Wood-decay inverts	Biomass of foliage inverts	Leaf litter	Pollen and nectar	Fruit and seeds	Epiphytes
Rowan	*Sorbus aucuparia*	***	**	*	*	****	****	****	***
Lime	*Tilia* (2 spp.)	****	***	**	***	****	****	*	**
Hornbeam	*Carpinus betulus*	***	**	**	*	***	*	***	**
Field Maple	*Acer campestre*	***	**	**	*	***	****	*	***
Sycamore	*Acer pseudoplatanus*	***	***	***	*****	*****	****	*	*****
European Larch	*Larix decidua*	*****	**	*	***	*	*	****	*
Juniper	*Juniperus communis*	-	-	-	-	-	-	-	-
Sweet Chestnut	*Castanea sativa*	***	***	****	*	*	*	*****	*
Holly	*Ilex aquifolium*	***	*	*	**	*	*****	****	**
Horse-chestnut	*Aesculus hippocastanum*	***	**	***	*	**	****	*	*
Walnut	*Juglans regia*	***	**			***			*
Yew	*Taxus baccata*	***	**	*	**	*	*	****	*
Holm Oak	*Quercus ilex*	***	***	*	*	*	*	*****	*
False-acacia	*Robinia pseudoacacia*	***	**	***	*	***	****	*	*

Data from Alexander *et al.* (2006).

What would go extinct if we removed trees?

If all the world's forests and trees were to be removed or were to die, a very large number of species would be liable to go extinct. The world's rainforests alone hold around 60 per cent of the world's terrestrial plant and animal species, and all the world's forests hold more than 80 per cent of species. Many species, such as saproxylic beetles living on dead wood, would be very unlikely to survive without trees. (*Saproxylic* comes from the Greek *sapro*, meaning 'dead', and *xylo*, meaning 'wood'.)

A much more likely scenario is for individual species to go extinct or at least become much rarer. We have seen this in Europe with the Dutch elm disease pandemic starting in the 1970s, which killed an estimated 28 million mature and

20 million young elm trees in Britain, removing 90 per cent of mature Field Elms *Ulmus minor* (Thomas *et al.* 2018). While many elms have remained alive to sucker from the roots, or in the case of Wych Elm *U. glabra* to produce new shoots from the base, large mature elm trees have almost disappeared from the landscape and elms survive as shrubby growth. This has had some consequences for insects that live on elms. As Table 5 shows, 124 species of insects and mites have been recorded on 'two species' of elms (presumably Wych Elm and English Elm *U. procera*). This is comparatively low, certainly compared to our native oaks, but the loss of mature elms has affected the White-letter Hairstreak butterfly *Satyrium w-album*, whose larvae feed on elm flowers and leaves, normally at the top of trees. This is possibly the insect most strongly associated with Wych Elm, and it has declined significantly in Britain following Dutch elm disease. It is predominantly found in England and Wales but fortunately since 2018 has been found breeding in Scotland where the elms are yet to fully succumb to Dutch elm disease.

Two moths are particularly associated with Wych Elm: the Clouded Magpie *Abraxas sylvata* found locally throughout Britain, feeding mainly on buds and ripening seeds, and the Dusky-lemon Sallow *Xanthia gilvago*. These are also likely to be declining in numbers. But there are some gains. The Comma butterfly *Polygonia c-album* has expanded its range due to climate change and has altered its host plant from the Hop *Humulus lupulus* to include other hosts, particularly Wych Elm and Nettles *Urtica dioica*. Dutch elm disease is also having a mixed impact on beetles. The rare Longhorn Beetle *Rosalia alpina*, normally found in beech wood, has been found in dead Wych Elm trees in Poland and Germany as populations expand into dying stands of elm. Conversely, threatened beetles such as *Quedius truncicola* (whose larvae primarily feed on fly larvae in very rotten wood) and the European Red Click Beetle *Elater ferrugineus* (the larvae of which live in old hollow deciduous trees, particularly oaks but also old elm and ash) are undoubtedly declining in mainland Europe as old, diseased elms gradually rot away. Since the elm has not completely gone from our landscape, the effects on other species have been mixed.

Potentially more damaging is the impact of ash dieback caused by the fungus *Hymenoscyphus fraxineus* (see Chapter 15), which has the potential to kill around 95 per cent of Ash trees in Britain and Ireland. Ash is the second most abundant tree species in small woodland patches (less than 0.5 hectares) after the native oaks (Maskell *et al.* 2013), and the third most abundant in large areas of high forest (Forestry Commission 2003). It has been estimated that there are 2.2 million individual Ash trees outside of woodland in Great Britain, most in England (1.8 million), second only again to native oaks. Ash is the commonest standard hedgerow tree in Great Britain (Thomas 2016). It has been estimated that we have over 1.7 billion Ash trees with a trunk more than 4 cm in diameter and something like 2 billion seedlings smaller than 4 cm, both inside and outside of woodland (Jon Stokes, Tree Council). To lose 95 per cent of these would have a huge effect on British wildlife, more so if the Emerald Ash Borer *Agrilus planipennis* arrives in Britain (Chapter 15), which would remove the remaining 5 per cent.

What species would be lost if the Ash tree largely disappears? Southwood (1961) recorded just 41 species of insects associated with Ash and concluded that Ash, along with Hazel, Holly *Ilex aquifolium* and Yew *Taxus baccata*, have 'remarkably few associated insects'. However, a more recent and wider review found 1,058 species associated with Ash or Ash woodland, including 12 birds, 55 mammals, 78 vascular plants, 58 bryophytes, 68 fungi, 239 invertebrates and 548 lichen species (around a quarter of all British lichens), many of which are threatened or endangered (Mitchell 2014a). For example, 31 per cent of the lichens (169 species) are threatened or endangered (Woods & Coppins 2012). Of the species identified by Mitchell and colleagues, 44 are considered to live only on Ash trees (living and dead), including 4 lichens, 11 fungi and 19 invertebrate species, and would be likely to become locally extinct. Another 62 species are 'highly associated' with Ash, including 13 lichens, 19 fungi, 6 bryophytes and 24 invertebrates, which would be likely to decline in number if they can't find a home on other surrounding tree species. In conclusion, ash dieback inevitably means that species living on Ash trees will be increasingly threatened or lost as the Ash declines, despite some moving to live on other tree species.

Comparisons have been made between Ash and other tree species on which the species associated with Ash could live. This considers their ecological similarity in terms of such things as tree height, bark pH, type of fruit and the habitat that the trees create (such as readiness of decomposition of the litter type). The conclusion is that no one tree species could ecologically replace Ash in the British landscape. A mixture of other tree species could partially make up for the loss of Ash, such as Alder *Alnus glutinosa*, Small-leaved Lime *Tilia cordata*, Rowan *Sorbus aucuparia*, and especially our native oaks *Quercus robur*/*Q. petraea*, Beech and, perhaps surprisingly, the non-native Sycamore. Indeed, Scottish Natural Heritage has taken the quite radical step of deciding to 'accept sycamore (a non-native tree species) within [ash woodland], as it shares many of the functions of ash' (see https://www.nature.scot/landscapes-and-habitats/habitat-types/woodland-habitats/ashwood).

Sycamore becomes even more important, since these Scottish Ash woodlands also contain Wych Elm. Dutch elm disease has been moving north in Great Britain and at the time of writing is causing the first loss of elm trees in southern and eastern Scotland. In time it is highly likely that this will carry on into the north and west, causing the loss of these trees. In many ways, Sycamore is also ecologically the best likely replacement for Wych Elm, and it readily moves into the space left by Wych Elm in the canopy of mixed lowland woodland (Thomas *et al.* 2018). For example, in one study in central Germany, Wych Elm made up 27 per cent of stems before declining rapidly through Dutch elm disease, at which point Sycamore made up 81 per cent of the seedlings (around 52,000 seedlings per hectare), whereas Wych Elm contributed just 3 per cent (Hüppe & Röhrig 1996).

SHOULD WE BE PLANTING MORE TREES?

As has been seen in this chapter, trees have tremendous value to us and to wildlife. This has understandably led many people and organisations to plant trees by the million. But views on whether we *should* be planting trees have become very polarised – some suggesting that it is the most important way to improve our environment and sense of wellbeing, and some suggesting that it is not just a waste of time and money but positively detrimental. As in many debates, good points are being made at both ends of the spectrum.

Planted trees are often difficult to establish, the more so the bigger they are. Richard Mabey pointed out in 2019 that planted trees need a lot of care, and we are in danger of creating needy 'arboreal pets' rather than creating a self-maintaining wooded area. Planted trees *do* need care; the Arboricultural Association is an advocate that if trees are planted, then they need to be looked after (Fig. 28). Planting also raises the question of where they would all go. There is conflict between land to be used for farming, conservation and tree planting. Sometimes, to again quote Richard Mabey, 'poor quality land' apparently suitable for tree planting is often prime conservation land. Oliver Rackham has also bemoaned that 'Too much attention, and too much money goes into the automatic and unintelligent planting of trees' (Rackham 1986, p. 29). He also comments that near his house, Ash trees were planted 'where there are already perfectly good ash saplings'. It is true that there are a number of cases where planting trees is not necessary and that all is needed is to protect a site from grazers and let trees establish themselves and grow (Fig. 29). This after all is what happens naturally around the world where conditions allow trees to grow. Nature abhors a vacuum, and trees readily plant themselves even after major events such as forest fires (Fig. 30).

There is, of course, the question of whether we should be planting trees to solve our climate change problem by soaking up carbon dioxide from the atmosphere – this is considered in the next section.

FIG 28. Planted trees need a lot of care, as this Arboricultural Association poster highlights.

Sometimes, however, there is a need or desire to establish trees different to those that are already in the area for conservation value; for example, such as planting elm in new places to foster the White-letter Hairstreak butterfly. In more urban areas and gardens, specific varieties or sizes of tree may be planted for aesthetic or ornamental value (Fig. 26). Planting may also be necessary in a harsh environment where small self-sown seedlings would face difficult odds – such as the verge of a road and in busy urban areas. Planting trees is also a superb way of getting people involved in their surroundings, taking ownership of 'their' trees as described above. The bottom line is that there are many very good conservation

FIG 29. Was the planting of trees necessary? (a) The green planting tube holds a dead birch seedling that was planted several years before the photograph was taken. In the meantime, 'volunteer' Downy Birch *Betula pubescens* seedlings have appeared and are doing far better than the planted seedling. In this case planting was not needed. (b) Here, young Common Lime *Tilia × europaea* and Silver Birch *Betula pendula* seedlings have been planted under a canopy of oak. The Lime seedlings are fairly shade tolerant and will likely survive, but the Birch, being shade intolerant, will have a harder time. The Lime planting will diversify this amenity woodland, but the Birch probably did not need planting.

FIG 30.
Self-sown
Lodgepole Pine
Pinus contorta
growing up
below adult
trees killed by
a fire 13 years
before the
photograph
was taken. This
slowness of
regeneration
can prompt
people to
want to plant.
Vermilion
Pass, Kootenay
National Park,
Canadian
Rockies.

and social reasons for planting trees, but, if it is feasible, then letting nature do its own thing will result in a stronger set of trees, better adapted to the growing conditions.

There are examples where mass planting of trees on a large scale has proved useful. For example, during the dust bowl of the American prairies in the 1930s, President Franklin Roosevelt hired men left unemployed by the Great Depression in the Civil Conservation Corps. They planted 220 million trees as shelterbelts in a successful ploy to reduce soil erosion. By the time the corps was disbanded in the 1940s, they had planted an estimated 3 billion trees across America (Williams 2008). It is also worth pointing out, however, that planting the wrong species in the wrong place can cause problems, especially if they are invasive, as is the case of Lodgepole Pine *Pinus contorta* in New Zealand (Fig. 31).

FIG 31. An example of a tree that should not have been planted where it was. (a) Lodgepole Pine *Pinus contorta* was introduced from western North America into New Zealand in 1927 as a timber tree. Pines are not native to the southern hemisphere. Unfortunately, Lodgepole Pine is so much at home that by the 1950s seedlings had rapidly spread into nearby Tongariro National Park (b) and were taking over the park's subalpine and grassland habitats, since they can tolerate conditions up to 1,800 m altitude, close to the maximum altitude that plants can grow here.

OXYGEN PRODUCTION, CARBON SEQUESTRATION
AND CLIMATE CHANGE

Oxygen

It is often quoted that a large deciduous tree like a beech will provide enough oxygen for 10 people each year (Thomas 2014). Based on such figures, it has been calculated that urban trees in the USA produce 61 million tonnes of oxygen per year, enough for two-thirds of the American population. So what would happen to oxygen levels in the atmosphere if we removed all of the world's forests – would we rapidly asphyxiate? Are forests the 'lungs of the Earth'? First of all it is worth pointing out that around half of the world's photosynthesis and thus oxygen production is carried out by phytoplankton in the oceans, so we're not quite so dependent upon trees as we might think. Moreover, an insightful article by Duursma and Boisson in 1993 pointed out that the worldwide production of oxygen since photosynthesis began 1.8 billion years ago, is 18 Pt (peta tonne, 1 with 15 noughts after it, or quadrillion tonnes), which is 15 times what the atmosphere holds; currently the atmosphere is 21 per cent oxygen, equivalent to 1.2 Pt of oxygen. Much of the excess oxygen has been used by decomposers in rotting dead material. In effect, what keeps oxygen in the atmosphere is the burial of organic matter in sediments, which is not decomposed as a result. Around half of this organic matter is produced on land and half in the oceans but almost all ends up buried in ocean sediments (Belcher *et al.* 2021), so this is a slow, stable cycle with a molecule of oxygen passing through an organism once every 9,000 years. In the short term, it has been calculated that each year all the world's living organisms, including humans, use just 0.011 per cent of what is in the atmosphere. Since the atmospheric stock of oxygen is so huge, it would take hundreds if not thousands of years to make much of a change to the oxygen level in the atmosphere even if all the world's forests disappeared overnight. The conclusion is that forests are important for all sorts of reasons but are not really the lungs of the world as far as oxygen is concerned.

Can we plant enough trees to cure climate change?

What about carbon dioxide? Surely here the forests are 'lungs' sequestering – capturing and storing – huge quantities of carbon from the atmosphere and helping us with climate change? Forests do indeed have great potential to help mitigate climate change (Anderegg *et al.* 2020). Much of the current emphasis is on tree planting to store carbon. The Intergovernmental Panel on Climate Change (IPCC 2018) has suggested that an extra 1 billion hectares of forest is needed to absorb enough carbon dioxide to keep global temperatures to less than 1.5 °C higher than pre-industrial levels by 2050. There is globally around 4.2 billion hectares of wooded land, so this is adding an extra quarter to what we already have. An analysis by Bastin *et al.* (2019) has suggested that there is room for another 0.9 billion hectares which would store (*sequester*) 205 Gt of carbon[1] over the next 30 years. But this calculation is not without its problems and is undoubtedly over-ambitious and very unlikely ever to be feasible. A more realistic figure is 40–100 Gt of carbon – a large amount but still only around a decade's worth of our carbon emissions (Waring *et al.* 2020).

In Britain, the Zero Carbon Britain campaign has recommended that the area of trees should be doubled by planting 3 million hectares (CAT 2013) – one and a half times the area of Wales. Various political parties have readily jumped on the tree-planting bandwagon as a partial solution to our climate change problems, including pledges such as planting 30,000 hectares of trees across the UK by 2025. This would be difficult but not unfeasible. However, as noted above, planted trees need extensive aftercare, and usually a good deal of replanting is required in later years, which adds to the cost and infrastructure needed. One cannot just plant and walk away.

Another significant problem to take into account is that planting trees into open habitats like grasslands (which is mostly what we do) causes disturbance to the soil, which triggers decomposition of the soil's organic matter and thus the release of carbon dioxide. In many cases this release is initially more than is being absorbed by the young trees, so planting trees can make climate change worse, not better, in the short term. This loss of carbon from soils can carry on for a number of years after planting, and on wet grasslands and moorlands, where there is plenty of organic matter in the soil, it can take over 40 years before more carbon is being stored in the planted area than in an area left undisturbed (Friggens *et al.* 2020). It is not just the disturbance from planting that causes this. Even grasslands that are naturally invaded by self-sown trees show some change because the tree roots alter the soil microbial community, which increases decomposition in the soil (Jackson *et al.* 2002).

Trees are also a source of methane released into the atmosphere, although just how much and the mechanisms behind the release are still being investigated (Putkinen *et al.* 2021). Methane is 80 times more effective as a greenhouse gas than carbon dioxide in the first 20 years after production, and

---

1 Calculations are nearly always done on the basis of carbon rather than carbon dioxide (1 tonne of carbon is equivalent to 3.67 tonnes of carbon dioxide).

so release of methane by planted trees will act to counter their effectiveness in solving climate change.

Putting these problems aside, in the short term – over the next few decades – tree planting can indeed make an important contribution to absorbing carbon from the atmosphere even though small trees are less effective in taking up carbon than larger trees, and plantations store 28 per cent less carbon than a natural forest (Liao *et al.* 2010) because they are simpler ecosystems. But what is the bottom line – can we plant enough trees to solve climate change? It has been calculated that if we globally do everything that is feasible to slow forest loss to ensure carbon that is stored stays stored, and plant trees where it is realistically possible, then we can expect to soak up around 12–15 per cent of our global carbon emissions. This is not inconsequential, and planting trees to store carbon is certainly worthwhile, but it is not the silver bullet that will solve our climate change problems.

There is another, longer-term, issue with planting trees that needs to be considered in the equation. In temperate woodland, around 10 tonnes of new biomass is grown in each hectare per year, which is 5 tonnes of *carbon* per hectare per year (Table 7). If we know how much carbon an area of woodland can soak up per year and we know how much carbon the UK releases per year, then it is a deceptively simple equation to work out how much extra woodland we need to solve climate change. However, there is a large flaw in this argument.

In the long term it is important to remember that the net *absorption* of carbon by a plantation or woodland actually *declines* with age until a point is reached when no extra carbon is being taken up. The reason for this is that while individual trees will keep getting bigger (as more wood is added to make the trees fatter) and store more carbon, and more carbon is stored in the soil organic matter, there comes a point when the woodland is mature and the amount of biomass held in the trees and in the soil approaches a maximum. At this point, the release of carbon by the rotting of fallen wood, leaf litter and soil carbon, added to plant respiration, matches the uptake of carbon by photosynthesis, and the woodland is effectively carbon neutral. If this were not so, the forest would keep on getting bigger and the soil would keep on getting deeper as more carbon is stored in it. In old woodlands, the release of carbon from decomposition will be approximately the same as the amount stored in new material.

The growth of new biomass is highest in the tropics (Table 7). Tropical forests cover just 10 per cent of global land area but are responsible for 35 per cent of global biomass growth per year and contain 60 per cent of terrestrial carbon (Pan *et al.* 2011). Thus, the lush, productive forests of the Amazon basin, which hold 40 per cent of tropical biomass found above ground, are thought of as the lungs of the world. In reality, however, since they are largely mature forests, they are carbon neutral, circulating huge amounts of carbon dioxide into and out of the atmosphere but not actually taking any great net amount from the atmosphere. On the positive side, they hold very large reserves of carbon. This reaches 500 tonnes of biomass per hectare, half of which is carbon (Table 7).

For businesses and communities trying to become carbon neutral by planting trees, this has important implications. Planting trees works well until the trees reach maturity, which for fast-growing poplars can be half a century or less. At this point there is no more net uptake of carbon. Hence, to keep up the momentum, another area of trees has to be planted while keeping the original area of woodland intact with all its carbon. The result is that we have to plant more woodlands in the long term just to stand still. The alternative is to fell the original area and replant it while keeping all the carbon stored in the felled timber, perhaps by making it into furniture or other structures that will not rot quickly. It has been estimated that all the UK's homes in 2009 stored 19 million tonnes of carbon (0.019 Gt) in their structural wood. This is really quite small compared to the 0.8 Gt of carbon stored in UK woodlands. Almost three-quarters of that is stored in the soil, the other quarter (c.0.2 Gt) being in the trees themselves (Read 2009). If we have to keep felling plantation trees, how do we keep all this carbon stored up? There have been all sorts of imaginative solutions suggested, including that trees from such areas should be posted down into disused coalmines where the wood will not rot and thus keep its stored carbon.

TABLE 7. Biomass and productivity of different global forest types. The figures in the first column are a measure of how much biomass there is in a hectare of forest at any one time. The figures in the second column are a measure of how much extra biomass is added to the forest in each hectare every year. Almost exactly half of the biomass and productivity is carbon. So the average amount of new carbon absorbed by a temperate forest per year is 5 tonnes per hectare.

	Biomass (t/ha)	Productivity (t/ha/yr)
Northern conifer forests	150	8
Temperate forests	200	10
Tropical forests	500	25

The effects of climate change on carbon storage

By increasing the amount of carbon dioxide in the air, we are giving trees more of the main raw ingredient for photosynthesis, so surely existing forests should be able to grow a little quicker and eventually larger. This works in a greenhouse where you can grow plants faster if you pump in extra carbon dioxide – called *carbon dioxide fertilisation*. However, most experiments out in the world's forests have shown that this may work just for a year or two, and in the longer term growth slows back to what it was before (Walker *et al.* 2020). This is most likely because other things, particularly nutrients such as nitrogen and phosphorus, become limiting and slow growth. Indeed, it has been calculated that by the end of the century the global growth of trees will be reduced by 19 per cent due to nitrogen limitation, and by 25 per cent if both nitrogen and phosphorus are considered (Wieder *et al.* 2015).

Nevertheless, at the moment, the increased temperatures in northern forests and long-range nitrogen pollution are helping otherwise mature forests to grow a little larger. Moreover, the longer growing season caused by climate change (Chapter 3) is helping forests to accumulate more carbon, simply because they have more days for growing. It has been worked out that in temperate forests in North America (and likely elsewhere as well), each extra day of spring leads to an extra carbon uptake of 45 kg of carbon per hectare of forest, and each day by which autumn is delayed adds another 98 kg of carbon per hectare – more than spring, since the tree already has a full crown of leaves (Richardson *et al.* 2012). Adding all this together, it is likely that by the 2080s the forests of North American will be locking up an additional 22 per cent more carbon than they are now (Zhu *et al.* 2018). However, this may be much less due to *sink limitations*; once the tree's storage reserves of sugars are full, extra photosynthesis does little to improve growth, since the sugars cannot be stored (Rollinson 2020).

Trees can help reduce climate change in other ways. Trees evaporate water from their leaves, which helps form clouds, which reflect light back into space, reducing the amount of heat in the atmosphere. They also release volatile organic compounds (VOCs), such as isoprene from deciduous leaves and monoterpenes from conifers, which also play a role in helping cloud formation to reflect even more light away.

Nevertheless, trees can also act to make climate change worse. It was noted in the section above that planting trees can cause more loss of carbon dioxide from soils than is absorbed by the young trees. But they can increase global warming in other ways. Trees absorb more light than grasslands and crops, due to having a lower reflectance (albedo), and the trees then release this into the atmosphere as heat. Replanting all grasslands and crop areas with trees would increase global temperature by an estimated 1.3 °C, nearly 60 per cent of that produced by doubling atmospheric carbon dioxide levels (Gibbard *et al.* 2005). Moreover, because of higher temperatures, trees are increasingly threatened with overheating which slows photosynthesis, so they soak up less carbon dioxide (Pau *et al.* 2018). On top of this, drought and resulting fires in a number of years such as 2005 and 2010 (Lewis *et al.* 2011) means that areas of the world such as the Amazon basin have been *releasing* carbon. In 2010, 2.2 Gt of carbon (equivalent to 8.1 Gt of carbon dioxide) was released, increasing by a quarter the quantity of carbon that we humans are releasing every year (8.3 Gt carbon). This raises the issue of what the future holds for our forests, since climate-induced death of forests is already occurring and could produce a 'dangerous carbon cycle feedback' (Anderegg *et al.* 2020).

Protection of forests is more important than tree planting
Woodlands and forests have a hugely important part to play in climate change. Their importance lies not so much in how much carbon they absorb per year, but how much carbon they have accumulated over millennia (Mackey *et al.* 2013). Forests cover 30 per cent of the world's land area and hold around 45 per cent of the carbon stored on land, which amounts to 2,780 Gt of carbon. This is about 3.3 times the amount already in the atmosphere (829 Gt). Carbon dioxide in the atmosphere has increased from 280 ppm (parts per million) in pre-industrial times to over 410 ppm, a current increase of over 45 per cent. If all the world's trees died and decomposed to release their carbon into the atmosphere, the atmospheric level of carbon dioxide would rise to around 1,700 ppm (more than 600 per cent above pre-industrial levels), with catastrophic effects on our world (UNEP 2009).

Such a wholesale collapse of the world's woodlands is unlikely to be so complete, but nevertheless it highlights that reducing deforestation, and stopping the decline in quality of forests, particularly in the tropics where most carbon is held (Table 7), is more important than planting new trees. Similarly, the UK's woodlands hold 213 million tonnes of carbon above and below ground, but 36 per cent of this is stored in ancient woodlands although they make up only 25 per cent of woodland area (Reid *et al.* 2021) – ancient woodlands cannot be easily replaced by planting new woodlands. There is a danger that while feeling good about planting trees, we forget about the impact of deforestation elsewhere. Carry on planting trees by all means, but our long-term survival is more dependent on reducing deforestation. There are currently around 3 trillion trees globally and yet around 15 billion are removed by us every year, and we have lost 46 per cent of our trees since human civilisation began (Crowther *et al.* 2015), something that is not in our long-term interests. Deforestation already accounts for 13 per cent of total carbon emissions (IPCC 2021). Fortunately, there is good news in the REDD+ initiative (United Nations' Reducing Emissions from Deforestation and Degradation), described in Chapter 16, which is acting to slow deforestation.

CHAPTER 3

Starting the Year:
Getting the Tree Growing

TRIGGERS FOR SPRING

As winter ends, a tree needs to know when to start its growth. It could use day length (as it does in the autumn – Chapter 10), but by itself this would be problematic, since good growing conditions start on different dates each year, so a set starting date linked to day length could mean vulnerable new growth appearing when nights are still frosty, or in an early spring, wasting valuable growing time. So the timing of bud burst is set mostly by temperature, with some influence of day length. Species intolerant of shade that tend to grow in the open, such as Ash, respond mostly to temperature, while trees that grow in denser, shaded woodlands, such as Beech, tend to rely a little more on day length (Basler & Körner 2012).

Roots do not have any special mechanisms for surviving winter – when it gets cold, they just stop growing and start again in spring when the surrounding soil warms to 5–6 °C. Even the fine roots (less than 2 mm diameter, going down to the smallest at 0.07 mm) can survive the winter, although there are some exceptions: the Walnut *Juglans regia* loses more than 90 per cent of its fine roots, which may explain its unusually slow leaf development in spring.

Shoots need it a little warmer and will stir into opening when exposed to temperatures around 10 °C and above. The lower temperature threshold for roots usually means that roots near the soil surface start growing and exploring soil for water a week or two before the shoots start appearing. Deeper, colder roots may only become warm enough to stir into action just as the shoots start, but the surface roots will be sufficient to provide water as the first shoots open.

The exact temperature needed for buds to open varies between species. The result is that species of trees and shrubs tend to produce leaves in the same order each year, spread over 2–3 months, although the whole sequence may be shifted by several weeks depending upon how warm the spring is. However, in some years there is a change in the normal order of leaf emergence due to the different emphasis that each species places on temperature and day length (Lechowicz 1984). This is seen in the old British rhyme 'Oak before Ash, in for a splash. Ash before oak, in for a soak'. This, of course, has nothing to do with how much rain will fall or indeed has fallen. The real cause is that oak is more sensitive to spring temperatures, so in warmer years, the buds open earlier than Ash. In colder years, the oak is more delayed than the Ash. If warmer years are drier, then the old rhyme makes sense. Interestingly, in Norfolk, oak came out first in 60 per cent of the years between 1750 and 1958, but the warmer springs over the last 40 years have led to oak coming out first in more than 90 per cent of these years (see http://www.naturescalendar.org.uk).

There is also some variation in the temperature needs of different individuals in the same species. In Pedunculate Oak *Quercus robur*, leaf appearance can vary by up to 20 days between individuals (Puchałka *et al.* 2017), since each tree has its internal thermostat set slightly differently. As an example of this, there are two Sycamores of a similar age that I pass regularly, and every year the same individual will just be putting out a few leaves while the other is grandly in full leaf. Age can also make a difference, since saplings are usually 1–3 weeks ahead of mature individuals of the same species. This is an obvious help to young trees growing in the shade of their parents, since the young trees receive a significant portion of their year's light supply in the early spring before the parents acquire their leaves and cast their offspring into shade.

In the tropics, where temperature is the same throughout the year, new leaves can be formed at any time of the year, usually in tune with losses to herbivores and competition with neighbours. In tropical areas with more of a seasonal climate, new leaves are produced in response to the very small changes in their environment. In dry tropical forests, many trees are deciduous and grow a new set of leaves at the start of the rainy season. The leaves appear just before the rains start and are actually induced by slight changes in temperature.

As spring progresses, bud opening starts with the terminal bud at the end of a branch and progresses back along the branch. If the tree is in the open, then buds tend to open first at the top of the crown, but in a woodland position, the branches at the bottom of the canopy open first, presumably to ensure they are not shaded while developing.

BUD BURST AND CLIMATE CHANGE

Phenology is concerned with the timing of recurring natural events such as leaf and flower production. With global warming, winters are becoming less cold, and warmer spring temperatures arrive earlier. Since temperature is important in triggering the start of growth, it is inevitable that trees and other plants are starting growth earlier in spring. In southern England, oak produced its first leaves around 2 May in 1950, but by 2000 they were coming out in the first week of April. For many temperate trees, bud burst is around 7–8 days earlier now than it was 30 years ago (Chmielewski & Rötzer 2001). The opening of flowers is also primarily governed by temperature, and many trees are showing a similar advance in their flowering phenology. Over a 15-year period from 1980 to 1995, apples, cherries and forsythia began flowering 1–3 weeks earlier (Fig. 32), and this trend is still continuing. Autumn is also getting later in terms of temperature,

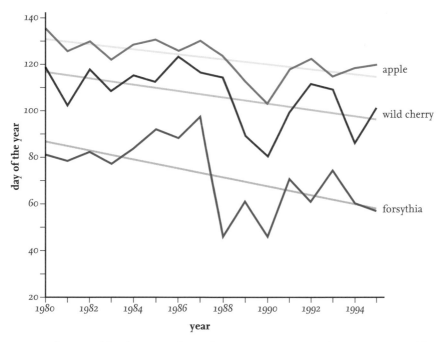

FIG 32. The time of first flowering of Forsythia *Forsythia* species in Cologne, Germany, Wild Cherry *Prunus avium* in Zurich, Switzerland, and Apple *Malus domestica* in Berlin, Germany, between 1980 and 1995. It is clear that, over time, flowering is starting earlier. From Thomas (2014) redrawn from Roetzer *et al.* (2000).

and although there are fewer data, what there are suggest that such things as leaf fall are being delayed, but not always! More on this in Chapter 10.

As described in Chapter 2, earlier leaf production does mean that temperate trees have longer growing seasons, which helps them take up more carbon to the tune of 45 kg of carbon per hectare of forest. However, the earlier leafing out and production of flowers is not without its problems. Earlier leaf flush of oaks is being matched by the earlier appearance of moth caterpillars, which are also responding to the warmer spring conditions, so leaves do not escape being eaten. However, a number of birds such as Blue Tits, Great Tits *Parus major* and Pied Flycatchers *Ficedula hypoleuca* time the appearance of their chicks to coincide with the abundance of caterpillars to feed their young. But while the leaves and caterpillars are coming earlier, the chicks are not, especially in the case of the Pied Flycatcher, which is migratory, arriving in Britain in the spring. This mismatch happens more often in the warmer south (Burgess *et al.* 2018).

Larger animals are also affected. The numbers of dormice in Britain have declined by 72 per cent between 1993 and 2014 (Goodwin *et al.* 2017). This has a number of potential causes but one of them is lack of food due to changes in spring phenology of the Hazel. Hazel is one of the first trees to flower very early in spring and is responding as other trees by producing its flowers even earlier. In the Midlands of England, flowering has moved from March and April through to January and February. Although daytime temperatures are high enough to trigger flowering this early in the year, the nights are still long and so the risk of frosts is high, killing the delicate flowers and reducing Hazel nut production, to the detriment of the autumn food supply of the dormouse.

GROWING NEW WOOD IN SPRING

Trees become larger by the buds on the branches opening to grow new branches complete with leaves and/or flowers. This elongation of branches makes the woody framework of the tree larger (called *primary growth*), helping it to grow tall to reach the sun. However, for trees to become tall, the stems need to be reinforced, and a key step in the evolution of trees was the development of *secondary growth*, allowing trees to become progressively wider and able to support the growing tree. This widening of the stem is carried out by the *vascular cambium* underneath the bark (Fig. 33). This cambium is a thin, seemingly insignificant layer of tissue, a millimetre or so thick, that coats the whole tree – trunk, branches and woody roots. Although the cambium is very thin, it is the tissue which makes the tree fatter by growing new woody cells on the inside (*secondary xylem*). This

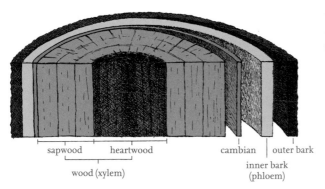

FIG 33. Cross-section of a tree. From Thomas (2014).

new wood is involved in moving water from the roots to the leaves and also in holding the tree up. Looking at a felled tree, the new wood produced each year makes up the annual growth rings. These rings are usually 2–4 mm wide in European trees, so the trunk becomes around 4–8 mm fatter each year (more on growth rates in Chapter 12).

The cambium also adds new cells to the inner bark (*secondary phloem*) which is made up of living cells that are involved in conducting sugars and other compounds around the tree. The cambium normally produces 4–10 times the number of new wood cells compared to inner bark cells, so the bark does not get anywhere near as thick as the wood. Moreover, the inner bark is used to create the outer bark, the waterproof skin of the tree, which is progressively lost from the outside of the tree, ensuring the whole bark stays comparatively thin – this is discussed further in Chapter 8.

At the start of winter, the cambium becomes dormant and is not capable of producing new cells even if favourable conditions are given. This coincides with the start of leaf fall in deciduous species. The cambium is said to be *at rest* and, like the buds, will not respond to warmth until the chilling requirement has been met, at which point the cambium is *quiescent* and is capable of starting cell division as soon as conditions are suitable. As temperatures become warmer and the buds begin to open, a hormonal signal is sent from the new shoots to the cambium to start producing new xylem and phloem. The signal is a hormonal cocktail made up primarily of auxins, mainly indole acetic acid (IAA), with a number of other hormones such as cytokinin, abscisic acid and ethylene undoubtedly also involved (Aloni 2010). Since the hormones are coming from the opening buds, the cambium starts to produce new cells beginning at the tips of the branches, moving down the trunk and into the roots. In conifers and most broadleaf trees, inner bark (phloem) is produced first to ensure that sugars stored over winter can be moved to the expanding buds. After this, new wood (xylem) begins to form, taking several weeks to reach the base of the trunk and another 4–6 weeks to reach the end of the woody roots, depending upon soil temperature.

In some broadleaf trees, notably oak and elm, the timing of new wood production is slightly different because they conduct most of their water in the ring of wood formed at the beginning of the year (these are described as having *ring-porous wood*; the reasons for this are explained in Chapter 6). Since these need to produce a new ring of wood before the leaves open, wood formation occurs rapidly through the whole tree, up to 2–3 weeks before bud break. They can do this because the cambium is very sensitive to the very low levels of auxins produced by the buds at the very beginning of their spring development.

In the tropics, where environmental conditions are fairly constant through the year, the cambium can be active all year round and keep growing wood constantly, stopped only by extreme events such as droughts or floods or by some internal signal that is not related to the time of year. In these cases, there may be no growth rings in the wood, and any rings that are produced are unlikely to be annual as they are in temperate areas. This, of course, makes the ageing of tropical trees more difficult.

Root and stem pressure

In late winter and early spring, between the thawing of soil and the opening of buds, a positive xylem pressure can develop in a number of trees, including birches, maples and walnuts. In birches, this positive pressure seems to originate mainly in the roots and is created by pumping sugars into the vessels, or tubes, of the xylem. This causes water to be pulled into the vessels by osmosis, creating a positive pressure which squeezes water up through the vessels of the roots and into the stems (Hölttä *et al.* 2020). This may also be supplemented by the active transport of water itself into the vessels (Wegner 2014). In birches this results in a pressure of 0.02 MPa (0.2 atmospheres) and can move water around 5 m up into the stems. This pressure is useful in refilling vessels blocked by winter-induced gas bubbles and thus restoring hydraulic conductivity before the start of the growing season (Schenk *et al.* 2020) – see Chapter 6 for more details. But it is also useful in getting sugars stored in the roots up to the rest of the tree. Sugars are normally transported just in the phloem, but by putting some extra sugar into the xylem in early spring, the supply to fund the spring growth is increased.

In maples and walnuts, the internal positive pressure is generated within the stem rather than in the roots. This was discovered because a cut stump shows no sap exudate from the intact roots while cut stems can show abundant sap exudate. But in reality, many maples can produce both stem and root pressure. Stem flow is generally dependent upon sunny days above 4 °C with hard frosts at night and temperatures below −4 °C, but there is still some debate as to exactly how stem pressure works. In Sugar Maple *Acer saccharum*, this could be entirely a physical process where gases in the wood fibres around the water-filled vessels are compressed on cold nights as ice builds up inside the fibre walls, giving the gas less room. In the morning, the gas warms and expands, pushing the melting water back into the vessels under pressure (Cirelli *et al.* 2008). Others have suggested that creating stem pressure is more an active process requiring energy and works in a similar way to the development of root pressure (Améglio *et al.* 2001). However it is created, xylem pressure in maples can be high, reaching 0.1–0.3 Mpa, enough to move water 10–20 m up a tree. In addition to refilling the vessels, this has the benefit of moving extra sugars even higher within the tree than root pressure alone.

If a hole is made into the xylem of a birch, maple or walnut (favourite of which is the Butternut *Juglans cinerea* of New England and Canada), the sugary sap (2–3 per cent sugar in maples) can be collected and boiled down to produce syrup of around 60 per cent sugar (Fig. 34). A birch may yield 20–100 litres and a sugar maple 50–75 litres during the season. Both make good syrup but that from birch tends to be very bitter.

FIG 34. Maple sap being boiled over fire to concentrate the sugar. As it loses water it is moved to progressively smaller pots until it arrives at the one on the far left (covered by a board) as maple syrup. This was a re-enactment of 'pioneer days' in New Brunswick, Canada – more modern methods, such as reverse osmosis, are now used to remove the excess water.

DEFENDING THE YOUNG SHOOTS FROM HERBIVORES

Young shoots are a sitting target for being eaten because they are nutritious and have not yet become tough by acquiring the mass of less palatable, lignified tissue. Think of the difference between eating a succulent young lettuce leaf compared to an older tougher one from the outside of the plant. Consequently, over the lifetime of a leaf, it has a 60–80 per cent chance of being eaten while still young and tender but is relatively safe thereafter. To try and avoid this loss, large plants like trees have evolved to mature leaves quickly and produce them synchronously across the crown to minimise damage by swamping herbivores with too much to eat. In many trees it is also worthwhile to add extra protection to young growth. This can be in the form of extensive hairs coating young twigs and leaves, which fall away as the shoot matures. These hairs can be a physical deterrent by making it more difficult for an aphid or caterpillar to reach the living tissue, especially if these hairs are dense and long or branched. But the hairs can also contain chemical deterrents, such as the stinging, nettle-like hairs found on young leaves of stinging trees, *Dendrocnide* species, found in the rainforests of Southeast Asia and northern Australia. These trees belong to the nettle family (Urticaceae) and in humans are renowned for causing pain that, while no worse than that of a European Nettle, lasts for much longer, sometimes for months. They are also known to have killed dogs, horses and even humans (Schmitt *et al.* 2013).

Young leaves can also have a thickened margin, as seen in Holly (Fig. 35). If you feel the pale edge of a holly leaf between the prickles, it is as if a thin rope has been added around the perimeter. This acts as a defence against most caterpillars, which start feeding at the leaf edge. It raises the question of why other plants don't do the same. The answer lies in a cost-benefit analysis. If it costs more to grow a thickened margin than it costs to grow new leaves lost to caterpillars, then there will be no natural selection for the margin. In the case of Holly, it is evergreen, so the investment works over the seven-year lifespan of an average leaf.

In the arms race of plant defence, some herbivores will inevitably overcome the defences. Even the powerful chemicals of the stinging trees are not proof against all herbivores, since the leaves are often seen with holes; this is to the advantage of the insects, since they have a food source without competition. In Holly, this is seen by the presence of Leaf Miners *Phytomyza ilicis*. The eggs of this fly are laid directly into the leaf tissue, avoiding the thickened margin, and the larvae feed within the leaf, leaving their distinctive 'mines' as they move. Other long-lived evergreen species, particularly in the tropics, tend to go a different

FIG 35. Leaves of Holly *Ilex aquifolium* with spines that deter larger herbivores such as deer and a thickened, pale leaf margin that deters leaf-feeding insects.

route and use chemical defences, described in Chapter 8, perhaps because the threat of tissue loss is greater.

Protecting the developing leaves

As young leaves expand, they develop into the main photosynthetic organs of the tree. For this they need chlorophyll. If you look at freshly opened tree leaves, they are often quite pale green or even pink deepening to red in maples and northern birches and even brown in cherries, beeches, poplars, *Catalpa* species and the Katsura tree *Cercidiphyllum japonicum* and Tree-of-heaven *Ailanthus altissima* (Fig. 36). Delaying putting expensive chlorophyll into leaves until the leaf is fully expanded and ready to photosynthesise saves it being wasted in leaves that are eaten. That explains pale leaves but not the expense of producing other pigments, predominantly red anthocyanins and yellow carotenoids, for highly coloured leaves. Part of the answer is that these pigments help prevent damage to young tissues by ultraviolet light. Certainly, young birch leaves in the long hours of sunshine in the Arctic tend to be red, and evergreen trees in hot, dry Mediterranean summers protect their leaves with yellow pigments against high light levels at midday and when it is too dry for photosynthesis. Another part of the answer is subterfuge. Nathaniel Dominy and colleagues (Dominy *et al.* 2002) noted that leaves on tropical trees were just as red whether they were at the top of the crown in full sunlight or were in deep shade lower in the canopy.

FIG 36. (a) Young leaves of this Henry's Lime *Tilia henryana*, named after the Irish plantsman Augustine Henry, and native to China, show the pink colour of the youngest leaves before chlorophyll is added. It is a stunningly attractive lime with the bristle-like hairs that fringe the edge of the leaf. (b) Colourful young leaves can also be selected for, such as in this unusual variety of Norway Spruce *Picea abies* 'Smultron', which has bright red young leaves that turn green when mature.

They concluded that the red pigments help to hide the leaves from insects. Most insects do not see into the longer wavelength light at the red end of the spectrum and so to them red leaves look very dark or possibly dead and so unattractive. This may also be important in the autumn colours of trees (Chapter 10).

Some trees produce such an abundance of red anthocyanins throughout the year that it is sufficient to mask the green chlorophyll, resulting in the copper-coloured leaves of the Copper Beech and Sycamore *Acer pseudoplatanus*

'Purpureum'. Anthocyanin production requires light, so the reddest leaves are on the outside of the canopy and the inner shaded ones can be almost a normal green. In trees such as the Red Hazel *Corylus avellana* 'Red Majestic', the red colour gradually fades over the summer, resulting in ordinary green leaves. The covering up of the chlorophyll in these red or copper trees does not appear to reduce growth.

Leaves can also be protected from herbivores by staying tightly rolled up until just before they start being used. This is seen particularly in tropical rainforests where many trees keep their leaves tightly folded or rolled up until they are more than half their final length. This does come at the cost of lost photosynthesis, but since these leaves tend to be long-lived and evergreen, the small loss at the beginning of their lives is outweighed by avoiding loss of leaf area at the vulnerable young stage. For a similar reason, palms keep new leaves tightly rolled. They have very few but large leaves so each one is important, and any small loss in early photosynthesis is outweighed by survival.

BEGINNING PHOTOSYNTHESIS

The chlorophyll in leaves is contained within small chloroplasts just 4–6 μm (thousandths of a millimetre) long. Chloroplasts are concentrated inside the cells just underneath the top skin, or epidermis, of the leaf. These cells (forming the *palisade mesophyll*) are at the top of the leaf, exposing the chloroplasts to as much light as possible. A study on Aspen *Populus tremula* found that an average leaf has around 30 million cells that each hold 40 chloroplasts, giving the whole tree an estimated 10^{15} chloroplasts, or a thousand, million, million chloroplasts (Keskitalo *et al.* 2005). The result is that each square millimetre of a tree leaf has up to 400,000 chloroplasts in it, giving a large broadleaf tree a total area exposed to the light of more than 350 km²! Is it any wonder that although forests cover around 30–35 per cent of the world's land surface they are responsible for 65 per cent of terrestrial productivity amounting to 150–175 Gt of carbon removed from the atmosphere each year (Welp *et al.* 2011), the equivalent of 550–640 Gt of carbon dioxide? (Chapter 2 discusses how this helps us with climate change.)

The raw material of photosynthesis is carbon dioxide and this needs to get into the leaf. The problem is that the outside of the leaf is covered by a waterproof skin. The cells of the outer epidermal skin secrete a waxy layer to form an almost airtight cuticle, designed to keep the inside of the leaf from drying out. So, to get carbon dioxide into the leaf there need to be holes through the cuticle. These are the stomata (*stoma*, singular). Stomata in tree leaves are similar in size and number to other plants, typically around 0.01-0.03 mm long and one-third that across, with densities usually in the range of 4,000–35,000 stomata per square centimetre (cm^2), although it can be as low as 1,400/cm^2 in larches to over 100,000/cm^2 in Scarlet Oak *Quercus coccinea*. Since the stomata are individually so small, even with such high numbers they still take up less than 1 per cent of the leaf surface area.

Each stoma has two specialised, sausage-shaped 'guard cells' along the edges, which can swell and shrink, allowing the stomata to be open or closed. The reason this is necessary is that as well as allowing the movement of gases in and out of the leaf, stomata let out large quantities of water vapour, so much so that more than 95 per cent of the water taken up by the roots is lost through the stomata. While this water loss is useful in carrying nutrients dissolved in the water to the top of the tree, and keeping it cool by evaporation, it is risky for the tree, especially during dry periods. The ability to close the stomata and reduce water loss when water is short is a valuable asset. It also means that the stomata can close at night when carbon dioxide is not needed and so save even more water.

The internal plumbing of the leaf is contained in the leaf veins. Within these veins, the xylem tubes conduct water from the roots to where it is needed in the leaves. Water movement is discussed in more detail in Chapter 6, but it is worth saying here that the density of veins delivering water in a leaf is finely balanced with the need for water as determined by stomatal density. Too few veins per stoma would leave the leaf short of water, and so the stomata would be closed more often to stop the leaf dehydrating, which would reduce the amount of carbon dioxide getting in, slowing photosynthesis. Too many veins would seem to be a better gamble but would mean extra costs of growing these veins, which has to be paid for by the leaf, and less room for photosynthetic cells so there would be less sugar left over to export to the rest of the tree. A fine balance is needed.

To reduce water loss, most trees have stomata only on the underside of the leaves to keep them away from the direct heat of the sun. However, trees habitually found in damp areas, such as poplars, have stomata on both leaf surfaces – this helps carbon dioxide uptake, and the extra water loss is not usually a problem, since it is plentiful. Stomata are usually spread evenly over the surface of the leaf but this is not always so. In Ash, stomata are most plentiful in the middle of the underside, while in Horse-chestnut they are predominantly around the edge of the broadest parts of the leaflets. This is tied to the physics of gas exchange between flat leaf surfaces of different-sized leaves and the atmosphere.

Once inside the leaf, carbon dioxide dissolves in the water bathing the cells and diffuses into the cells to the chloroplasts. As photosynthesis progresses, the carbon dioxide is combined with the hydrogen produced from splitting water molecules to create sugars. The oxygen produced as a by-product of splitting water diffuses out of the stomata into the atmosphere (just how much is produced was discussed in Chapter 2). The sugars are transported through the phloem tubes in the veins to the rest of the tree to be used in growth and respiration, or to be stored for later. In a 14-year-old Scots Pine in Sweden (2.8 m tall with a trunk 6.2 cm in diameter), the annual sugar production was estimated at around 4 kg by Ågren *et al.* (1980). Of this, 10 per cent was used in respiration and 90 per cent was used in growth – 17 per cent in growing new wood, 17 per cent in growing new needles and a huge 57 per cent in growing fine roots and their associated mycorrhizas (see Chapter 14), plus a small amount (around 9 per cent) put into storage in case of need. These sugars are moved through the phloem of the leaf veins and inner bark, which, unlike water movement in the xylem, takes energy. But the benefit is that sugars can move both ways through the phloem, away from the leaves when photosynthesis is going well but also towards the leaves if they are deeply shaded or otherwise impaired, and towards buds in the spring.

Water shortages and photosynthesis: CAM and C_4

Water loss from leaves can be controlled by closing the stomata. But even greater water savings can be made if carbon dioxide is taken up at night, when evaporation is at its lowest, and the stored carbon dioxide is used during the day, when the stomata can be kept tightly closed. This process is called *Crassulacean Acid Metabolism*, or CAM, so named because it was first found in the succulent family Crassulaceae, and the carbon dioxide is stored inside the plant as organic acids, mostly malic acid. These acids are broken down in the day to release carbon dioxide inside the leaves, allowing photosynthesis to happen. CAM is very common in desert plants and succulents but rare in trees and is known only from two tropical genera: *Euphorbia* (spurges), particularly common in Hawaii, and *Clusia*, a genus of small trees and vines found from Florida down through South America. Even in these few trees, the conversion to CAM is not complete, since they may use CAM in the dry season and revert to normal photosynthesis during the wet season. Seeds of *Clusia* usually germinate in the canopy of a host tree, and during this phase they can be short of water until their roots reach the ground, and so seedlings use CAM all the time (Ball *et al.* 1991). To make life even more interesting, different leaves on a tree, and possibly even parts of the same leaf, can undergo normal and CAM photosynthesis at the same time.

Another variation is C_4 *photosynthesis* (as opposed to the normal C_3 photosynthesis). The names come from how many carbon atoms are in the compounds made by photosynthesis. In C_4 plants the internal structure of the leaf is modified, allowing carbon dioxide to be very strongly absorbed even at low concentrations so that photosynthesis can proceed at a high rate when the stomata are nearly closed, hence reducing water loss in hot, dry environments. Around 3 per cent of the world's plants are C_4, and they are particularly common amongst tropical grasses, but this is even rarer in trees than CAM and found only in a few desert shrubs and a few trees in the families of Chenopodiaceae and Polygonaceae in the Middle East. It is most common in the *Euphorbia* trees of Hawaii mentioned above that can be CAM, C_3 or C_4, depending upon the growing conditions (Young *et al.* 2020). Just why CAM and C_4 are so rare in trees is still open to speculation.

Photosynthesis outside leaves

Some trees do away with leaves completely or almost completely. Since the flat leaf blade is the part that loses most water, trees in dry areas, such as many acacias, have lost the leaf blade, and the petiole, or leaf stalk, is flattened out to make a leaf-like organ (a *phyllode*) that is green and photosynthesises, but being thicker, it is better at resisting water loss (Fig. 37a). Other trees and shrubs have leaves when water is readily available, but when conditions become drier, they lose the leaves and use just their green photosynthetic branches. This is seen in the brooms *Cytisus* of Europe, the Creosote-bush *Larrea tridentata* (Fig. 37b) and even the larger trees of North American deserts (Fig. 38). In desert shrubs up to half of the annual sugar production can come from the green branches. As well as keeping the shrub alive, this gives a supply of sugar that can help in growing new leaves when conditions improve, supplementing stored sugars (Saveyn *et al.* 2010).

Even temperate deciduous trees that have normal leaves can use stem photosynthesis to gain extra sugars, especially during the shoulder seasons of spring and autumn when there are no leaves but it is still warm. In the European Spindle *Euonymus europaeus*, up to 17 per cent of the annual production of sugars can come from the bark (Thomas *et al.* 2011). In American Aspen *Populus tremuloides*, the chlorophyll content of one-year-old twigs can be similar to that in leaves (Aschan & Pfanz 2001), and stem photosynthesis can increase the growth of stems by 30 per cent (Bloemen *et al.* 2016). It is also worth pointing out that flowers and young fruits can produce more than half of their own sugars from photosynthesis in their green parts.

Other trees have permanently abandoned photosynthetic leaves. This is seen in the she-oaks *Casuarina* species of the southern hemisphere. Here, all the

FIG 37. Losing leaves. (a) Leaves of the Australian Coastal Wattle *Acacia sophorae* have lost the leaf blade and are reduced to just a broad, flattened petiole (a phyllode). (b) The desert shrub Creosote-bush *Larrea tridentata* will shed the whole leaf if conditions become too dry. (a) Barren Joey Head near Sydney, Australia, and (b) Death Valley, California, USA.

leaves are reduced to toothed sheaths in whorls around the twigs, creating trees which look like giant, branched horsetails. A similar reduction of leaves to small brown scales is found in the Japanese Umbrella Pine *Sciadopitys verticillata* and

FIG 38. Blue Palo Verde *Parkinsonia florida* (previously *Cercidium floridum*), a desert tree which is drought-deciduous, losing its leaves in dry periods but still capable of photosynthesis using the conspicuous green branches and trunk. Palo Verde means 'green stick' in Spanish. Sonoran Desert, Arizona, USA.

the celery-topped pines *Phyllocladus* species native to the southern hemisphere (Fig. 39a, b). In all these cases photosynthesis is carried out by the green branches. These are not as efficient at producing sugars as flat leaf blades, but the branches are better at holding water and it is a compromise that works well in areas that have long dry periods. In some cases, photosynthesis is increased by the branches becoming flattened so they can catch more light. The celery-topped pines (Fig 39b) are superb examples, with flattened stems called either *phylloclades, cladophylls* or *cladodes*, all meaning essentially the same thing. To us in Europe these may look rather exotic, but phylloclades can be found in some of our own woody plants, such as the Butcher's-broom *Ruscus aculeatus* (Thomas & Mukassabi 2014) native to southern England and the Mediterranean (Fig. 39c). Accidental phylloclades can

FIG 39.
Photosynthesis in modified stems. (a) Japanese Umbrella Pine *Sciadopitys verticillata*, where the leaves have been reduced to small scales and photosynthesis occurs in the cylindrical green branches, looking like the spokes of an umbrella. (b) A branch from Celery-topped Pine *Phyllocladus aspleniifolius* which uses flattened branches called phylloclades for photosynthesis. These flattened stems look like miniature sticks of celery, hence the name. (c) Butcher's-broom *Ruscus aculeatus* also has phylloclades. The flowers and fruits appear to grow out of the middle of the 'leaves' and appear very strange, but really the flowers are growing on a stem just like in any other plant.
(a) Keele, Staffordshire, (b) Tongariro National Park, New Zealand, and (c) Blanes, Spain.

also be found in trees such as Holly that occasionally grow an anomalous flattened or *fasciated* stem that looks like a ribbed green ribbon with leaves. It looks strange but is harmless.

Stem photosynthesis benefits from the high carbon dioxide concentration trapped in gas and liquid under the bark – up to a concentration of 26 per cent compared to 0.4 per cent in the atmosphere (Teskey *et al.* 2008). This carbon dioxide originates from respiration within the living tissues of the stem and also from respiration in the roots and the transport of the resulting carbon dioxide up into the stem, dissolved in the water in the xylem. Some of this carbon dioxide will diffuse out of the stem through the breathing holes in the bark (lenticels), but this is slow, allowing a build-up inside the stem. Using this internal supply of carbon dioxide allows some photosynthesis to continue even in the worst drought when the stomata in the leaves are tightly shut, starving the leaves of carbon dioxide. This effectively decouples photosynthesis and water shortage (De Roo *et al.* 2020), allowing the tree to carry on with some photosynthesis until the plant eventually dies of drought.

DISPLAYING ALL THOSE LEAVES

Looking up into a tree crown in early spring before the leaves appear, it is usually possible to see that the base of the crown is made up of large 'framework' branches that hold increasingly smaller branches towards the outside of the crown, with the finest holding the leaves when they appear. Branching has obviously evolved in trees and other plant groups as a way of the plant catching more light and stealing light from shorter neighbours. There are, of course, extra supportive costs in growing a wide spreading crown, and so there is a compromise between being large to increase photosynthesis and the costs involved. Some, such as palms and tree ferns, are unbranched and increase photosynthesis by having large spreading leaves, while an oak in the open has a huge spreading woody crown with comparatively small leaves. These compromises and how they affect the shape of the trees are explored further in Chapter 11.

Leaves are costly to produce in spring and, for the tree to survive, leaves need to pay back both the cost of making and running them (respiration) and make a profit in sugars to keep the rest of the tree going. This means that leaves need to be effectively displayed to avoid self-shading. For the first leaves in spring this is fine, as they have plenty of room, but a large deciduous tree such as an oak may eventually hold up to 700,000 leaves by the end of spring, so how can a tree with no brain work out how to optimally display all these leaves?

One way to do this is to treat each leaf as a short-term investment by displaying it in full sun at the end of a branch. It will be very productive, but the high light also leads to a rapid fall in the capacity to photosynthesise (called photoinhibition) and in effect it rapidly burns out. This is despite the aid of having yellow pigments which have a protective role in shielding the chloroplasts from too high a light intensity. A problem of this strategy is that high photosynthetic ability requires high nitrogen in the leaves to supply the necessary components of photosynthesis, which makes them vulnerable to herbivores and also less robust to physical damage (Wright *et al.* 2004). There rapidly comes a point where the oldest leaves start failing or have been eaten, and the resources held would be more valuable if moved to a new leaf. At this point, the next leaf up the branch expands and takes its place. This leads to a short leaf lifespan and a quick turnover in leaves. This successive leafing out is found in alders, where individual leaves might last less than a month, compared to other trees, where a leaf survives all summer long.

Leaves in high light can last longer if they avoid too much light. This can be done by leaves that can adjust their angle to be more vertical, 'spilling' excess light. Some eucalypts are very good at this, creating largely shadeless forests, since all their leaves hang down. Another adaptation is seen in Silver Lime *Tilia tomentosa*, where in hot, dry weather the leaves turn over, exposing the white underside of the leaves, helping to reflect sunlight away to reduce heating and water loss (Fig. 40).

The alternative is to produce all the leaves more or less at once (synchronous leafing) and arrange the leaves to avoid self-shading as much as possible. One way is to produce the leaves in a *monolayer* (single layer) so that the leaves form a two-dimensional mosaic with no overlapping. Beeches and most maples take this approach to ensure most leaves get the maximum amount of light. This strategy works very well in the deep shade of the understorey where light is at a premium. Saplings of Small-leaved Lime in deep shade produce a monolayer of leaves and can stay alive for many years, growing very slowly, waiting for a gap to appear in the canopy.

Another strategy is to have multiple layers of leaves – a *multilayer* tree – but arranged to minimise self-shading. *Phyllotaxy* (the arrangement of leaves along a stem) in a spiral or opposing pairs of leaves helps put leaves into the gaps of the leaves above (Fig. 41). The petiole can vary in length – longer further down a shoot – and it can bend to put a leaf in the best position to catch light. Leaves will repeatedly go through slow movements to orientate themselves in the best light. Inevitably, leaves further down in the crown will be increasingly shaded as the leaves above them develop. This is dealt with by having sun leaves and shade leaves.

FIG 40. The leaves of Silver Lime *Tilia tomentosa* have a dense white covering of hairs on the underside of the leaf. On hot, dry days, the tree reduces water loss by (a) turning the outermost leaves over to reflect excess sunlight. (b) In sunny conditions, the leaves turn mostly on the top of the tree. (a) Keele, Staffordshire, and (b) Istanbul, Turkey.

FIG 41. The spiral arrangement of leaves along a stem helps prevent leaves from shading each other. Sequoia National Park, California, USA.

Sun leaves and shade leaves

In multilayer trees, some leaves will spend their life growing in shaded conditions, even in a fairly open crown such as in birches. Leaves in the sun will look and operate differently from those in the shade. Sun leaves perform well in bright light by being thicker, with many layers of chloroplast-rich cells. By contrast, shade leaves compensate for lower light levels by being larger in area, thinner with a thinner cuticle, darker green with more chlorophyll per gram of leaf and fewer of the shielding yellow pigments, and less lobed (Fig. 42), with half to a quarter the density of stomata. Also, the chloroplasts are concentrated in a thin layer just under the epidermis of the leaf, allowing them to work more efficiently at low light intensities.

Shade leaves are also able to open their stomata and start photosynthesis very quickly and so can make use of the brief moving sunflecks that penetrate the leaves above. Sun leaves put in this position tend to be sluggish and unable to use the sunflecks. Moreover, the high respiration rates of sun leaves require more sugars to keep them productive and they are unlikely to survive in the shade. By the same token, shade leaves cannot handle bright light for long periods and if suddenly exposed to light will fare poorly. This is readily seen when severely pruning a hedge of Hazel or Hornbeam *Carpinus betulus*, where the shade leaves exposed by cutting can be scorched and damaged by the sudden bright light conditions. A similar consequence is seen in saplings of rainforest trees suddenly exposed to light by the fall of an overhead giant.

FIG 42. Leaves from a Sweet Gum *Liquidambar styraciflua* collected from the sun leaves at top of the tree (bottom left) and progressively lower in the crown to the lowest shade leaves (bottom right). The shade leaves have a larger area and are thinner and less lobed. As importantly, it can be seen that the change from sun leaves to shade leaves is gradual such that each leaf is adapted to the amount of light it receives, and the terms 'sun leaves' and 'shade leaves' represent the two ends of a continuum. Although this is a tree we are familiar with in Europe, it also grows in high-altitude cloud forest in Honduras where this photograph was taken.

As each leaf emerges in the spring, it reacts to the conditions it experiences and may be a shade leaf or sun leaf, or somewhere along the continuum between the two. Shaded saplings may thus be all shade leaves, and a monolayer tree in the open may be all sun leaves. For many trees there is usually a mix of sun leaves on the outside of the crown passing down to full shade leaves at the inner bottom of the crown. It is important to note that individual leaves are able to change. In our pruned hedge discussed above, the shade leaves that survive will gradually become thicker and take on the role of sun leaves. Conversely, in a tree progressively shaded by competing neighbours, its sun leaves can gradually become larger and thinner, changing into shade leaves.

Shade leaves are obviously useful, and since they have a longer lifespan than sun leaves, they can dribble some sugar into the system through the whole growing season. But since they receive comparatively little light, they actually add comparatively little to the overall growth of a tree. The trunks of cherry trees are valuable for producing veneers, but this involves removing the lower branches to create a long trunk. In a study using 13-year-old Wild Cherry trees, 6–7 m tall, it was found that removal of the lowest branches to reduce the depth of the crown by a quarter did not affect height growth, and the trunk diameter growth was reduced by only 4 per cent the next year. Even removing half the crown height did not reduce height growth and reduced the diameter growth by 5 per cent in the first year and 9 per cent in the next (Springmann *et al*. 2011). This is excellent news if you are growing veneer logs in plantations and need long trunks with no branches. It also shows that the shade leaves at the bottom of the crown contribute relatively little to the growth of the tree. For trees in the wild, however, even the small contribution of shade leaves will help them in competition with their neighbours and may make all the difference between survival and death.

As branches grow longer in spring and the tree's crown expands, there will come a point where a leaf at the bottom of the tree is just not receiving enough light. Most leaves need around 20 per cent of full sunlight to reach their *compensation point*, where photosynthesis is greater than respiration and they are producing more sugar than they consume. Yet beneath large forest trees such as beeches, light levels can be down to 2–3 per cent of full sunlight and is one reason why so little grows on the floor of a beech forest. Leaves at the very bottom of the crown in late spring can thus find themselves in very deep shade and operating below the compensation point. The solution to a leaf that is consuming more sugars than it is producing is simply to shed the leaf. It is like investing on the stock exchange: if a particular company is doing poorly, it may be better to get rid of those shares quickly before they become too much of a drain. This is seen in trees as a loss of some leaves in late spring, which represents those leaves too shaded to continue and which are a drain on the system.

Loss of leaves in spring and later in the summer is not just caused by too little light but also by other things that the leaf needs being in short supply. The most obvious of these will be water (Fig. 43). Leaves can normally control water loss by closing the stomata, as described above, but if water shortage becomes critical, a tree will lose some or even all of its leaves. In effect the leaves act as *hydraulic fuses*, as it is more cost-effective to lose leaves than for the stem to suffer irreversible damage (Wolfe *et al*. 2016). In Walnut trees, for example, Tyree *et al*. (1993) found that in times of drought, the petiole of the leaf lost 87 per cent of its water conductivity while nearby stems lost only 14 per cent, so the leaf would be critically short of water, inducing the tree to shed the leaves before the stem developed a damaging lack of water. This shedding of leaves from shade or drought is a deliberate, planned process (Chapter 10) – they don't usually just wither and die as in Figure 43. The evidence for this is that the leaves usually

FIG 43. (a) California Buckeye *Aesculus californica* copes with the Mediterranean climate of its native habitat by starting growth in the wet winter, and by late spring, when water is in short supply, (b) the leaves die off and hang dead on the tree. If grown on the Californian coast, the leaves stay alive until the autumn. Foothills of the Sierra Nevada Mountains, California, USA.

FIG 44. Leaves on one branch of a Pear *Pyrus communis* going yellow due to a combination of shade and drought. In normal years the leaves on the right, shaded by a hazel, will survive, but in the year the photo was taken, the soil was also very dry and the combination of shade and closed stomata due to a lack of water meant that photosynthesis was low, and the leaves were a burden on the tree and thus shed. The leaves are going yellow as the useful components, including chlorophyll, are broken down and taken back into the tree before the shedding process starts.

go yellow before being shed, as the useful components are broken down and reabsorbed before the leaves are finally dropped (Fig. 44).

A similar case is met in evergreen trees. There comes a point where the oldest, most shaded leaves will eventually be shed. In some this happens throughout the year, often with a midsummer peak when the soil is driest, as in *Cupressus* species. Others drop their oldest leaves in one go, as seen in Holly *Ilex aquifolium* in spring and early summer.

WHY ARE TREE LEAVES DIFFERENT SIZES AND SHAPES?

Even in temperate trees, leaves come in a huge range of sizes and shapes. Then add in the even wider range from other biomes around the world and we are left with a bewildering array of different leaves. It can be difficult to pick out the reasons why any one leaf or that of a species is the shape and size it is, since many factors will affect these. But some general principles can be seen based on the conflicting needs to photosynthesise (intercepting light and taking up carbon dioxide) while at the same time conserving water.

Large leaves tend to be found in tropical areas where it is hot, wet and sunny. They are good at catching light, resulting in high sugar production. The downside is that the high light intensity will also heat the leaves, but since the environment is wet, as in rainforests, there is plenty of water to evaporate to keep the leaves cool. Evaporation of water is most effective at the edges of leaves, since a thick still layer (*boundary layer*) of moist air builds up over the main part of the leaf, which thermally insulates it. So tropical leaves in full sun tend to be highly lobed to increase the amount of edge. Large leaves also work well in more shaded tropical conditions, acting as large, thin shade leaves that do not have the problem of being overheated by the sun. In this case, the boundary layer helps the leaf by reducing heat loss, leading to a temperature 3–10 °C higher than the surrounding air, aiding photosynthesis in the cooler, moist inner regions of the forest. Inside the forest, wind speed is also lower, so the big, thin leaves are not at risk of being damaged. So why don't leaves keep on getting bigger and bigger? The answer appears to be that bigger leaves need to be supported by heavier duty branches which require more investment, until eventually the extra leaf area is not covering the cost of the bigger branches. Very large leaves would also be much more difficult to protect against wind damage, as seen in the leaves of bananas that start intact but are extensively shredded by the end of their life.

Tropical rainforest leaves can be quite small, generally becoming smaller with altitude (Fig. 45), but whatever the size, they mostly tend to be glossy with a long drip tip (Fig. 46a). These features encourage water to rapidly drain from the leaf surface, collecting at the end and forming a drop big enough to fall off. This helps reduce water sitting on the leaves and leaching minerals from the leaf and also reduces the growth of epiphytes, particularly algae, which would block light. Look ahead to Figure 217 in Chapter 15 to see what can happen if water remains. The exception is seen in leaves emerging from the top of the rainforest canopy; these do not have drip tips because they rapidly dry in the sun. Although drip tips are common in the tropics, they are not only found there; even in lower-rainfall temperate forests, drip tips can still be found in trees such as limes and birches (Fig. 46b) for the same reasons.

FIG 45. Variation in size of rainforest leaves, with my foot as scale. The two largest leaves of *Elaeagia auriculata* are several orders of magnitude larger than the small *Clusia* species leaf at the top, and these are several orders of magnitude larger than the smallest found at high altitude such as *Toxicidendron striatum* – not shown but smaller than half of the nail on my big toe. High-altitude cloud forest, Honduras.

FIG 46. (a) Tropical leaf with a drip tip (*Miconia* species) from the cloud forest of Honduras. (b) Trees of wet temperate areas, such as Henry's Lime *Tilia henryana*, can also have a drip tip.

Moving away from the tropics to the deciduous forests of Europe, conditions become drier and the leaves become smaller and usually thinner. These leaves are only on the tree for a few months (compared to the years of an evergreen tropical rainforest leaf) and in effect they are short-lived, cheap-to-build, disposable leaves that can be thin, and it does not matter if they are easily damaged. Like the rainforest leaves in the open, large-leaved temperate trees often have lobes and big teeth to create turbulence to reduce the boundary layer, making the leaf effectively smaller, aiding heat loss by evaporating water. Some go further, such as in Aspen (*Populus tremula* in Europe and *P. tremuloides* in North America), where the petiole is laterally compressed, causing the leaf to shake in the gentlest of breezes, further removing the boundary layer. This is like shaking our hands to dry them, which may account for poplars being renowned for being high water users and doing best on moist alluvial soils.

In more arid areas, such as the Mediterranean, leaves become even smaller but thicker. The thick leaf with a deep cuticle reduces water loss. The small size also perhaps counterintuitively helps reduce water loss because although the boundary layer is thinner, the air moves more easily past the leaf, and heat is lost more by convection than by water loss. Photosynthesis in these leaves can be quite poor, partly because of water shortage and partly because inside the small, thick leaves the chloroplasts have to be stacked up, so they shade each other and compete for carbon dioxide. Low sugar production means that it can take a long time to repay construction costs, often more than a year, so the leaves have to be even tougher to survive the blazing hot summer and be low palatability in order to survive for a number of years. On the positive side, smaller leaves means that there are often more of them. Since each leaf has a bud in the axil, a tree with more leaves has more buds, which gives it more options in surviving environmental extremes and damage that might kill some of the buds.

Small evergreen broadleaf leaves are also found in cool, moist areas. In the nutrient-poor soils, nitrogen in particular is in short supply, so again it takes more than one year to pay back construction costs (Midgley *et al.* 2004). Here, the leaves need to be tough to survive the cold winter. Size helps, since small evergreen leaves are also better at coping with frosts because they radiate less energy into the sky at night than larger leaves and so stay warmer (Lusk *et al.* 2018). They also give less resistance to high winds and are good at coping with water shortages, particularly in spring when the air warms but the water in the soil is still largely frozen. Small leaves are also common on wetter soils in Europe, where root growth is limited and so water and nutrients are harder to supply to the shoot, resulting in the very small leaves of heathers such as *Calluna vulgaris* and the many *Erica* species. These small leaves can be summed up by saying that a longer lifespan of evergreen leaves is linked to a high leaf weight per unit area (g/m^2) – the longer lifespan needed to pay back costs requires tough, thick leaves, and so the same area of leaf is heavier than in temperate areas, particularly in low rainfall areas (Wright *et al.* 2004).

Reduction in leaf size with inhospitable conditions reaches its limits in conifers, which have needles and scales. Needles often occur singly, but in pines

FIG 47. Scale leaves of a variety of Hinoki Cypress *Chamaecyparis obtusa* 'Tetragona Aurea', which emphasises the individual scale leaves.

they are usually in bundles of two, three or five, which fit together to make a cylinder carried on dwarf branches (however, there can be up to eight needles in a bundle or, as in the Pinyon Pine, single, cylindrical needles). A needle contains everything that a flat leaf does but is adapted to harsh dry or cold conditions by having a thick cuticle and a very low surface area for the volume of the leaf, making it very good at resisting wind and water loss and good at shedding snow. Moreover, there are comparatively few stomata, which are set deep in sunken pits often covered in wax (creating lines of small waxy dots along the needle) to further reduce water loss. In most cypresses and in a few others, such as the red pines in the Podocarpaceae (*Dacrydium* spp.), and in the Tasmanian cedars (*Athrotaxis* spp.) and Japanese Red-cedar *Cryptomeria japonica*, both in the redwood family (Taxodiaceae), the leaves are even more reduced to photosynthetic scales just a few millimetres long, which clasp the stem (Fig. 47). In some, including many junipers, the free tip of the scale is elongated to form a needle, increasing the photosynthetic area.

Compound leaves: When are they beneficial?

Compound leaves resemble small branches with a number of leaves (Fig. 48). The way to tell them apart is that a branch with leaves has a bud at the base of each leaf, whereas the only bud on a compound leaf is right at the base of the petiole. If a compound leaf resembles a small branch, why not just grow a small branch? The answer is that a compound leaf is cheaper to build than a branch with leaves, simply because the petiole and central stalk (rachis) have only to survive one year

FIG 48. Compound leaves. (a) Chinese Sumac *Rhus chinensis* (previously *R. javanica*) from China and Southeast Asia, showing a pinnate leaf made up of leaflets joined along a winged stalk, or rachis. As with simple leaves, the leaf stalk below the rachis is called the petiole. (b) Palmate compound leaves of the Red Buckeye *Aesculus pavia* native to southeastern USA on the left and Horse-chestnut *A. hippocastanum* on the right, showing stalked and unstalked leaflets, respectively. (a) Deogyusan National Park, South Korea, and (b) Thorp Perrow, Yorkshire.

and are held up mainly by water pressure. It is thus structurally less costly than a similar-sized branch full of fibrous material that is reused over a number of years.

Compound leaves are useful where rapid height growth is important. A tree that is growing rapidly can display cheap throwaway compound leaves rather than branches with leaves which would be rapidly shaded and only used for a short time. The saved energy can be invested in more rapid height growth, which could make the difference in out-competing neighbours or not. Moreover, the rachis of compound leaves can have a higher hydraulic conductivity than a branch with simple leaves, and therefore better water flow into the compound leaf, which supports higher photosynthetic capacity (Yang *et al.* 2019). This use of efficient, disposable 'branches' is why many species that invade open areas, such as tropical trees in rainforest openings, have compound leaves. They are also found in temperate trees such as Manitoba Maple *Acer negundo* and some ash trees which are native to floodplains and benefit from the ability to grow quickly after damage. In extreme cases the use of compound leaves means that a tree can grow for some years with no or few woody branches, using large compound leaves instead. In the tropics this includes mahogany trees (various species of *Swietenia*) and in temperate areas trees such as the Tree-of-heaven, Kentucky Coffeetree *Gymnocladus dioicus* and the Japanese Angelica Tree *Aralia elata*.

Some trees with simple leaves compete in the same habitats using large single leaves in place of compound leaves and, like the species above, grow few if any branches for the first few years. This includes catalpas *Catalpa* and the Foxglove-tree *Paulownia tomentosa* (Fig. 49) but not many other species, presumably because large simple leaves are more vulnerable to damage and costlier than compound leaves. Other trees have found their own answer. Poplars, for example, invade gaps using small simple leaves held on thin and inexpensive twigs that are soon shed, doing almost the same job as the petiole and rachis of a compound leaf.

FIG 49. Foxglove-tree *Paulownia tomentosa* showing the lack of branches when young. Only when the tree is many metres tall does it start to produce woody branches. This tree at Keele University was planted to celebrate a member of staff retiring, and I think they were a bit perturbed when all that was planted was a thin stick!

Compound leaves are also useful in arid areas for reducing water loss after leaf loss. Small twigs, which have a large surface area to their volume as well as thin bark, are the biggest source of water loss for the tree once leaves have fallen. Thus, in areas with very dry summers, such as savannahs, thorn forests and warm deserts, having, in effect, disposable branches in the shape of compound leaves is beneficial. This is seen in the Baobab *Adansonia digitata* of African savannahs and the horse-chestnuts *Aesculus* of various dry parts of the world that have palmate compound leaves (looking like a palm with fingers – Fig 48b). It is also seen in trees with pinnate leaves, as shown in Figure 48a, and found in mesquites *Prosopis* spp. and palo verdes *Parkinsonia* spp. of North American desert grasslands (Fig. 38), as well as in acacias of Australia and Africa. Temperate trees are unlikely to be affected by such water loss in winter, so we have fewer species with compound leaves, but Ash trees growing on dry limestone screes in central England show that compound leaves have their advantages even here.

STORAGE OF RESERVES

In deciduous trees, the first leaves in spring are grown using reserves stored in previous years. Most evergreen conifers (but not all – for example, Norway Spruce) have the advantage that the needles from previous years can grow sugars to fund the new spring growth and so are less reliant on food storage in spring. But all trees, once the leaves are up and running, will start to store food as starch or fats as a reserve for the future. In the late nineteenth century, trees were catalogued as either starch trees (most ring-porous broadleaf trees – see Chapter 6 – and a few conifers like firs and spruces) or fat trees (most diffuse-porous trees and pines), depending on what they were thought to store. In reality, although the proportions vary, most trees will store both. Starch provides quick energy, while fats are needed in the production of volatile organic compounds (VOCs) such as jasmonates (Chapter 14). Fat storage is generally favoured by low temperatures and is also particularly important in extreme conditions such as in drought-stressed trees (Hartmann & Trumbore 2016). Fats are more difficult and energy-demanding to store than carbohydrates but they have more energy per gram.

These stores are very important in the long-term survival of a tree. Stored food is required daily, since the tree needs to respire in the night when photosynthesis has stopped. It is also needed seasonally, to kick-start spring in deciduous trees but also as an insurance should anything happen to the new leaves, and over the decades, to replace any major losses of parts of the tree or to

FIG 50. Sugary sap from flower stalks of the Palmyra Palm *Borassus flabellifer* being boiled to create palm sugar. Siem Reap, Cambodia.

cope with insect epidemics removing generations of leaves. How these are used is discussed in Chapter 12.

The reserves can also act as a savings account used to fund years of abundant fruit production, especially in trees that show masting – periodic high numbers of seeds (Chapter 4). Some trees can store enormous quantities of reserves and are used by humans. For example, the Southeast Asian Sago Palm *Metroxylon sagu* stores food in a readily available cache in the centre of the stem. Trees are felled and the central pith is pulverised in water to release the large grains of starch, used as sago, which when cooked resembles frogspawn! The flower stalks of palms can be cut and the abundant sugary syrup collected to make palm sugar (Fig. 50) and a fiery alcoholic drink. In Cambodia, fruits from various palms are pressed and the sugary liquid used to make palm wine.

CHAPTER 4

Spring and Early Summer Activity: Flowers and Seedlings

I once had a student in an adult education class who was certain that most European trees did not flower. This misconception is quite understandable, partly because many of our native trees are wind-pollinated and so the flowers can be fairly inconspicuous. It is compounded by many trees being comparatively long-lived compared to animals and other plants (see Chapter 12) and so can spend their early years getting established in the canopy before giving over resources to reproduction. By this point many of the flowers are way above head height, adding to their general invisibility. Having said this, some trees do have flowers that are instantly recognisable as flowers, with all their component parts, as seen in Figure 51.

AGE OF FIRST FLOWERING

The start of reproduction in a young tree is the usual marker of the tree changing from being a juvenile to becoming mature. Many European conifers and broadleaf trees are 10–40 years old before they start flowering. Pioneer species that invade gaps, like birches and willows that need to grow and produce seeds before they are shaded out, may start earlier, even as young as five years old. By contrast, long-lived forest trees such as oaks may be closer to 60 years old before they start producing seeds, or may be approaching a century in heavily shaded trees. Exactly when a tree begins to flower is affected by many factors but, generally, the better the growing conditions the younger they begin flowering. Tropical trees start flowering at a similar age to temperate trees. For pioneer trees

FIG 51. Flower terminology using the ornamental flowering cherry *Prunus* 'Umineko' as an example. The outside of the flower is made up of five green sepals, or bud scales (collectively called the calyx) – seen opening at bottom left. Inside these are five white petals (the corolla, which with the sepals form the perianth). The female part of the flower (the carpel) is in the centre of the flower. This is composed of the green pinhead-like stigma in the centre of the flower that catches the pollen. The stigma is supported on a style, which leads down to the ovary hidden below the sepals. The ovary contains one or more ovules, which will become seeds. Around the carpel are the many male stamens. Each stamen is made up of a white 'stick' (the filament) and a tiny terminal yellow anther that produces the pollen. The centre of the 'Umineko' flower becomes pink with age.

such as Teak *Tectona grandis*, this is often less than five years, and in well-grown nursery stock, flowers can appear on trees less than six months old! For most the norm is 10–30 years old, although the Dipterocarp trees of Malaysia may delay this until they are 60 years old.

WHY DO MANY TEMPERATE TREES FLOWER IN THE SPRING?

When trees do get around to flowering (and they all must, at least in their native lands, or they would have disappeared long ago), it is noticeable in temperate areas that many flower in early spring before the leaves appear, sometimes called precocious flowering, *hysteranthy* or *proteranthy*. In the Great Lakes area of North America, 30 per cent of hardwood trees flower before the leaves appear (Gougherty & Gougherty 2018). This includes species of ashes, poplars, elms, alders, birches, and redbuds/Judas trees *Cercis* species. Some large genera such as maples *Acer* and cherries *Prunus* contain a mixture of species that flower before the leaves or with them, perhaps just because of the sheer number of species and the different habitats in which they live. Silver Maple *Acer saccharinum*, Red Maple *A. rubrum* and Box Elder *A. negundo* flower before leaf emergence, but the Sugar Maple *A. saccharum*, Moosewood *A. pensylvanicum* and Mountain Maple *A. spicatum* flower after leaf emergence. A similar story could be told about most temperate regions.

Early flowering comes with the risk of flowers or pollen being damaged by frost, but since many trees do it, there must be an advantage. A logical explanation would be that early flowerers have more time through the growing season to produce large seeds. This turns out to be true for herbaceous plants but not for trees (Bolmgren & Cowan 2008), probably because the large amount of food that can be stored in a tree makes large seeds possible whenever the flowers appear. However, early flowering may allow trees to disperse seeds earlier in the growing season, giving them more time to reach suitable habitats and maybe even germinate and become cold-hardy before the first frosts of winter. In the Great Lakes region, Gougherty & Gougherty (2018) found that species which flower before the leaves appear tended to be tolerant of low rainfall, so maybe this avoids having flowers and leaves demanding too much water at the same time. Another possible explanation is that producing flowers before the leaves may have a similar advantage that the flowers and new leaves are not competing for stored sugars at the same time.

While all these reasons may be true, the most compelling reason for early flowering is linked to wind pollination. Globally, 85 per cent of flowering plants (trees and herbaceous species) are pollinated by animals (Ollerton *et al.* 2011), but spreading pollen by wind is common in higher latitudes where there are just a few dominant tree species. Wind is very good at moving pollen long distances – hundreds of kilometres – but is very unspecific in where the pollen ends up going. This is less of a problem if there are many trees of the same species in the same area, since whichever way the wind is blowing, some pollen is likely to be shared between the trees. The downside is that it does take a lot of pollen. Many trees concentrate their flowers in dangling catkins – the name meaning 'little cat', since the catkin looks like a kitten's tail. Each catkin of a birch (Figs. 52 and 53) and Hazel produces around 5.5 and 4 million pollen grains, respectively. Conifers can also produce prodigious amounts of pollen (Fig. 54) – no wonder wind-pollinated trees are a major cause of hay fever in spring!

a
b

FIG 52. (a) Male catkins of *Betula luminifera*, native to the Himalayas and noted for its long catkins. In birches, both male and female catkins are long and pendulous. (b) Long male catkins and very short female catkins of Sweet Chestnut *Castanea sativa*.

In some trees, such as Sweet Chestnut, the female catkin is much shorter than the pollen-producing male (Fig. 52b), but the long male catkin still produces abundant pollen. With this amount of pollen flying through the air, it makes evolutionary sense for deciduous wind-pollinated trees to produce their pollen while the canopy is bare of leaves in early spring to reduce the surrounding surfaces that 'compete' with the stigmas for pollen. Other adaptations help this along. Pollen tends to be released primarily when the air is dry, to further reduce the risk of pollen sticking to wet branches and being washed out of the air by rain. Flowers also tend to be on the ends of branches, where it is windiest. Catkins generally produce their pollen from behind scales (the bracts and bracteoles in Fig. 53) that touch each other, which means that most pollen is released when the catkin is being bent by the wind, ensuring that the pollen is released only when there is a chance of it blowing further.

The pollen from wind-pollinated trees is smooth, rounded and small; 20–30 μm in diameter in broadleaves and 50–150 μm in conifers, compared to up to 300 μm – 0.3 mm – in insect-pollinated plants. While all pollen grains have

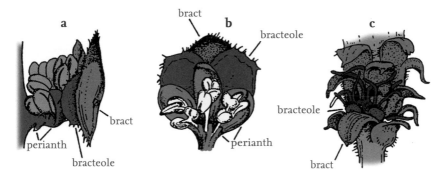

FIG 53. Flowers of Silver Birch *Betula pendula*. (a) Side view of a group of male flowers borne on a single catkin scale made up of a bract and two smaller bracteoles, which are all modified leaves. (b) A similar group seen from below. (c) Part of a female catkin showing the red stigmas. From Proctor & Yeo (1973).

the same settling rate (dropping in still air at around 3–12 cm per second), being small, smooth and round helps the pollen from wind-pollinated flowers to be blown further through the air before they drop to the ground. Small pollen, however, can be a problem in that the wind can whisk the pollen straight past the stigma – the female part of the flower that catches the pollen. Nature, of course, has overcome this problem! Part of the solution is that the stigma and the style upon which it is borne create turbulence in the air as the pollen passes, which can be enough to allow the pollen grains to fall out of the slipstream onto the waiting stigma. Leaving nothing to chance, this is made even more likely because pollen flying through the air develops a strong positive charge, much like a hand moving across nylon develops a static electricity charge. At the same time, the stigma develops a negative charge, allowing the pollen grains to be plucked electrostatically from the air by the stigma as the grains blow past, particularly in lower wind speeds created by turbulence (Bowker & Crenshaw 2007).

This explains why deciduous trees flower in early spring, but what about evergreen conifers? Conifers are gymnosperms rather than angiosperms and so do not have flowers as such – strictly speaking, they have male and female strobili (singular *strobilus*), or cones, that act as flowers. However, they still have pollen to move from the male to the female strobilus. Evergreens obviously don't have a leafless period but partly get around this problem by ensuring that the male and female strobili are held on the tips of branches as far away from the leaves as possible. Nevertheless, evergreen conifers have less to gain from reproducing in spring, and while pines produce strobili in spring, others such as the cedars

Cedrus produce them in the autumn or, like the northern incense cedars *Calocedrus* and the Coastal Redwood *Sequoia sempervirens*, reproduce in winter when the wind speeds are highest.

Instead of a flower with a stigma to catch pollen, the female strobilus has the unfertilised seeds buried deep between a series of scales (which will become the scales of the cone). Being wind-pollinated, conifers produce abundant pollen (Fig. 54), and most conifers, including conifer relatives such as the Maidenhair Tree and cycads, produce a sticky 'pollination drop' at the tip of each scale, which traps the pollen as it flies past (Fig. 55). As the drop dries, the pollen is pulled down towards the young unfertilised seeds. This is not just a passive process, since the drop will dry much quicker once pollen is added, typically within 10 minutes with added pollen compared to days or even weeks with no pollen. Some conifers do not have pollination drops, including firs, cedars, larches, hemlocks *Tsuga*, the southern kauris *Agathis* and monkey puzzles *Araucaria*. Instead, these have slippery scales that funnel the falling pollen into the depths of the cones, safe from being blown away by the wind.

FIG 54. A cloud of pollen released from a Silver Fir *Abies alba* after someone brushed past the tree. The many yellow male strobili can be seen along with some taller, purple female strobili on the left.

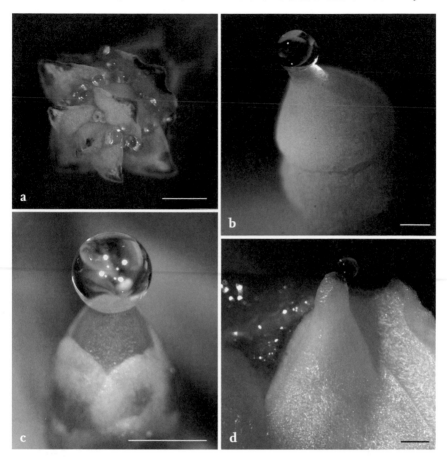

FIG 55. Pollination drops in conifers. (a) Lawson Cypress *Chamaecyparis lawsoniana*, (b) Maidenhair Tree *Ginkgo biloba*, (c) a hybrid Yew *Taxus* × *media*, and (d) Douglas Fir *Pseudotsuga menziesii*. Scale bars are 1 mm long. From Coulter *et al.* (2012).

INSECT POLLINATION

In tree species that tend to be scattered at low density through a woodland, wind pollination does not work as well, since the chance of wind blowing pollen from one individual to another can be very low, especially in the low wind speeds inside a dense woodland. This is particularly the case in tropical rainforests made up of huge numbers of tree species such that 30–65 per cent of tree species grow at densities of less than one individual per hectare. In these cases, pollination by insects attracted by sight and scent tends to dominate. Insects are abundant – globally, three-quarters of all animals are insects – and they are well suited to the role. Insect pollination is clearly a very effective mechanism since, for example, in Costa Rican rainforest 90 per cent of the trees are insect-pollinated. In temperate areas, insect pollination is indeed found in trees that tend to grow as solitary, widely spread individuals such as whitebeams *Sorbus*, hawthorns *Crataegus* and Crab Apple *Malus sylvestris*.

Using insects for pollination means that flowering does not need to be concentrated in spring, and flowers can be produced at any time in the growing season, with the advantage that they are open in weather that is likely to be better for pollinators than the fickle frosts of spring. Spreading their flowering through the growing season also has the advantage that it reduces competition for pollinators. However, flowering does need to be synchronous within a species to ensure that pollen can be spread between trees – cross-pollination (see below). In temperate areas there are large enough seasonal changes in temperature and other climatic cues to ensure this synchronicity within a species. In the tropics, however, with little variation in climate through the year, this is more of a problem and has to rely on quite subtle changes. For example, in the Amazon rainforests on the equator, day length is the same throughout the year. However, the time of sunrise and sunset moves forwards and backwards by 30 minutes throughout the year due to the elliptic path of the Earth around the sun and the tilt of the Earth's axis relative to its orbit. Although this translates into a change of just 5–7 minutes over a 20-day period, many trees use this minute signal as a trigger for flowering. This results in flowering being concentrated at the spring and autumn equinoxes, when rate of change is greatest, although different species tend to be staggered from each other. It is probably the time of sunset that is crucial (Borchert *et al.* 2005). To detect these small changes obviously needs a very sensitive internal clock, which links to the tree's internal calendar, described in the next chapter.

Insect pollination can be problematic for trees due to their often large size. Trees need to provide enough reward in their flowers to attract pollinators but not so large that the insects can do all their foraging on one tree, which would result in mainly self-pollinated seeds. Some tropical trees solve this problem by having only some parts of the crown in flower at any one time, even just a few flowers open at once, while other parts are in bud or in fruit. This gives the appearance of a series of small trees and will encourage insects to keep moving between trees. As long as different trees of the same species are still in synchrony with each other, this works fine. Insect behaviour can also help. A number of hawkmoths but especially solitary euglossid bees, along with some bats, hummingbirds and monkeys, have evolved to exploit these extended blooming trees by *traplining* (Ohashi & Thomson 2007). The pollinators build an internal map of these trees that produce a reliable source of food over a long period, and repeatedly fly complex, constantly changing feeding routes, visiting widely spread trees to optimise their foraging (Kembro *et al.* 2019). This is similar to how an Arctic trapper would visit their spread-out line of traps, hence the name. Traplining bees can travel long distances between trees, covering more than 20 km in a day. In comparison, temperate bees move pollen comparatively shorter distances: bumblebees up to 1.5 km, while honeybees routinely travel just 600–800 m. By providing a reliable food source, a tree benefits from having pollen moved more reliably to individuals of the same species than it would with randomly flying bees, and the pollinators benefit from spending less time looking for nectar. Euglossid bees may also benefit by using scents taken from plants to establish territories, ensuring themselves a more exclusive food supply.

Temperate trees can use similar strategies to persuade pollinators to visit different trees by encouraging pollinators to visit only those flowers that have not yet been pollinated. A classic example is seen in Horse-chestnut. When the individual flowers open, the petals have yellow markings which change to red once the flower has been pollinated (Fig. 56). Since bees are not receptive to the red end of the spectrum, the red markings would appear an unattractive black to a bee, which therefore avoids these flowers (Thomas *et al.* 2019). To reinforce the message, as the flowers change colour, the scent disappears and nectar production is also greatly reduced. This is beneficial to the tree, since the effort of pollinators is not wasted on flowers that are already pollinated and they are more likely to visit different trees, and the bees benefit from a signal that indicates which flowers hold no reward for them. Surely, though, it would be cheaper for the petals just to be shed from pollinated flowers? Indeed, this is what happens in most trees. The difference for the Horse-chestnut is that it grows naturally in dry habitats of Greece and other Balkan countries where trees are widely scattered. By keeping the pollinated flowers, the tree maintains a large visual signal to attract pollinators from faraway trees. If the petals from pollinated flowers were progressively lost, it is likely that the last few flowers would go unpollinated, since bees and other insects would not be able to detect them from a distance. Horse-chestnuts hedge their bets because pollination is supplemented by wind; significant amounts of pollen are detected in the air of urban areas, enough to contribute significantly to hay fever.

In a similar way, woody plants such as European dogwoods and Bougainvillea *Bougainvillea spectabilis*, native to South America, use large petal-like bracts (which are modified, coloured leaves) to help make small, inconspicuous flowers

FIG 56. Horse-chestnut flowers *Aesculus hippocastanum* which have yellow markings on the petals that change to red once the flower has been pollinated.

more attractive to pollinators. The European Guelder Rose *Viburnum opulus* and *Hydrangeas* native to the Americas and Asia produce similar results using large, sterile flowers around the outside of the flower head to advertise the smaller and less showy fertile flowers in the centre.

A different strategy in insect pollination is to be more of a generalist that attracts a wide range of flies and beetles. This includes trees like magnolias, apples, Rowan, European Spindle, some maples and hawthorns, and a long list of others. The strategy is more similar to wind pollination and involves producing lots of pollen in the hope that some will be transported to another individual of the same species. Like wind pollination, this works best if there are many trees of the same species in close proximity. Common features of these generalist trees are flat, upward-facing flowers with a strong scent and with many stamens and easily reached nectar that requires little effort or specialisation for the insects to forage (Fig. 57).

FIG 57. Hawthorn flowers *Crataegus monogyna*. These flat, upright flowers are perfect for attracting a wide range of pollinating insects that can walk from flower to flower to feed and pollinate.

Other trees take a middle approach and attract a more limited number of specific pollinators that are more likely to go straight to another tree of the same kind. This reduces the amount of pollen required but does mean that the flowers need to conform to what will attract the insects. Bees, butterflies and moths fall into this category, often attracted to specific colours of flower, particularly yellows and blues and including red for butterflies. This means that butterflies will also visit predominantly bee-pollinated flowers such as privets, limes and Alder Buckthorn *Frangula alnus* in Europe. Pollination can be made more specific by only allowing some insects access to the pollen, seen to perfection in many orchids. But trees can do the same; for example, many shrubs in the heather family (Ericaceae) rely on *buzz pollination*, where the anther is tubular and the insect must grasp the flower and vibrate its flight muscles at a specific frequency to trigger the explosive release of the pollen (Vallejo-Marín 2019). Honeybees cannot do this, so pollination is by a more select group of solitary bees. Buzz pollination can be seen nearer to home in tomatoes.

BIRDS AND OTHER VERTEBRATES AS POLLINATORS

It is commonly accepted that the first flowering plants (angiosperms – see Chapter 1) were insect-pollinated and even now they remain the principal animal pollinators. Nevertheless, birds can also be important pollinators, especially in the tropics, with hummingbirds and honeycreepers in the New World and honeyeaters and lorikeets in the Old World. Bird-pollinated flowers are often vivid red, with tubular flowers to hide the abundant nectar from the wrong pollinators. They are also usually odourless, and this has contributed to the idea that birds have a poor sense of smell, but there is growing evidence that this is not so (Steiger *et al.* 2008). Examples are the *Fuchsia* species commonly grown in Europe. There are other types of bird-pollinated flowers, including Australian trees in the Proteaceae family such as the bottle-brush trees *Banksia* with masses of protruding stamens to dust the birds' feathers as they land to drink nectar. Eucalypts are even more specialised, since the petals have been lost and the bud scales (sepals) are modified into a cap over the flower, called the operculum, which is pushed off as the bud opens, leaving tufts of yellow, red or white stamens acting as a landing platform for bird pollinators (Fig. 58). Bird pollination also happens to an extent in temperate areas. In Europe, small warblers and Blue Tits

FIG 58. The Mottlecah or Rose of the West *Eucalyptus macrocarpa* is a eucalypt endemic to southwestern Australia. The large flowers can be up to 10 cm in diameter, and they then produce large fruits, hence the species name. The cap (operculum) has fallen from the flower on the right to expose the many stamens, pollinated primarily by hummingbirds. The flower on the left is fading, revealing the young fruit behind. Perth Botanic Gardens, Australia.

FIG 59. Catkins of Goat Willow *Salix caprea* from female (left) and male (right) bushes. Although wind-pollinated, there is a nectary at the base of each flower within both male and female catkins, which attracts Blue Tits.

Cyanistes caeruleus will take nectar from cherries, mahonias, Gooseberry *Ribes uva-crispa* and Almond *Prunus dulcis*, and at the same time transport pollen on their feathers.

The best example is seen in the catkins of willows (Fig. 59). These have separate male and female catkins on different trees (dioecious – see below). Each catkin is made up of classic wind-pollinated flowers with the stamens and stigma exposed to the air with no distracting petals – perfect for spreading pollen by wind. However, each male and female flower within a catkin also has a large nectary, producing a large glistening drop of nectar. Willows are thus hedging their bets by being wind- and animal-pollinated. Moths certainly use willows for food in early spring but so too do Blue Tits, since the nectar is readily accessible to their small beaks (Kay 1985). As they feed, the face and chest feathers pick up enough pollen that a yellowish tinge can be seen from a distance, and the birds act as pollinators, moving pollen from male to female trees. Willows produce enough nectar for Blue Tits to get their daily energy requirement in less than four hours of feeding, despite being much larger than insects and warm-blooded as well (and so needing a lot of energy), although they may need to balance their diet by also eating insects. To illustrate how abundant nectar can be in bird-pollinated flowers, aboriginal peoples in Australia collect nectar from various bottle-brush trees *Callistemon* as food. Producing so much nectar is obviously expensive for the tree, but where insects are in short supply, as in Australia or in Europe in very early spring, or highly competed for, as in the tropics, then the investment is worthwhile. There is the extra advantage that birds will carry pollen further than bees, which is beneficial in dry, open vegetation where trees are further apart, and they also show considerable constancy to flowers of one species and can visit many thousands of flowers in a day.

A number of trees are also pollinated by mammals. Bats are attracted to flowers that obviously open at night, produce abundant nectar and have a sour or musty smell. The flowers tend to be rather dull in colour, partly because they are pollinated at night and because bats are colour-blind. Bat-pollinated trees include a number of tropical trees, including Kapok *Ceiba pentandra*, African Baobab, Balsa *Ochroma lagopus* and the Durian *Durio zibethinus* from Southeast Asia that has very strong-smelling fruit – once smelt, never forgotten (Chapter 9)! Some bats take nectar while hovering like hummingbirds but many land to drink, requiring flowers to be strong. To make access easier for these comparatively large pollinators, flowers are often produced when the trees are leafless or the flowers are borne on the trunk (known as cauliflory – see below) or ends of branches, away from the leaves.

Non-flying mammals tend to be poor cross-pollinators, since it is harder for them to reach the flowers across several trees. However, a number of monkeys and possums that are good tree climbers are undoubtedly involved in the pollination of trees. Even smaller mammals can play a part in shorter trees. Some proteas *Protea* are pollinated by rodents in South Africa, and Traveller's Palm *Ravenala madagascariensis* in Madagascar is pollinated by the abundant lemurs. Banksias and eucalypts in Australia that are known to be pollinated by birds are also visited by marsupial mice and Honey Possums *Tarsipes rostratus*, which have long tongues for feeding on nectar, and fur which picks up pollen. Large land-based animals may also play a role in pollination but this may be more apocryphal than actual. It was thought for a long time that the Knobthorn Acacia *Acacia nigrescens* that grows in semi-arid savannahs of Africa was at least partially pollinated by giraffes *Giraffa* sp. The flowers form up to 40 per cent of the giraffe's annual diet, particularly at the end of the dry season, even though the flowers and leaves are guarded by dense thorns. Giraffes can travel up to 20 km per day with pollen and even whole flowers on their noses and so would seem to be capable of acting as pollinators. However, they will eat around 85 per cent of the flowers they can reach, and so even if the remaining flowers are pollinated, there will be overall fewer seeds. Moreover, most seeds are produced at the tops of trees out of the reach of giraffes and likely result from insect or bird pollination. Sadly, therefore, it seems unlikely that the giraffe is a good pollinator and if it has any role in pollination, it is in moving pollen further than the birds and insects can (Fleming *et al.* 2006).

WHY SO MANY FLOWERS AND FRUITS?

Trees, in common with other plants, produce far more flowers than can ever turn into fruits or cones. Pines typically produce mature cones from over three-quarters of their female strobili, but broadleafed trees like Mango *Mangifera indica*, Teak and Horse-chestnut may produce only 1 fruit from 15–20 flowers down to only 1 in 1,000 in the Kapok in tropical Africa. Part of the reason for this apparent waste can be down to poor pollination. As seen above, pollination can be a fairly random process, so it makes evolutionary sense to have some extra flowers to allow for those that don't make it.

However, this is not the whole answer, since some young fruit with pollinated seeds are often aborted. In some cases, there is physically not enough room for all fruits to develop. A flower head may have many flowers in order to attract pollinators, but only some can develop normally, since fruits are usually larger than the flowers. Another important reason is that it allows a plant to hedge its bets. If some young fruit are lost due to frost or drought, there are others that can mature. Also, it is a way of allowing the tree to produce as many fruits as possible in a good summer when there is an abundance of sugar and water. This is a similar process to that seen in birds, which produce an optimistic number of eggs; the smallest chick(s) will die if there is not enough food and survive if there is. The same thing happens in trees, since a young fruit is competing with its neighbours for sugars, and rather than all partly starve, the weakest will fail and be aborted. Flowers at the base of a flower head often get first claim on food and water. Those above may or may not get enough sugar and water to survive, and how far up the shoot fruits are produced depends upon just how good the summer is.

Fruit is costly for the tree to produce, so the ability to discard fruits also allows resources to be invested in the highest-quality fruit. Most fruit need one or more fertilised seeds to ensure the fruit develops normally, and it is not uncommon for fruits and cones to be dropped if they contain too few seeds. The threshold number can vary annually in any one tree. The tree can also detect fruits heavily infested with seed-eating larvae and selectively abort them. I have an apple tree affected by the Codling Moth *Cydia pomonella*, the larvae of which burrow into the young fruit and devour the core, including the developing seeds (see Chapter 15). The first small fruits falling from the tree (the so-called June drop) are likely to be those infested, as the tree sheds them early to avoid wasting resources on fruit with no viable seed.

Cross-pollination, or outbreeding – fertilisation by pollen from a genetically different plant – is better for a species in the long term, since it produces genetic variation and suppresses the expression of harmful recessive genes that are often detrimental. *Inbreeding depression* can lead to poor germination, reduced survival of seedlings, chlorosis of seedlings (lack of chlorophyll) and reduced height growth. However, self-pollination, or inbreeding, is a useful insurance policy for producing some seeds, albeit not the best in the long term but better than none at all. There is some evidence that cross-pollinated flowers begin fruit growth quicker than self-pollinated flowers, and so have first call on resources and are the ones most likely to survive. In this way the tree will grow the preferred cross-pollinated seeds but will fill up with self-pollinated flowers if there are enough resources available. Darwin (1876) also noted that cross-pollinated flowers of Broom *Cytisus scoparius* (called *Sarothamnus scoparius* in Darwin's day) produced twice as many seeds as self-pollinated flowers.

HOW DO TREES ENCOURAGE CROSS-POLLINATION?

If cross-pollination is better than self-pollination, how do trees favour this when reproduction involves the use of a proxy agent like wind or an animal, and there is little direct control of where pollen comes from or goes to? This is a particular problem for a large wind-pollinated tree that is surrounded by a cloud of its own pollen, and for animal-pollinated trees where the animals can do most of their foraging in one crown. A solution is to be *self-incompatible*, so that a tree cannot be pollinated by its own pollen. This is particularly common in wind-pollinated trees such as Beech, where self-pollination leads to empty nuts, and in isolated trees, where self-pollination is most likely to produce fewer nuts than those in woodlands. Self-incompatibility is also found in insect-pollinated trees such as apples *Malus*. This can result in problems for orchard owners, since most apple varieties are cloned by grafting, and so an orchard of one variety is, in effect, all the same tree. The solution is to plant two or more other varieties, and there are readily available charts showing which varieties will pollinate which others. This explains why a lone apple tree may be barren unless it has several varieties grafted onto one stem, while a single Victoria plum (self-compatible, or self-fertile) fruits well.

The majority of trees, however, are not completely self-incompatible and have at least some capacity for self-pollination as an insurance policy, allowing self-pollination if cross-pollination fails. When pollen lands on the stigma of a flower, it germinates and grows a pollen tube down through the style to reach the ovary, where it can fertilise the ovules or young seeds. Cross-pollination can be given priority by slowing down the growth of the self-pollen tube through the stigma and style, allowing pollen from another plant to reach the ovules first. This means that pollen from another tree will fertilise the seeds if it is available, and if not, the tree's own pollen will eventually do the job.

Another mechanism for favouring cross-pollination is to mature the male and female parts of the flower at different times, called *dichogamy*. The norm is for the female stigma to be receptive to pollen before the stamens start producing pollen (*protogyny*), since the stigma can catch pollen from another tree before being besieged by its own abundant pollen but can be self-pollinated if other pollen is absent. The opposite situation, where the male stamens produce pollen and wither before the stigma becomes receptive (*protandry*), is more common in animal-pollinated flowers. Self-pollination here is more difficult but can be achieved by bending or curling the stigma to touch parts of the flower still dusted in its own pollen or by having another late-maturing set of anthers do the job.

A step further is to move from hermaphrodite flowers with male and female in the same flower to having separate unisex male and female flowers on the same tree (*monoecious*). This is particularly common in wind-pollinated trees and is often accompanied by the male flowers being on a different part of the branch or the crown to reduce the female flowers being swamped by the tree's own pollen. An added advantage of this is that female flowers and subsequently the fruit can be on stronger branches or even directly on the trunk (a condition called *cauliflory*) in species with particularly large and heavy fruits such as Cocoa tree *Theobroma cacao*, Durian (Fig. 122 in Chapter 9), a number of figs (Fig. 60) in the tropics and Judas Tree *Cercis siliquastrum* in temperate areas.

An even further step is to have the male and female flowers on *separate* trees (*dioecious* species), as found in willows (Fig. 59), yews, Maidenhair Tree, Holly and

FIG 60. Cauliflory – fruits that grow directly on large branches or the trunk. (a) Cocoa pods *Theobroma cacao* growing in Honduras. (b) Elephant Ear Fig *Ficus auriculata* native to Southeast Asia growing in the Brisbane Botanic Gardens, Australia.

many tropical trees. This ensures cross-pollination but prevents self-pollination, and only half the trees (the ones with female flowers) contribute to seed production. To overcome these disadvantages, dioecious species tend to flower at a younger age and use insects for pollination rather than relying on the vagaries of wind (Ohya *et al.* 2017). Large fleshy fruits are also more common in dioecious trees, perhaps by allowing the female trees to concentrate on fruit production and help ensure good seed dispersal (Walas *et al.* 2018). Nevertheless, only about 6 per cent of all flowering plants (angiosperms) around the world are dioecious (Queenborough *et al.* 2009), but in the temperate forests of North Carolina, dioecy is found in 12 per cent of trees (26 per cent if shrubs are included), rising to about 20 per cent (32 per cent with shrubs) in Costa Rica and 40 per cent in Nigeria. We find dioecy useful in urban areas in that we can avoid planting female poplars (which produce huge amounts of annoying fluffy seeds) and, in warmer areas, the female Maidenhair Tree (which produces fruits smelling strongly of rancid butter).

TREES THAT REASSIGN SEX

Being male or female in trees and other higher plants is not as tightly controlled genetically as it is in higher animals. The default position of higher plants is being hermaphrodite, and single-sex flowers are formed by the suppression of one sex or the other within those flowers. This allows some fluidity in the sex of individual flowers and of whole trees. A dioecious male or female tree may have a few hermaphrodite flowers (for example, in the shrubby Butcher's Broom), or flowers of the opposite sex (yews can produce a few flowers or a whole branch of the opposite sex), or even change sex completely, once or repeatedly. Male trees of the normally female Irish Yew *Taxus baccata* 'Fastigiata', a clone originating from a single female tree, and female trees of the normally male Italian Poplar clones *Populus* × *euroamericana* 'Serotina' are known. European Ash and some maples take a bit of beating in this fluidity, since individual trees can be all male or female, some male with one or more female branches or the other way round, some branches male one year and female the next, some with hermaphrodite flowers; and this can all vary from year to year (Thomas 2016).

Part of the rationale for this changing of sex could be that female flowers and fruits are more expensive to produce than male flowers, so parts of a tree can switch to female if resources allow. The cost of being female is particularly seen in dioecious species. Yew, Maidenhair Tree and poplars show 'male vigour', where the male trees are taller, grow faster, flower at an earlier age and live longer because of lower reproductive effort (Walas *et al.* 2018).

The proportion of female flowers on monoecious trees usually increases with better growing conditions. Moreover, these trees have a greater proportion of cheaper-to-make male flowers when young, increasing the proportion of female flowers as they mature and have greater resources available. Many trees end up entirely female before they die, particularly in pioneer trees that invade gaps. Moosewood, which grows in gaps in forests around the Great Lakes, is progressively shaded as the gap fills, and it responds by changing from male to entirely female, so putting all its remaining stored resources into seeds before it dies. There are other species that do the opposite and are mainly female when young, including the Cook Pine *Araucaria columnaris*, North American Bigtooth Maple *Acer grandidentatum* and a few pines. What these have in common is being wind-pollinated, and the advantage appears to be that male trees are taller and more exposed to the wind to help move pollen, while shorter trees are perhaps better at accumulating pollen falling in the calmer air and so do better as females.

SEEDS AND FRUITS WITHOUT POLLEN: APOMIXIS AND PARTHENOCARPY

We discussed above the need for seeds to be pollinated to stimulate the development of fruit. However, some trees and other plants can form seeds without pollen, a process called *apomixis*. The rose family (Rosaceae) is particularly good at this, including apomictic genera such as *Amelanchier* (mespils, Saskatoon berry), *Crataegus* (hawthorns), *Sorbus* (whitebeams, rowans) and *Malus* (apples). Strictly speaking, all but the apples physically need pollen to be present for the seed to develop properly, but all the genetic material comes from the mother. This results in all offspring from a mother tree forming a clone of similar-looking individuals, genetically identical to each other and the mother. Since these may look slightly different from nearby clones derived from a different mother tree, they are usually referred to as 'microspecies'. There is always an ongoing debate between 'lumpers', who call these all the same species, and 'splitters', who see these as valuable and discrete species. As discussed in Chapter 1, in Britain we have up to 160 native woody species (leaving out hybrids), but another 309 microspecies can be added to this. The choice is yours!

Apomixis is common in angiosperms but not in conifers. However, a strange case of apomixis, called *paternal apomixis*, has been found in a rare Mediterranean conifer, *Cupressus dupreziana*. Pollen and ovules normally contain a single DNA strand of each chromosome (described as haploid) that come together in the embryo when the pollen fertilises the ovule (referred to as diploid). In this conifer, the pollen remains diploid and when it lands on a female cone of any other *Cupressus* species, it grows into an embryo inside the seed without fertilising the host embryo and so with no genes involved from the maternal plant. In a sense, the male is acting like a cuckoo, parasitising the female. The new seedlings look identical to the male parent regardless of what the maternal species might be (Pichot *et al.* 2001). Sadly, this species is on the edge of extinction.

Parthenocarpy (from the Greek *parthenos*, meaning 'virgin') is similar to apomixis (there is no fertilisation from pollen, and the fruit still grows), but in this case no seeds develop. Many common temperate trees (including maples, birches, ashes, elms, hollies, beeches, firs and junipers but not oaks) are known to produce fruit that are empty of any seeds. Parthenocarpic fruit are usually smaller than normal, but this has been exploited to produce seedless varieties of fig, clementine, oranges, apples, pears and bananas, although the last, of course, do not grow on trees – see Chapter 1. The black flecks along the middle of a banana are all that are left of the ovules; wild bananas can be 90 per cent seed with dozens of large seeds and very little but strong-flavoured flesh.

THE FIRST SEEDLINGS IN SPRING

Most temperate trees shed their seeds in the autumn and begin to germinate sometimes in the autumn or early winter but mostly in the spring. For seedlings in deciduous forests, germinating in autumn or spring gives them an advantage, since this is when the canopy is bare of leaves, giving them more light to help them establish before shade conditions begin. In most woodlands the leafless canopy casts some shade such that light levels are likely to be around 70–80 per cent full sunlight. Once leaves are out, however, in an open birch woodland there may still be 20–50 per cent of sunlight at ground level, but this drops to 2–5 per cent under the dense shade of Beech. In evergreen forests, which cast shade all year long, summer light levels are around 11–13 per cent under Scots Pine in natural woodlands, down to 2–3 per cent under Norway Spruce, and as low as 0.2–2.0 per cent in tropical rainforests. Anything below 20 per cent sunlight is usually below the light compensation point where the sugar production of photosynthesis just matches the sugar usage by respiration – the break-even point at which the seedlings can grow.

Large seeds (and tree seeds tend to be large) readily attract the attention of herbivores once on the ground and so are more likely to begin to germinate in the autumn than are small seeds. However, germinating in spring is generally more common because it avoids the tender seedlings having to survive tough winter conditions and gives the seedlings the whole growing season before winter returns. This can obviously cause problems if spring is too wet or too dry for shallow-rooted seedlings, which accounts for the high rate of loss in the first weeks of life. Oaks have the best of both worlds, with the young root (the radicle) growing out in autumn, followed by the above-ground shoot in spring. The vulnerability of large seeds to being eaten explains why very few forest trees have persistent *soil seed banks* – seeds stored in the soil for years or even decades. These are found usually only in woody plants with small seeds such as heathers.

Around 8 per cent of the world's plants have seeds that die if desiccated below 30–40 per cent moisture content, called *recalcitrant* seeds in contrast to *orthodox*, or desiccation-tolerant, seeds. This rises, however, to 33 per cent in the world's trees and can be as high as 52 per cent in evergreen rainforest trees (Wyse & Dickie 2017, Wyse *et. al.* 2018). Trees with recalcitrant seeds include Avocado *Persea americana*, Cocoa, Rubber Tree *Hevea brasiliensis* from South America, and Mango from South Asia. Most recalcitrant species grow naturally in dense forests where establishment is in gaps and growth has to be rapid if the seedling is to survive. In these species, the seeds are shed while still completely hydrated, unlike orthodox seeds where the water content is reduced to less than 20 per cent before they are shed. Being full of water allows the recalcitrant seeds to grow on their mother plant until the last minute to become as large as possible before being shed. It also means that when they hit the ground, they are primed and ready to germinate straight away and, in effect, barely stop growing. But this speed, helping them to successfully invade gaps, comes at the expense of being sensitive to drying (Tweddle *et al.* 2003). In temperate areas, recalcitrant seeds include species of oak *Quercus* and chestnuts, including *Castanea* (e.g. Sweet Chestnut) and *Aesculus* (e.g. Horse-chestnut). Most of these can germinate in the autumn (at least producing the first young root) before rapid growth of the shoot in spring. This benefits the Horse-chestnut in its native Balkan Peninsula, since the seeds can germinate in the autumn as soon as they fall, taking advantage of the autumn rains (Thomas *et al.* 2019), and all are ready for growth in early spring. Orthodox seeds are able to survive for longer periods in the soil due to their being drier and are more likely to form part of the soil seed bank.

The majority of seeds have some form of dormancy when they land on the ground. Even recalcitrant seeds that produce the root in the autumn will not produce shoots until the dormancy is broken. This ensures that the vulnerable new shoot does not start growing in a warm autumn or midwinter spell. There are exceptions such as tropical species which obviously have no winter worries, and a few temperate species such as elms which release their seeds in the spring, ready to germinate straight away. For most tree seeds, dormancy is broken by a period of winter chilling, which can be mimicked by *stratifying* the seeds by storing them in damp sand in the refrigerator. Seeds that get little cold stratification in a warm winter will eventually be induced to germinate by spring warmth but it will take longer.

In a few trees, such as Ash, dormancy is caused by the embryo of the seed being too immature to grow at seed fall. The spring after seed fall, around 5 per cent of Ash seeds may germinate but most will need the following summer to mature, leading to most germination the following spring, 18 months after they have fallen from the tree. This would seem to make them vulnerable to being eaten, but judging by the forest of Ash seedlings in my garden, it cannot be much of a problem. In others, dormancy is caused by a very hard seed that is impermeable to water, so germination cannot happen until the seed coat is broken. This may be by the fluctuating temperatures of hot days and cool nights or the intense heat of fire (such as in gorse *Ulex*) causing the seed coat to expand and contract, leading to cracking. Or the seed coat can be weakened by fungal decay or partial digestion by passing through an animal's gut. A good example is Yew, where the seeds germinate quickly upon passing through a bird's gut as the seeds are moved in the autumn, followed by winter chilling. By contrast, seeds that drop to the ground may take 1–2 years to germinate once the seed coat has partially rotted away. Gardeners can short-cut the process by scarifying the seeds by abrasion with sandpaper, nicking the seed coat with a knife or soaking them in dilute acid. Even something as simple as removing the seed coat of an acorn leads to quicker and more complete germination in a batch of seeds.

Once the seed begins to germinate there are differences in what happens to the cotyledons, or seed leaves. With small seeds, as in pines, beeches, maples and ashes, most conifers and many tropical species, germination is *epigeal* (*epi* meaning 'above'), and the cotyledons are brought up above the ground. For this to happen, the shoot below the cotyledons expands rapidly, pushing the cotyledons above ground, often initially still enclosed in their seed case, where they will expand, turn green and start photosynthesising (Fig. 61a, b). This allows the seedling to supplement the small amount of food stored in the seed by allowing the cotyledons to start producing sugars before the first true leaves can be produced, usually some weeks but sometimes up to 1–3 months after germination. Once the true leaves are unfurled, the cotyledons wither and fall.

For large-seeded species, such as oaks, walnuts, cherries, hazels, Horse-chestnut and the tropical Rubber Tree, germination is *hypogeal* (*hypo* meaning 'below'). Here, the shoot *above* the cotyledons grows, drawing the shoot and

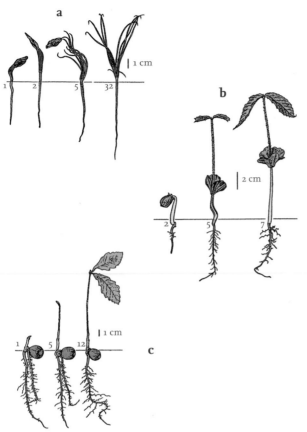

FIG 61. Seedlings of (a) Longleaf Pine *Pinus palustris*, (b) American Beech *Fagus grandifolia* and (c) Bur Oak *Quercus macrocarpa* at different numbers of days after germination. Germinating pine and beech bring the cotyledons above the ground surface (epigeal germination), whereas oak keeps them below ground level (hypogeal). Pines have up to a dozen or even more cotyledons, while the broadleaf trees have two. However, it is not uncommon to find different numbers, and Sycamore *Acer pseudoplatanus* will frequently have three or four cotyledons. From Anon (1948).

eventually the first true leaves above ground, leaving the large cotyledons safely underground (as in oaks, Fig. 61c). The downside of this is that the cotyledons cannot, of course, photosynthesise and this has to wait until the first true leaves are opened. But the shortfall in sugars produced by early photosynthesis is funded by the large reserves of food stored in the large seed. The advantage of hypogeal germination is that the long-lived, food-rich cotyledons are safely in the ground below the reach of most herbivores. Moreover, the cotyledons of large seeds tend to be heavy things to balance on a thin shoot and young root system, and the seedling is less likely to be damaged if they stay safely in the ground.

Some trees go a stage further in sinking the contents of the germinating seeds deeper into the ground – termed *cryptogeal* germination. This is found in Monkey Puzzle *Araucaria araucana* and some other species in the same genus, and also in a few African trees which use this to shield the young embryo from frequent fires (Jackson 1974). At germination, food is moved from the seed into the base of the young root (the *hypocotyl*), making a deeply buried tuber. This seems to be a way of getting valuable resources in the large seed sitting on the soil surface underground as quickly as possible.

MAST YEARS AND SUCCESSFUL SPRING GERMINATION

An additional way in which large-seeded species can increase the chance of successful establishment is by producing a superabundance of seeds in *mast years* with years of very few seeds in between. Masting species are not that common globally; Pearse *et al.* (2017) identified 363 species, just over 0.1 per cent of the 300,000 vascular plants. Although rare globally, masting is found in a number of important temperate trees such as beeches, ashes and oaks, and a number of conifers. In the last century, Beech produced mast years every 2–3 years and Pedunculate Oak every 3–4 years, with exceptionally high seed production occurring every 5–12 and 6–7 years, respectively. This abundance of seeds in a mast year is not without cost; growth in Beech, measured by annual ring-width of wood, is less in mast years (Hacket-Pain *et al.* 2017).

The advantage, however, comes because large seeds are readily eaten by herbivores such as insects, small mammals, squirrels and birds – wood pigeons can remove 100 acorns per day. In mast years the seed-eaters are swamped with more seeds than they can consume (called *predator satiation*), leaving some alive to germinate. These herbivores (predators) are then starved in years of low seed production, keeping their populations low, which helps more seeds survive in mast years. Studies of Beech in Britain have shown that up to 100 per cent of Beech seed is eaten by mice and birds in years when there is a poor crop, but over 50 per cent of seed may be left at the end of a winter following a mast year. This is despite the large flocks of Bramblings *Fringilla montifringilla* and Great Tits *Parus major* attracted from mainland Europe in good Beech mast years. Another benefit of masting is that the abundant flowers in mast years makes wind-pollination more successful, resulting in even more seeds in mast years. Masting works: most seedlings of Beech and oak result from mast years.

There has been much speculation about what triggers a mast year synchronously across many trees. The most likely candidates are stores of food within the tree (a tree needs to rebuild its stores after a mast year, so two consecutive mast years are unlikely) and weather cues such as warm weather. The trees are responding to a weather cue that triggers mass seed production, so they are not coordinating with each other via chemical or mycorrhizal signals but with the weather. Trees that respond most strongly to the weather cue will be most successful, and so natural selection acts to increase the variability in seed numbers between years and ensures the synchronicity between trees (Janzen 1971). This sensitivity to weather is, however, leading to problems due to climate warming.

A long-term study on Beech was started in 1980 by Prof John Packham and Dr Geoff Hilton, following the seed production of 139 trees across 12 English sites. From this the effects of global warming are becoming apparent. There are some good aspects, in that Beech being a warmth-loving tree has gradually produced more seeds over the study in response to a warming climate. However, over the past few decades synchronicity between individual trees began to falter, with different trees within the same woodland becoming out of synch, and the year-to-year variation in seed numbers began to decline, so the difference between a mast year and a non-mast year was much less than it had been, and there have been far fewer 'famine' years (Bogdziewicz et al. 2020b). The result is that there is now a more constant number of seeds in a woodland each year as some trees are masting and others are not, and so herbivores are eating more of them.

The main seed predator of Beech are larvae of the Large Beech Piercer *Cydia fagiglandana*, a micro-moth that feeds on the developing Beech seeds. The young larva burrows into one of the two seeds within the woody cupule and spends the summer feasting on the rich food, usually moving onto the other seed in the cupule when food runs short. By the time the nuts fall from the tree, taking the larva with them to overwinter on the ground, the seed is dead. With the breakdown of masting, the proportion of seeds eaten by the larvae has increased from around 1 per cent in the 1980s to 40 per cent in recent years. Moreover, pollination efficiency also declined, since there were no longer big mast years when huge numbers of wind-pollinated flowers were being produced at once, that made pollination more certain. The net result is that the cost to the tree of producing a viable seed has more than doubled over the last 40 years. The extra herbivory by the moth larvae is continuing to drive natural selection and so, in the long term, the benefits of masting will be selected for again, but selection will likely be over centuries while climate warming is occurring over decades (Bogdziewicz et al. 2020a). Over the lifetimes of our children and grandchildren, they may see fewer oak, Ash and Beech seedlings in our woodlands as a consequence of climate warming.

Masting is highly beneficial to some but not all trees. Trees with small seeds that are not sought by herbivores, such as birches and willows, have more to gain by producing as many seeds each year to ensure some germination and establishment whenever conditions allow. Similarly, in trees that have seeds moved using animal-eaten fruits, it would be counter-productive if fruits were left uneaten on the tree. Masting is also more unusual in tropical forests, presumably because the herbivores can move and change diets in enough numbers to eat the increased quantities of seed. But in the animal-poor dipterocarp forests of Southeast Asia, there is masting. Here, the seed fall of different species is tightly coordinated to drop large numbers of seeds at the same time so that even with movement of herbivores not all the seeds will be eaten.

HOW IS SEED SIZE RELATED TO SPRING?

If you are a gardener, you will know that the seeds of herbaceous plants are often quite small and difficult to sow individually. Some tree seeds that rely on being blown by the wind can be equally as small or even smaller, so for Heather, there can be 33,000 seeds to the gram and for rhododendrons around 11,000 per gram. Other wind-spread seeds – which tend to be small and light – such as birches, are a little larger and have two small wings to help them fly, but even here there are 6,000 seeds per gram, so each seed weighs less than 0.0002 g, or 0.2 mg each. In comparison, there are around 1,000 lettuce seeds in a gram. However, most tree seeds are somewhat larger. Ash seeds weigh 0.07 g each and Beech seeds 0.2 g. But many are still bigger, with Pedunculate Oak weighing in at 3.5 g each, Horse-chestnut at 11 g, up to Coconuts at 250 g each. Even these are surpassed by the Seychelles Double Coconut *Lodoicea maldivica* (Fig. 62) which can weigh 18–20 kg each, and a female tree can carry over 100 kg of fruits!

FIG 62. Seychelles Double Coconut *Lodoicea maldivica* growing in a bed in the Singapore Botanic Gardens. To give a scale, the coconut is about the same size as a human pelvis. Normally the new shoot would be some way from the seed, but this specimen was presumably constrained within a pot when young.

Why are many tree seeds so large? In many cases, seed size within a species is a compromise between a number of competing factors. Nor is this static as the evolutionary pressure is never constant, so seed size will be constantly changing as their surroundings hone the seed size for maximal survival. Having said this, for some trees the reasons for seed size are fairly clear-cut. For woodland trees the classic answer is that the extra food and nutrients stored in a large seed produce larger, quicker-growing seedlings in spring that are able to grow above the shade of the litter and woodland herb layer. Oak seedlings can grow 5–10 cm before the first leaves appear, and just a few centimetres of extra growth can make the difference between being too shaded to survive against surviving above the competing vegetation (Harper *et al.* 1970). The extra food reserves can also help seedlings to be more tolerant of low light and competition for nutrients and water once the trees in the canopy above start producing leaves, usually a few weeks after the seeds have germinated. Seedlings of Sycamore are germinating in my garden a month or more before the adults begin to produce their leaves, giving the seedlings a light-rich head start. Larger seedlings are also more resistant to the physical assault of falling debris inside a woodland, which improves their survival (Leisham *et al.* 2000).

There are also benefits below ground. Large seeds can grow roots downwards more quickly, penetrating below any loose layers of undecomposed leaves found in woodland which are notorious for rapid drying, leaving a shallowly rooted seedling prone to drying out. As an example, larger acorns of Cork Oak *Quercus suber*, weighing up to 8 g, are better able to produce seedlings under drought conditions than smaller acorns, which can weigh as little as 3 g (Ramírez-Valiente *et al.* 2009). This is matched in temperate areas by woodland species that produce smaller seeds flourishing in climates that are more constantly wet. Large seeds may also be beneficial on nutrient-poor soils and so explains the production of large seeds typical on the impoverished soils of dense tropical forests (Fig. 63).

FIG 63. A large fruit of a dipterocarp *Dipterocarpus* species. Penang, Malaysia.

The very large size of coconuts – the largest seeds in the world – has a slightly different explanation. They float in sea water (how far is described in Chapter 9) and will hopefully be washed up on a

FIG 64. Coconut *Cocos nucifera* seedling that has moved itself some 2 m from below the high tide line by growing up under the beach to escape the sea. Penang, Malaysia.

beach. The problem here is that unless they are hurled inland by a large wave, they end up just below the high tide line, which is not ideal. But the coconut can solve this problem. As the coconut germinates, the young root penetrates into the freshwater that flows seawards through the sand of the beach and turns to grow against the freshwater flow and hence up the beach. This can move the growing 'bud' of the embryo (the *plumule*) up to several metres before it produces a new shoot and root system. This very nicely moves the young plant from below the high tide level up above the high tides (Fig. 64). A similar thing happens in the Seychelles Double Coconut but for a slightly different reason. The huge nuts are not renowned for their aerodynamic properties, and when ripe they do nothing else but fall straight downwards. This leaves the germinating seedlings in the shade of their mother, so early root growth puts the young plumule a half-metre or so underground and up to 3–4 m sideways from the coconut, hopefully putting the new seedling in higher light. Enough energy remains to help grow the first leaves up through the 10 m tall canopy for up to four years. This is not without a cost: the Seychelles Double Coconut can bear fewer than 20 fruits at a time and these take 6–7 years to develop (Edwards *et al.* 2003).

As in all things biological, seed size is always a compromise. Large seeds may have an advantage in shaded wooded habitats, but they can be difficult to protect from herbivores, which may then require the production of expensive defensive chemicals. Conversely, small seeds may escape herbivores just by being overlooked or not worth the effort of finding, giving them more freedom to germinate when conditions are best suited to survival. Pioneer species such as birches and heathers that invade open areas, where the vegetation cover is sparse, can afford to have small seeds, since there is little competition with tall neighbours. Small seeds are also likely to disperse more widely so that some find new open sites. This compromise in size is nicely illustrated by a group of woody legumes in Central America that are plagued by the larvae of bruchid beetles which live inside the seed. Different species adopt different compromises; some species have large seeds (an average of 3.0 g per seed) which have to be heavily protected by expensive toxins, others have smaller seeds (average 0.26 g) which are readily attacked and rely on some being missed, and one species has very small seed (average 0.003 g), which has the advantage that it is too small for a beetle larva to grow in but the disadvantage that developing seedlings have a small food supply to aid establishment (Janzen 1969).

In some cases, large seeds have evolved to escape the effects of herbivores, since they can afford to lose more of the seed before they are killed. The seeds of Oriental White Oak *Quercus aliena*, a common tree in eastern Asia, can be up to 4.3 g each, compared to 3.5 g in Pedunculate Oak. The large size seems to have evolved to cope with weevil infestations of *Curculio* species and still live (Yi & Yang 2010). There is evidence that partially eaten acorns of Pyrenean Oak *Quercus pyrenaica* nibbled by rodents produce seedlings just as well as intact seeds, and even experimentally removing two-thirds of the length of the acorn (but leaving the embryo) did not affect dispersal distances. In addition, germination rates were higher and faster in damaged seeds (probably due to the seed coat removal removing physiological barriers), and early root growth was also faster than from intact seeds (Perea *et al.* 2011).

Having explored the nuances of the ecological role of seed size, it's important to note that in some cases seed size may have little to do with its ability to germinate and survive as a seedling. For example, in northern Europe, our common oak species are dependent on jays for dispersal of the acorns. Jays collect acorns and cache them in the ground to eat in the winter. Some will be left uneaten or forgotten and these are the ones that are ready buried and most likely to germinate and grow, away from the shade of the mother tree. Jays tend to go for large acorns but only so large, usually around 17–19 mm wide, because they can fit 1–5 seeds snuggly within their crop. Smaller and larger ones are rejected although jays may occasionally carry a larger one in their beak. Seed size in oaks has therefore been selected for by the jays rather than the needs of establishment (Pons & Pausas 2007a). Proof of this comes in spring when the uneaten buried acorns start germinating. When a jay finds a young oak seedling, it gives it a sharp pull upwards, exposing the acorn containing the cotyledons, which are nipped off and eaten. Small seedlings may be uprooted but most survive despite losing their food reserves. Survival and growth of North American oaks is lower if the cotyledons are removed (Yi *et al.* 2019), but seedling growth of Pedunculate Oak is unaffected by the removal of the sugar-rich cotyledons even on nutrient-poor soil. This underlines the fact that the large food supply is not crucial to survival (Sonesson 1994), and large seed size is selected for by the jays.

Similarly, many pines around the world have evolved to be spread by other corvid birds that form buried caches, including Swiss Stone Pine *Pinus cembra*, the North American Whitebark Pine *P. albicaulis* and the pinyon pines of the American Southwest – especially Pinyon *P. monophylla* and Colorado Pinyon *P. edulis* (Lanner 1996). These pines have large wingless seeds (0.15–0.25 g), making them of interest to the birds, compared to the small (0.005 g) winged seeds of the wind-dispersed Scots Pine.

HOW MANY OF LAST YEAR'S SEEDS PRODUCE SEEDLINGS?

Plants, including trees, are at a disadvantage in that they can't control where their offspring will end up. By clever dispersal techniques (Chapter 9) they can certainly improve the chance of a seed getting to a place where it can germinate but it is still a chancy process. If we were to collect acorns and plant them in a nursery, we might get 95 per cent of them to germinate and almost as many can be grown through to saplings. Yet in the wild, the chancy processes mean that the likelihood of a seed landing where it can germinate is slim, and seed mortality is high, often around 95 per cent. Of the 5 per cent that germinate, around 95 per cent are likely to die within the first year due to issues of water shortage, too much shade, being eaten or being attacked by a pathogen. Add up these various hurdles and the probability of an acorn becoming an oak seedling is something like one in a thousand, and of becoming a mature tree more like one in a million. With these low probabilities of success in early life, and adding in the further ravages of pathogens and herbivores, the need for such large numbers of seeds to get just a few trees becomes a pragmatic reality. Fortunately, once this vulnerable young stage is passed, survival tends to be very good (see Fig. 171 in Chapter 12). Moreover, we should look at trees on their own timescales. If an oak can live for a thousand years, a few decades with few seedlings, as happened in the second half of the last century when jay numbers were very low, is unlikely to be significant.

CHAPTER 5

The Lazy Days of Summer: Growth Above and Below Ground

WHY DO TREES COMPLETE THEIR SHOOT GROWTH SO EARLY?

Catching light is so important for a tree's survival that leaves are produced rapidly in spring, as seen in the last chapter. What is perhaps more surprising is that once the first flush of leaves is produced, many trees stop the production of leaves and the growth of branches and produce their over-wintering buds early in the growing season, called *summer dormancy*. Why stop growth of the branches so early in the summer when growing conditions are just reaching their best and extra valuable height could be gained by continuing a little longer?

Determinate and indeterminate growth

To answer this, we need to address how a branch grows. A bud contains the next year's shoot preformed in miniature, consisting of the twig and leaves (and maybe flowers or just flowers) ready to be expanded. The opening of the bud and growth of the new leaves is thus rapid, making the most of the spring and summer sun. How much of the shoot is preformed varies greatly. Trees such as ashes, beeches, hornbeams, oaks, hickories, walnuts, horse-chestnuts and many maples and most conifers show *fixed* or *determinate* growth. Here, the whole of the shoot is preformed, so spring growth occurs in a single, rapid flush and is complete in a matter of days to a few weeks, followed by the formation of the next set of buds that may not open until next spring.

Other trees, such as elms, limes, cherries, birches, poplars, willows, alders, apples and Sweet Gum *Liquidambar styraciflua*, as well as conifers such as larches,

FIG 65. Mature leaves of various maples showing differences between shape of early (left of each leaf pair) and late (right) leaves. (a) Tatar Maple *Acer tataricum* from the Caucasus; (b) Grey Snake-bark Maple *A. rufinerve* from Japan; (c) Field Maple *A. campestre* from Europe; (d) Trident Maple *A. buergerianum* from China; (e) Montpellier Maple *A. monspessulanum* from central Europe to Southwest Asia; and (f) Cretan Maple *A. sempervirens* from the eastern Mediterranean. From Critchfield (1971).

junipers, Western Red-cedar *Thuja plicata*, Coastal Redwood and Maidenhair Tree, show *free* or *indeterminate* growth where only some of the leaves are preformed. These *early leaves* will expand in spring, after which the shoot will continue to produce and expand other *late leaves*. As an example, the Tulip Tree usually produces eight early leaves and between 6 and 12 late leaves. In some trees, such as maples, birches and eucalypts, this can lead to a difference in leaf shape (called *heterophylly*) between early and late leaves, such that it can be easy to see how many leaves were preformed in the bud. The differences in shape can be quite striking in many maples (Fig. 65). In Sweet Gum and Maidenhair Tree, early leaves may be less deeply lobed, and in Silver Birch *Betula pendula*, early leaves are proportionately wider. Growth in indeterminate species continues for longer than in those that are determinate but still normally stops well before the end of the growing season.

There is also a third strategy of *intermediate* or *rhythmic growth* where leaves are produced in a series of flushes over a summer, each followed by bud formation and then a new flush. This is common in fast-growing southern USA pines, such as Loblolly *Pinus taeda*, Shortleaf *P. echinata* and Monterey *P. radiata*, plus European Olive. Repeated flushing is also seen in tropical species such as Cocoa, Rubber, Avocado, Mango and Tea. Tea can produce a new flush of leaves every few weeks, which is fortunate for the tea producer, since it is the new shoot tips that are harvested. Sometimes the recurrent flushing can merge towards almost continuous *free growth* suited to the warmer conditions where these intermediate species grow.

So why stop growth in summer?

For most temperate fixed-growth species, stopping growth so early in the summer would appear to be a huge disadvantage, since they cannot take advantage of a good growing season by producing more leaves. The leaf buds were formed the previous year, and so the amount of growth this year is correlated to the weather of the previous summer. This could mean that the benefits of a good summer will be diminished if the tree is handicapped by having small buds with few preformed leaves as a result of a previous bad summer. The tree can compensate to a degree by making each leaf larger and the shoots longer (less shading of neighbouring larger leaves), but it still appears to be losing out. Indeterminate species do have a little more leeway in being able to produce more leaves later in the summer, which, since many of these grow in brightly lit openings in woodlands, would be a distinct advantage.

Preforming leaves does have the advantage in getting the leaves open quickly to make use of as much of the growing season as possible. This is important because each leaf needs to pay back the cost of its production before the autumn, otherwise it is a drain on the tree. If leaves open too late in the season, there will come a point in the summer where new leaves will cost more to grow than they can hope to recover in photosynthesis in the short time left in the season. This effectively means that leaves will normally only emerge in the early part of the summer. Moreover, if new outer leaves on vigorous growth start shading those produced earlier in the year, then those shaded will be less likely to repay their costs, so it pays for growth to stop. Indeed, those trees with repeated flushes of leaves are found in areas of long or continuous growing seasons and many are either typical of forest gaps and so less likely to suffer from self-shading or regularly shed their older leaves.

To every rule there is an exception! Leaves can appear later in the year than normal on a second flush of growth from terminal buds. Both trees with determinate and indeterminate growth can show this second burst of leaves, called *lammas growth* (Fig. 66). In oaks this tends to appear if the first set of leaves

FIG 66. Lammas growth in Pedunculate Oak *Quercus robur*. This is new growth from terminal buds late in the summer, usually produced to replace leaves damaged by insects. Called lammas because it usually occurs around Lammas Day, the 'bread-feast' harvest festival, traditionally 1 August or the seventh Sunday after Trinity.

produced in spring is extensively damaged by, for example, caterpillars. Lammas growth is also seen in elms, hickories, beech, alder and a number of other conifers, including Scots Pine and Norway Spruce. In oaks the leaves on lammas growth are longer and more deeply lobed, while in pines there may be more needles per bunch than normal but shorter, giving a tufted appearance. Lammas growth in evergreen trees can be susceptible to winter injury if it does not have enough time to harden off before the first frosts.

Lammas growth is also seen in a range of broadleaf trees and conifers if the top of a young tree is browsed by deer, helping the beleaguered tree to regrow some of the lost height. For example, one-year-old seedlings of Douglas Fir *Pseudotsuga menziesii* in Oregon were found to add an extra third to their height from lammas growth. But lammas growth can also happen in trees that have abundant light, water and nutrients, as if they are hedging their bets to do some more growing in good years, overcoming the problems of only producing leaves in the early spring.

WHEN DO FLOWER BUDS FORM?

If a tree flowers early in spring (Chapter 4), when do the flower buds form? In temperate broadleaf trees and conifers, flower buds are usually formed in late summer or autumn the previous year, which then overwinter and are ready to open the following spring or summer. This explains why early winter frosts and very cold winters can affect flowering the next spring. Bud formation the previous year is not always the case, however. There are, of course, exceptions: some warm temperate trees such as varieties of buddleia, fuchsia, hibiscus and lime *Tilia* form their buds in spring just prior to flowering. This produces the difference that those woody plants producing flower buds last year are flowering on 'old wood', whereas those producing flower buds in the spring are flowering on 'new wood', which, of course, affects when the tree should be pruned. A lime pruned hard in the winter will still flower in the spring, while a hazel flowering on old wood will have lost its flower buds.

Generally, the better the growing season when the buds are formed, the greater the number of flowers. Paradoxically, a really bad growing season, such as a drought, can also induce a large number of flowers, called a *stress crop*. The tree appears to react to extreme conditions by putting stored resources into a last-ditch attempt at producing flowers and seeds in case it dies. Partial ring-barking of the trunk, removing part of the inner bark, or phloem, has also been used to induce flowering in fruit trees and conifers. In this case, the partial breaking of the phloem connection between crown and roots creates a bottleneck, keeping more of the carbohydrates produced by the leaves in the crown, which are put into flowers. This is short-term thinking, since the roots are partially starved, which will weaken the tree. This may lead to death or at least to jeopardising future crops.

A few trees, such as some eucalypts, start their flowers two or more years before opening. In the more equitable climate of the tropics, flower buds are often produced more evenly over the year, taking just a few months to fully form, and these may open straight away, as in fig species, or be accumulated until environmental conditions are right for them to open, which may be sometimes three times a year as in strangler figs, or only once every 5–6 years as in the Dipterocarp rainforests of Southeast Asia.

The plant clock and calendar

Trees that flower in spring can use temperature to trigger their development – the buds begin to form when a threshold air temperature is reached. For most trees, however, producing flower buds at the right time of year requires a good internal calendar so that it knows where in the year it is. In turn, this depends upon a sensitive internal clock.

The internal clock of plants is run by a light-sensing pigment in the leaves called phytochrome. This exists in two forms which absorb light from different parts of the spectrum (Fig. 67). Our human eyes can see wavelengths of light between 390–700 nm (nanometres). The *inactive* form of phytochrome (often abbreviated as P_R) is converted to the *active* form (P_{FR}) when it absorbs *red* light around 660 nm. As there is a lot of this in sunlight, it really means that all the leaf's phytochrome becomes the active form at dawn. The active form, P_{FR}, can be converted back to the inactive form by longer-wavelength red light at 730 nm, which, since we humans cannot see it, is called *far-red* light (hence P_{FR}). More importantly, the active form created during the day will slowly convert back to the inactive form in the dark at a predictable speed. So the amount of active phytochrome in the leaf at dawn will tell the plant how long the night was – if there is a great deal of the active form at dawn, the night was short; if there is very little, the night was long.

This gives the plant a clock. It has a biochemical memory of how much active form of phytochrome there was at dawn yesterday so it can predict when dawn will begin today. Indeed, plants start getting ready for photosynthesis before

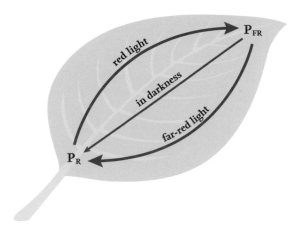

FIG 67. The plant clock and calendar depend upon two forms of phytochrome. The *inactive* form (P_R) is converted to the *active* form (P_{FR}) when it absorbs *red* light around 660 nm. The active form reverts to the inactive form when exposed to longer-wavelength far-red light at 730 nm, but this also happens at a slow, predictable rate in darkness. This allows the plant to measure the length of the night.

dawn because the clock tells the plant when dawn is approaching. Obviously daylength changes during the year so the clock at the start of each day will be a few minutes out. In the spring, as nights become progressively shorter, the plant will expect dawn a bit later than it occurs because the night yesterday was longer than it is today. But since it is only a few minutes out it is accurate enough, and the clock is reset every morning as all the phytochrome is reset to the active form.

Knowing how long the night is also forms the basis of the plant's internal calendar. Trees that produce their flowers in the autumn need the short days of approaching winter to initiate flower buds. This really means they need long nights, which is detected by having little active phytochrome remaining at dawn. In these plants a critical low level of the active form at dawn stimulates flowering. Conversely, trees that produce flower buds in midsummer need long days measured as really short nights. So these will start producing flowers when the amount of active phytochrome left at dawn reaches a critically high level.

SLOWING OF WOOD GROWTH

The growth of new wood (xylem) and new inner bark (phloem) by the cambium begins in early spring as the buds start to open (Chapter 4), but by early summer the inner bark production stops and by midsummer wood production also begins to slow, and at this point the annual growth ring (tree ring) is complete. In many deciduous temperate trees, wood formation stops a number of weeks before there is any sign of leaf coloration, although ring-porous trees like oaks (Chapter 6) might keep going a little longer, until the first sign of coloration (Dox et al. 2020). In common conifers such as larches, pines, firs and spruces, maximum wood growth rate coincides with the longest day length rather than the warmest weather and so begins to decline after the June solstice (Rossi et al. 2006). Having said this, the amount of wood grown in a year – i.e. the ring width – is related to summer temperature, described in Chapter 12. This comparatively early stopping to wood growth helps to ensure that all the new cells are complete before winter. Xylem cells are composed of a mixture of cellulose and hemicellulose that make up the main walls of the cell, which are later impregnated with lignin. This lignin makes the cells impervious to water, helping them in their role of water transport, and also makes them more rigid and increases their compressive strength and so makes the wood stronger. Lignin deposition is one of the last processes in making the xylem cells and needs time. This is illustrated by Black Pine Pinus nigra, growing 250 m above the tree line in the Apennines of Italy, where the growing season is so short that in many years a significant part of the annual ring fails to develop lignin before low autumn temperatures stop further development (Piermattei *et al.* 2015). This lack of lignin will inevitably make the wood weaker and less resistant to fungal decay.

When the cambium is most active in early summer (usually May to June), it is physically quite weak, and the bark is most easily peeled from the tree. Tanners exploit this by felling trees at this time, allowing the bark to be more easily peeled for use in leather tanning. This is not so good if you want decorative pieces of wood complete with their bark; better to fell these trees in the winter when the bark is more tightly and permanently attached.

ROOTS, WATER AND SUMMER

If you dig down below a tree, you quickly come across big woody *framework* or *coarse roots* radiating out from the trunk, and on shallow soils these may be seen up at the soil surface, as in Figure 68. These are long-lived roots that have two main roles – holding the tree up by forming a root plate (described in Chapter 11) and transporting water and dissolved minerals taken up by the smaller roots back to the rest of the tree.

The lateral coarse roots are those we see when a tree is uprooted. They can be very thick at the base of the trunk (they grow in diameter just as the trunk does) but rapidly narrow to just 2–5 cm in diameter towards the edge of the crown. As they pass the edge of the crown they are thin and easily cut or snapped and may seem unimportant. However, more than half of the weight and surface area of all a tree's roots may be out beyond the crown. Their role is not so much to hold the tree up, though they do help, but mainly to forage for water and nutrients. In temperate trees the total spread of roots away from the trunk is typically 2–3 times the radius of the crown, and even up to four times the radius on dry sandy soils. This usually approximates to a spread of from once to twice the height of the tree.

The radiating coarse lateral roots have smaller-diameter woody side branches that grow outwards and upwards to the soil surface, often reaching several metres in length. They may branch four or more times to end in fans of short, fine non-woody roots. In Red Oak, for example, there may be as many as 10–20 of these branches along a metre of lateral root below the crown, dropping to one branch every 1–5 m on roots out beyond the crown. These side roots tend to stay thin, seldom over 4–6 mm diameter in Red Oak and hold the *fine roots* typically 1–2 mm long and 0.2–1.0 mm in diameter but can be down to 0.07 mm in diameter – so fine as to be invisible to the naked eye. Reynolds (1975) estimated that in Douglas

FIG 68. Coarse woody roots growing away from the tree to form a shallow root plate. (a) Beech *Fagus sylvatica*. (b) A fig *Ficus* species producing many aerial roots around the trunk but still growing the flat, shallow root plate. Kuala Lumpur, Malaysia.

Fir, around half the total length of all the roots were less than 0.5 mm thick and 95 per cent less than 1 mm in diameter. The finest roots may also have small root hairs which act to further increase the contact of the root with the soil. These can vary from 1.0 mm long in blackcurrants *Ribes*, down to just 0.1 mm long in apples *Malus*. Being so small and delicate, root hairs may live for only a few hours, days or weeks before being replaced by new hairs as the growing tip elongates. Despite their usefulness, root hairs are often lacking, particularly where their role is taken over by mycorrhizal fungi (Chapter 14).

Roots not only spread over a large area, they are also very abundant within an area of soil and so very good at taking up water. Walter Lyford (1980), working at Harvard Forest, part of Harvard University in rural Massachusetts, painstakingly dissected cubes of soil taken from the forest floor below Red Oaks. In one cubic centimetre of this soil he found an average of 1,000 root tips, more than 2.5 m of root with a surface area of 6 cm^2 (so six times the area of the top of the cube), not counting mycorrhizal fungal or root hairs. It may sound as if the cube of soil must have been solid root, but in reality the roots made up just 3 per cent of the cube's volume. Scale this up, and a mature Red Oak will have something like 500 million live root tips, the majority of which will be very close to the soil surface.

The coarse roots of most trees are remarkably shallow, so that in temperate trees, 80 per cent of the roots are in the top 60 cm of the soil and 90–99 per cent are in the top 1 m. On very dry soils or fissured rocks, some roots can be much deeper (look ahead to Fig. 70), but these are the exception. Fine roots tend to be even more concentrated near the soil surface. In a North Carolina oak forest, it has been found that 90 per cent of the weight of fine roots less than 2.5 mm diameter was in the top 10–13 cm of soil. In tropical trees this can be even more extreme, with fine roots confined to the top 5 cm of soil or even forming a mat of roots over the soil surface that may contain more than half of a tree's fine roots. The result is that in temperate woodlands and gardens, the fine roots of a tree are normally up in the leaf litter, usually in the moist compacted litter below the most recent leaf fall, growing between the layers of leaves, branching and spreading as they go.

This complete root system makes an impressive water foraging system, best seen at work on shrinkable clay soils such as London clay and to a lesser extent on Oxford clays. On these soils, water uptake by trees causes the clay to shrink, which can lead to building subsidence. A survey on these soils by Cutler and Richardson (1981) found that oaks, poplars and some maples could damage buildings 30 m away and willows 40 m, showing just how far the roots spread (Table 8). This whole problem of subsidence is made worse by the pattern of water uptake through a year. Glenda Jones and colleagues (2009) looked at water uptake from the soil around a willow *Salix* species and an English Oak growing on highly shrinkable London clay. They found that the high density of roots, reaching out to twice the width of the crown, tended to keep the soil fairly dry all year, so changes in ground level were no more than 1.0–1.5 cm throughout the year. There was greater vertical movement at the drip line around the edge of the crown, presumably because of the concentration of roots here to take advantage of water running off the crown, but shrinkage and swelling tended to be fairly slow and consistent over large areas so that the soil went gently up and down, more or less as a single unit. The problem came at the outer edge of the root systems around three times the width of the crown. The soil here rapidly filled with water during winter and was equally rapidly emptied by root growth in this area responding to drought, especially at the extreme limit of their spread. This resulted in 3–6 cm of vertical movement of the soil over short distances of 3–5 m, which could have potentially disastrous effects on a building.

Sadly, this explains why insurance companies are often reluctant to accept the risk of trees on shrinkable clays even when they are tens of metres away from a building. As can be seen from this, it is safer if the trees are closer to the building! This might, of course, raise images of woody roots penetrating foundations and undermining walls. In reality, trees close to buildings usually do little damage, since the delicate fine roots cannot easily penetrate a building's foundations. The problem, of course, comes when there is a small crack that the roots can grow through, since once these roots are through the brickwork and become woody, they can exert great pressure on foundations, making cracks bigger as the roots grow in diameter. This is probably most common in old drains and sewers where fine roots penetrate cracks in the joints and then proliferate inside the warm, moist, nutritious contents.

Fortunately, not all trees have roots that spread as far as oaks and willows, as can be seen in Table 8. Large trees such as ash, elm, lime, maple and horse-chestnut caused damage to buildings more than 20 m away, while smaller fruit trees and *Prunus* and *Sorbus* species caused subsidence less than 10 m away. Every cloud has a silver lining, and while Common Leyland Cypress *Cupressus* × *leylandii*[2] spreads its roots over 20 m from the trunk, 90 per cent of the recorded damage has been within 5 m of the tree.

---

2 Leyland Cypress is an interspecies hybrid, indicated by the 'x' before the species name, between two North American West Coast species, Monterey Cypress *Cupressus macrocarpa* and Nootka Cypress *Cupressus nootkatensis*. However, Nootka Cypress used to be in a different genus, *Chamaecyparis nootkatensis*, so Leyland Cypress was originally classed as an intergeneric hybrid × *Cupressocyparis leylandii*, indicated by the 'x' before the genus name.

TABLE 8. Distances over which trees have been seen to cause damage to buildings by subsidence on predominantly clay soils in southeast England. Dashes indicate that data are not available.

Common name	Scientific name	Maximum tree-to-damage distance (m)	Distance within which 90 per cent of damage cases were found (m)	No. of trees
Willow	Salix	40	18	124
Oak	Quercus	30	18	293
Poplar	Populus	30	20	191
Elm	Ulmus	25	19	70
Horse-chestnut	Aesculus	23	15	63
Ash	Fraxinus	21	13	145
Lime	Tilia	20	11	238
Maple	Acer	20	12	135
Cypresses	Cupressus & Chamaecyparis	20	5	31
Hornbeam	Carpinus	17	–	8
Plane	Platanus	15	10	327
Beech	Fagus	15	11	23
False-acacia	Robinia	12	11	20
Hawthorn	Crataegus	12	9	65
Rowan & Whitebeam	Sorbus	11	10	32
Cherries, etc.	Prunus	11	8	144
Birch	Betula	10	8	35
Elder	Sambucus	8	–	13
Walnut	Juglans	8	–	3
Laburnum	Laburnum	7	–	7
Fig	Ficus	5	–	3

Based on a survey from 1971–79 conducted by the Royal Botanical Gardens, Kew. From Cutler & Richardson (1981).

Exploring the soil

The coarse roots and their smaller branches making up the *framework roots* are the part that we tend to be most aware of. However, the fine roots are equally important to the tree in their role of exploring the soil for water and minerals. Fine roots tend not to be able to grow towards areas rich in water or minerals; rather, they are opportunistic when they find a rich patch.

As a fine root grows it produces side branches, often in a spiral along the root (like a shoot producing leaves) so that the new roots spread in different directions, helping to explore the largest volume of soil. The tip of the root is protected by a hard root cap which is part of the root, but nevertheless the vulnerable tip can be damaged or killed, leading to new branches arising from behind the injury. In a similar way, if the root tip meets an obstacle such as a stone, the tip will tend to branch around the obstacle. The same result can be produced artificially by cutting roots with a spade to encourage the formation of a root ball – a useful preliminary before moving a sapling. Although the new root branches have a tendency to carry on growing in the same direction as the original one, all these incidents increase the number of roots exploring the soil. This is beneficial in helping to ensure that new volumes of soil are explored and not missed.

On top of this, when the roots encounter an area of soil rich in water or minerals, they will branch rapidly until the resources are exhausted, at which point most of these new roots rapidly die off. The loss is compensated for by new roots continuing to grow elsewhere in the root system, maybe even just a few centimetres away. This rapid fine-root turnover benefits the tree, since it is not keeping roots alive that are in unproductive parts of the soil. As a consequence, the lifespan of fine roots is usually measured in days, although on moist soils they can live for longer, ranging from typically 95 days in Aspen, with only 30 per cent of the fine roots surviving longer than 200 days, to 336 days in White Oak where 80 per cent of fine roots survived longer than 200 days (McCormack *et al.* 2012).

Fine-root turnover goes some way to explaining why the growth of fine roots of temperate trees tends to slow down and even stop in early summer once the buds in the crown have opened. In Pennsylvania, USA, McCormack *et al.* (2014) found that in 12 conifer and broadleaf trees, most root growth occurred in early summer to midsummer, slowing or stopping at the beginning of July. This may partly be due to limited sugary resources being commandeered by the active above-ground growth, leaving less for the roots, or due to soils getting a little too warm (especially near the surface where many fine roots are concentrated), but is most likely due to the near-surface soil becoming too dry for fine-root growth and survival. After the summer lull, root growth may speed up again in the moister, cooler autumn.

Stopping fine-root growth in midsummer could potentially leave a tree without sufficient water. The likelihood of this is reduced by having fine roots across the whole of the rooting zone taking up water. There are plenty of fine roots at the edge of the root system furthest away from the tree, but nearer the trunk the original fine roots will be long dead, leaving large volumes of soil with only large coarse roots passing through it that have a limited ability to absorb water and minerals. To avoid this, the coarse roots near the trunk are able to produce new, smaller roots that can re-occupy this otherwise lost soil. Some trees also have a tap root (Fig. 69) which can produce lateral roots lower in the soil to help with access to water (the value of these is discussed below under *hydraulic redistribution*).

Different tree species have evolved different ways of exploring the soil. Trees such as Ash and American Southern Magnolia *Magnolia grandiflora* commonly grown in Europe have long, fast-growing, fairly unbranched lateral roots designed to explore a large volume of soil very quickly. By contrast, others such as European Beech and American Green Ash *Fraxinus pennsylvanica* have shorter,

FIG 69. Side view of the root system of a young cherry *Prunus* species showing the wide-spreading lateral branches surrounded by smaller, shallow fibrous roots. Underneath this is a stout tap root (being held) which upon reaching the high-water table on clay soil has branched sideways to avoid the waterlogged soil.

slower-growing roots with many branches, which help to utilise a smaller volume more effectively. This may explain why beech suffers from drought more than ash, since it uses up the available water and cannot exploit new areas of soil quickly enough.

Some trees are more flexible in their rooting pattern than others and are better at adapting their root growth to suit soil conditions as these change. For example, willows and spruces, which are very good at coping with shallow or waterlogged soil, can root deeply on dry soils. Similarly, pines will produce a long tap root as well as widespread lateral roots on deep soils but can cope very well with very shallow soils, depending just on the laterals. Others, such as Silver Fir, Sycamore and oaks, are less flexible and depend strongly on a deep tap root from which laterals are produced and do not do well on shallow soils. The tap root can be very long if conditions allow or dictate (Fig. 70).

This flexibility of growth helps different tree species avoid competition with each other. As an example, a study in southern Germany found that Sessile Oak

FIG 70. (a) Western Sycamore (a plane tree in European parlance) *Platanus racemosa* growing on rock with a long tap root penetrating down through the crack in the rock, which may also be making the crack wider. (b) More prosaically, this Sycamore *Acer pseudoplatanus* growing in rubble under a concrete driveway has produced a formidable set of roots around a central tap root, all delving downwards to aid stability and find a reliable water supply. (a) Sequoia National Park, California, and (b) East London.

and Beech produce similar quantities of fine roots when grown on their own, but, when grown together, Beech will produce 4–5 times more fine roots than the Oak, even though the trees are similar in size and leaf area. To avoid being out-competed for water, the Oak produces fine roots equally in all directions from the trunk to explore a large volume of soil, while the fine roots of the Beech are more concentrated in areas of nutrient-rich leaf litter beyond their own crowns, so exploring smaller areas of soil in more detail (Leuschner *et al.* 2001). Such large variations in root abundance through competition have also been seen between pines and eucalyptus and maples and pines and doubtless occur to a greater or lesser extent in many mixed groups of trees.

Agroforestry

Flexibility of root growth enables trees to coexist with agriculture. A lone tree growing in a regularly ploughed field will have the bulk of its coarse and fine roots just below the maximum depth of ploughing. This is also exploited in *agroforestry*, where agricultural crops and trees are grown together, particularly in tropical areas. The trees provide shelter from a hot, drying sun and also themselves provide fruit and fuel. The system works on the assumption that the crops are using water near the soil surface, and the trees are using deeper water, 20–30 cm down below the crop roots. This does indeed happen in some systems. For example, shade Coffee *Coffea arabica* can be grown under the shade of a variety of trees (Fig. 71), and a study in Mexico showed that the Coffee took water out of the top 15 cm of the soil while the trees used water from 15–120 cm below the surface (Muñoz-Villers *et al.* 2020). However, the vertical zonation is not always quite as discrete (Bayala & Prieto 2020). It is true that in dry conditions trees can usually access water deeper in the soil than other plants, but they will often still have shallow fine roots that will compete with the crops. Many agroforestry trees get 30–40 per cent of their yearly water needs from the top 30 cm of soil, using progressively deeper water once the surface is fairly dry. As a consequence, crop yield is normally lower beneath the trees. In China, it was found that rubber trees acquired 40 per cent of their water from the top 20 cm of the soil, as did crops such as Tea, Cocoa and Galangal *Alpinia officinarum*, a relative of ginger. This was reflected by the top 20 cm of soil containing 60 per cent of the rubber trees' fine roots and 50–62 per cent of those of the crops (Yang *et al.* 2020). During the dry season, there was competition for water between the trees and crops at all levels in the soil. In the wet season, however, when the whole soil profile was moister, the trees coped with the presence of crops by taking water from deeper in the soil, supporting the traditional view of agroforestry. On the positive side, the crops helped *increase* overall soil moisture in the wet season by encouraging more

FIG 71. Shade Coffee *Coffea arabica* being grown under a canopy of trees in what was dense rainforest a few years before. In this case, the main value of the trees that have been left standing from the original forest is in providing shade for the Coffee rather than providing fruit or fuel. Nevertheless, the crops and trees act as an agroforestry system. Cusuco National Park, Honduras.

water to soak into the soil rather than running off the surface and, due to shading of the soil, also reducing evaporation, providing a bigger pot of water for both crops and trees.

What makes the system workable is that both crops and trees provide resources to the farmers, and some loss of crop production is worth the gain from the trees. The system can be improved by growing the trees and crops in alternate strips and digging a trench between them to help keep the tree and crop roots separate. When it is working well, the water issue is partly relieved because the crops will need less water, since they are shaded from the hot sun and will evaporate less water. Moreover, the crops may gain from hydraulic redistribution of water from deeper in the soil to the surface (Bayala & Prieto 2020), described below.

HYDRAULIC REDISTRIBUTION: MOVING WATER AROUND THE ROOTS

Most tree roots are remarkably shallow and near the soil surface. Perhaps counterintuitively, this helps the tree stand up (described in Chapter 11) and, since most water comes from rain and most minerals are released by decomposition near the surface of the soil, also puts the roots in the best place to absorb both water and minerals, often in direct competition with roots of herbaceous plants around the tree.

Having said this, if conditions allow, some roots will go deeper. Suitable conditions are soil that is not waterlogged and thus has enough oxygen and that has an open enough structure to allow the fragile finest roots to physically penetrate downwards. In some trees this begins as a carrot-like tap root, but is often quickly replaced by a number of 'sinker roots' that grow down from the lateral framework roots within a metre or so of the trunk (these are shown in Fig. 153 in Chapter 11). In temperate and tropical soils, this leads to a few roots penetrating down to around 2 m below the surface, but on well-drained soils they may reach down 4–5 m in deciduous trees and up to 7–8 m in Mediterranean conifers (Körner 2005). As an example, Beech in Germany will normally grow a few roots down to 1.7 m below the soil surface, but on very dry sandy soils or on rocks that have deep fractures, with more oxygen lower down, it can reach down 3.5 m (Leuschner *et al.* 2001). In very dry areas, where the soil is easily penetrated, the roots may go even deeper in search of water and have been found down to 12 m in an acacia dug up in the excavation of the Suez Canal, 13 m for the tap root of Cork Oak, 15 m in pines on the western slopes of the Sierra Nevada mountains, USA, and 10–30 m in eucalypts. In very unusual conditions, roots can be deeper still. Roots of a mesquite bush, a *Prosopis* species, were found down to 53 m in a gravel bed of an open-pit mine near Tucson, Arizona, and in the deep, open sands of the Kalahari in Botswana, roots of a local tree, *Boscia albitrunca*, were found more than 68 m down a borehole (with the water table at 141 m). Fig roots in South Africa have been reported at an astonishing 120 m depth.

Once the roots grow down to rock or soil too solid to penetrate, waterlogged soil or low oxygen concentrations, they often turn to grow horizontally to form a new set of roots similar to but smaller than the surface lateral roots. In some trees these deeper roots may instead repeatedly branch to produce a bushy end like a broom. As well as absorbing water, these lower roots catch nutrients washed down below the normal level of roots that would otherwise be lost to the trees and other plants. This can be seen with trees such as birches that reduce the acidity of the soil surface by catching nutrients with their deeper roots and dropping them back onto the soil surface in their leaf litter. Deeper roots also help to provide water from lower soil layers in dry periods that may be unavailable to plants with only shallow roots, as seen in the agroforestry systems above.

Gaining extra water can, however, be quite a subtle process through the movement of water around the soil. During the day, fine roots at all depths will take up water that will be used by the leaves of the crown, as in Figure 72b. At night, when the leaf stomata are closed, this upward flow of water (transpiration) largely stops, but water may still continue to move through the roots in a process called *hydraulic redistribution* (Burgess *et al.* 1998). On a hot day, the fine roots near the surface of the soil will absorb most water and at the end of the day the soil here is usually drier than deeper in the soil. This creates a gradient in moisture through the roots that causes water to flow from moister deeper levels up towards the drier surface roots, and even out of the roots into the surrounding soil (Fig. 72a) – which is why this is often referred to as hydraulic lifting. It might be in the best interests of the tree to hold this water inside the roots, but since this is purely a physical process, the tree cannot easily stop the water moving from the moist inside of the root out into the dry soil. Moreover, the root would not be able to hold all the water moved in a night. Fortunately the roots can reabsorb the water from the soil the next day, and so the system works.

a   b   c

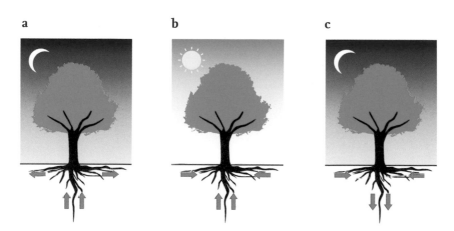

FIG 72. Hydraulic redistribution of water under a tree. In dry times (a) water is lifted at night from moist areas deep in the soil up to more shallow roots where it escapes into the soil. During the next day (b) the tree takes up water from the remoistened shallow soil and also from deeper into the soil. Following rain (c), water can move in the opposite direction at night from wet surface layers deeper into the soil. From Thomas (2014) based on Lee *et al.* (2005).

Water can also move downwards or sideways, since the roots are acting as passive wicks moving water from wetter to drier soil. If a tree is at the edge of a woodland and the roots in the open area have access to more rain, then water will flow across the roots into the drier soils of the woodland. Similarly, after rain when the surface of the soil is saturated, water can flow through the roots down into deeper layers, as in Figure 72c, which acts to lock water away where it is less likely to be lost by evaporation and below the roots of most competitors. Eucalypts in Western Australia have been seen to carry water through the tap root down to several metres below the surface.

Water movement from deeper roots can amount to 80–100 litres per night in dry areas and around 4–20 litres in temperate areas (Hafner *et al.* 2020a). Just how much is moved depends on the size of the tree and how much water it is transpiring in a day, but can represent anything from 10–80 per cent of the tree's daily water needs. This can make a very big difference to continued growth and survival during hot, dry summers and periods of drought.

It might appear that the roots have access to the same amount of water wherever it is in the soil, and so hydraulic redistribution is just something that happens because of the nature of the tubes in the roots and is of no real benefit or loss to the tree. But there are some distinct advantages for the tree in moving water up through the soil. Shallow roots offer a shorter pathway between roots and leaves than deeper roots so leaves can pull water more easily from the shallow roots than from those that are deeper, so it helps the tree if the surface is recharged at night. Moreover, the shallow roots are mixed in with the roots of other vegetation and when it rains these roots will have the best chance of competing and taking up water; deep roots will just get what is left. So it is advantageous to keep the shallow fine roots alive and working by keeping the soil moist. Moreover, some of the water moved into the shallow roots will leak out of the roots into the *rhizosphere*, the thin layer of soil surrounding fine roots where microbial and chemical reactions occur. If the rhizosphere is kept moist, this can help maintain contact of the root with the soil and help nutrient capture from the soil, since nutrients need to dissolve in water to be absorbed. It also increases the activity of soil microorganisms, such as mycorrhizal fungi that not only help with nutrient uptake from the soil but also help defend the tree against pathogens (Chapter 14).

This movement of water can also inadvertently allow trees to share water. Experiments have been done that allow water sharing between trees to be traced (Fig. 73) and they found that hydraulic redistribution from saplings of well-watered temperate trees can provide up to a quarter of the water used by neighbouring trees undergoing drought (Hafner *et al.* 2020b). Of course, this also has benefits for herbaceous vegetation, soil microbes and woodland animals in times of drought, underlining the value of trees within an ecosystem. Water movement can be particularly useful for trees growing in urban streets where soil compaction and small rooting spaces can limit amounts of available water but which can be supplemented by, for example, water from a leaking water pipe being shared amongst surrounding trees.

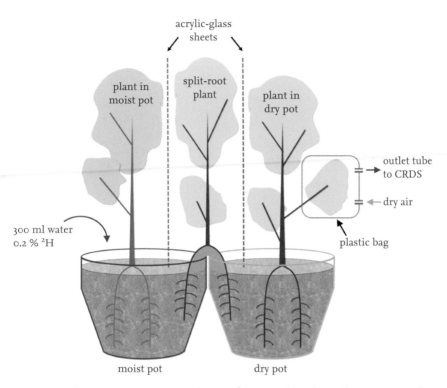

FIG 73. In this experiment, mixtures of two- to four-year-old saplings of temperate conifers and broadleaf trees were planted in two pots with the middle tree having half its roots in each pot. The moist pot was kept well-watered while the dry pot was given less water over a month-long period. At this point, the moist pot was given water containing a small amount of deuterium (^{2}H, or heavy hydrogen, which has an extra neutron in the nucleus doubling its atomic weight). Acrylic-glass sheets were then added to prevent the crowns of the trees from touching. Leaves of the tree growing solely in the dry pot were placed in a plastic bag inflated with dry air, and the amount of deuterium in the air leaving was measured in a spectrometer (labelled CRDS). It was found that water moved from the moist pot through the split-root plant into the dry pot and was being absorbed by the dry tree. From Hafner *et al.* (2020b).

Hydraulic redistribution is an example of trees being beautifully adapted to their environment. It is important to note, however, that this is not trees being altruistic and 'helping' each other – it is more an accidental outcome of a mechanism that has evolved to benefit individual trees. This is discussed further in Chapter 14. Moreover, hydraulic redistribution is not always beneficial. There are some seasonal parts of the Amazon basin that experience regular droughts, and in these cases the movement of water from deeper soil layers during the early stages of a drought can use up available water quickly and cause problems later. This is rather like having a larder full of food that needs to last a month, but everything is eaten within the first week because it is easily done, leading to starvation and possibly death (Thomas 2014). However, temperate trees appear to be better stewards of their resources, since hydraulic redistribution supplies water to the surface fairly slowly, eking out the supplies during a drought.

CHAPTER 6

Supplying Enough Water for the Summer

HOW MUCH WATER DOES A TREE USE?

An apple sapling 2 m tall may use around 7,000 litres of water in a summer (while it has its leaves), and a similar amount can be used by a more mature Mediterranean oak that, living in a dry environment, is more parsimonious in its use of water. In temperate areas the amount of water used rises to around 17,000 litres in an average-sized birch to 40,000 litres in a summer for a larger deciduous tree. If we assume that the growing season is six months long, the large tree will be using an average of 220 litres, or 48 gallons, of water per day. This fits well with measurements of less than 100 litres per day in conifers, to 20–400 litres per day in eucalypts and many temperate trees such as oak, reaching around 500 litres per day in a well-watered palm and as high as 1,200 litres per day in the Amazonian rainforest tree *Eperua purpurea* emerging above the canopy and bearing the full force of the hot sun and wind (Thomas 2014).

More than 95 per cent of the water ascending a tree is lost via *transpiration* – the evaporation of water from the surface of the tree, and primarily from inside the leaves (McElrone *et al.* 2013). This might seem very wasteful but it is an inevitable side effect of photosynthesis. The leaves have holes (stomata) that allow carbon dioxide and oxygen to diffuse in and out, but these also allow water vapour to escape. If the stomata are closed to reduce water loss, the tree would run out of carbon dioxide and photosynthesis would stop. On the positive side, if water is readily available, this loss will not adversely affect the tree and, indeed, can be beneficial, since the transpiration stream helps cool heated leaves – 98 per cent of the light energy reaching the leaves may be dispersed by evaporating water. Moreover, the transpiration stream is one of the main ways of delivering minerals like nitrogen and phosphorus that are dissolved in the water to the growing points of the tree. Although transpiration involves large amounts of water, most trees will carefully limit this to the minimum, since leaves in full sunlight have a narrow safety margin between working well and wilting.

How much water is lost by the tree is, of course, dependent upon conditions. Leaves in sunny, windy positions will lose more than shaded leaves. Consequently, trees in dense woodlands or between tall buildings in urban areas use about two-thirds the water of a solitary tree in the open. Trees will also use more water when it is available, such that Narrow-leaved Ironbark *Eucalyptus crebra* in eastern Australia has been found to use less than 25 litres per day in dry periods but more than 250 litres per day when more than 25 mm of rain fell (Zeppel *et al.* 2008).

Transpiration is driven by the sun's energy, but many trees continue to use at least some water at night, typically in the order of 10 per cent per hour of that during the day, although in the Coastal Redwood of California this can rise to 40 per cent. This night-time loss is counterintuitive, since the stomata should be firmly closed to save water. However, in Coastal Redwood, night-time transpiration may help to get nutrients to the top of the tree that are too short of water to allow the stomata to open much during the day. It may also help oxygen to diffuse into the leaves at night (needed for respiration) when there is no photosynthesis and so no oxygen production inside the leaf.

It is becoming clear that absorption of water by leaves is likely to be a common phenomenon in plants, helping to relieve water stress and maybe even refilling xylem tubes blocked by gas bubbles. The ability to absorb water through the leaves has been found in at least 124 plant species, representing more than 90 per cent of those tested and including woody plants as diverse as Paraña Pine *Araucaria angustifolia*, Douglas Fir, Holm Oak *Quercus ilex*, the Dragon Tree *Dracaena draco* of the Canary Islands (Fig. 74) and the heathland dwarf shrub Cowberry *Vaccinium vitis-idaea* – but interestingly not Olives or Ivy *Hedera helix* (Dawson & Goldsmith 2018). In the abundant fog of the Californian coast, water uptake by the topmost leaves of the Coastal Redwood from fog may be their main source of water, since it is so difficult for them to get water from the roots.

MOVING WATER THROUGH THE TREE

It has long been known that water moves up a tree through the wood, or xylem, which is made up largely of dead cells. In gymnosperms or conifers (Fig. 75a) the main cells in the xylem are *tracheids* that both conduct water and give strength

FIG 74. (a) Cloud building up on the north coast of Tenerife, Canary Islands, that brings foggy conditions to the laurel forest it covers. Fog condenses on the foliage and drips to the ground, supplying up to 20 times the amount of water received from rainfall and so very important for the development of the forest. (b) Some of this fog can be directly absorbed through the leaves of many of the species, including the Dragon Tree *Dracaena draco*. Shown here is the famous tree at Icod on the north coast of Tenerife.

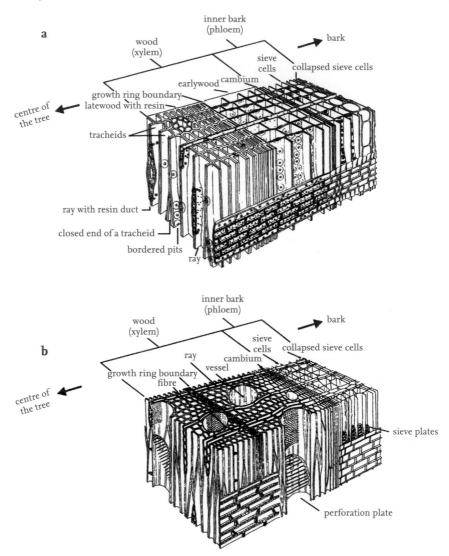

a

inner bark (phloem)

bark

wood (xylem)

sieve cells

collapsed sieve cells

earlywood cambium

growth ring boundary
latewood with resin

centre of the tree

tracheids

ray with resin duct

closed end of a tracheid

bordered pits

ray

b

inner bark (phloem)

bark

wood (xylem)

sieve cells

collapsed sieve cells

ray

cambium

vessel

growth ring boundary
fibre

centre of the tree

sieve plates

perforation plate

to the trunk. In angiosperms or broadleaf trees (Fig. 75b), the water conduction occurs through the *vessels* while the *fibres* surrounding the vessels give most of the strength to the wood. These tracheids and vessels start life as normal living cells but as they mature they die and become largely empty tubes full of water, joined together by holes between the individual tubes.

OPPOSITE: **FIG 75.** The detailed structure of wood of a (a) conifer and (b) broadleaf. Conifer wood is composed of tracheids that make up 90–94 per cent of the wood volume. They conduct water and give the wood strength. Individual tracheids are very short and are joined by bordered pits. The rays run from the centre of the tree towards the bark and are made up of living cells, and there are also resin ducts that run vertically through the wood. In broadleaf trees, water is conducted through vessels (c.30 per cent of the wood volume), and strength is provided by fibres in between that make up around 50 per cent of the wood volume. Rays are also present (20 per cent of volume). The vessels are made up of shorter vessel elements joined by perforation plates. From Schweingruber (1996).

For centuries it was known that water moved up through these dead drainpipes but exactly how this worked was not fully understood. Capillary forces that draw water up a narrow tube by surface tension could account for the rise of water by a metre or so, and sometimes roots produce a positive pressure that can squeeze water up to around 20 m (described in Chapter 3), but this is nowhere near the height of the tallest trees which are over 100 m. The first comprehensive view of how this happens was put forward by Dixon & Jolly in their 1895 paper 'On the ascent of sap', and while there have been a number of alternative ideas put forward, their *cohesion-tension theory* is still the best explanation.

The process begins with transpiration. The cells inside the leaf lose water to the comparatively dry air outside through the stomata in the leaves. The drier cells pull water from their neighbours, creating a suction, referred to as a *tension* or *negative pressure*. When this suction reaches the xylem tubes, it pulls water from within the tubes. In this way the evaporation of water creates a suction, or tension, that runs from the cells of the leaf through the continuous column of water in the trunk and on to the ends of the finest roots, which pulls water in from the surrounding soil. This is the *tension* in the cohesion-tension theory. For this to be able to work, the columns of water running through the tree need to hold together under these huge forces, which fortunately they do due to the very strong *cohesive* forces between the water molecules helped by friction between the water and the xylem cell walls. There has been debate about whether the cohesive forces between water molecules would be enough to allow these columns to hold together. Evidence for this came with a paper in 2008 by Wheeler and Stroock from Cornell University, who made miniature trees out of a hydrogel (a porous jelly filled with water) just 1 mm thick. They found that putting suction on one side created tensions larger than found in trees, without the water column snapping. It thus appears that the column of water in a tree is indeed strong enough to hold together under great tension.

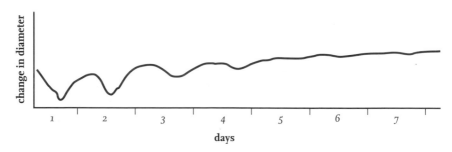

FIG 76. Variation in trunk diameter over several days in a hypothetical tree. The left-hand end of the line shows that the diameter of a trunk reduces in the afternoon when evaporation of water from the leaves causes great suction inside the trunk, pulling the sides of the trunk inwards. At night, the trunk relaxes outwards because, without the sun, less water is evaporated and the suction is reduced. The right-hand end does not show such pronounced shrinkage during the day because rain and high humidity are reducing water loss from the leaves, and all that is left is the gentle upward trend due to diameter growth. From Thomas (2014) based on data from Kozlowski (1968).

Perhaps the most striking visual evidence that such great negative pressures or tensions exist inside trees can be seen in Figure 76, where the tension is so great that it pulls the sides of the trunk inwards during the middle of a sunny day when transpiration is highest. Admittedly the change in diameter is small; measured under the bark (so only the wood is involved), this shrinkage is in the order of 0.01–0.25 mm on a tree 30–40 cm diameter, a change of just 0.01–0.07 per cent (Offenthaler *et al.* 2001). But how much pressure would you have to exert on a piece of lumber extracted from a tree to achieve this amount of deformation without crushing the cells? Further corroboration comes from the theoretical limit to the cohesion of water that would support a column of water 120–130 m at its longest, which neatly matches the tallest trees in the world at around 115 m (Chapter 12). In summary, the most supported view of how water is pulled up the tree is by the evaporation of water from the leaf (tension), the high strength of the water column (cohesion) and the ability of wood (xylem) to withstand these forces.

The remarkable thing about this way of getting water up a tree is that the energy to run it comes directly from the sun and not from the tree. If the tree had to provide the necessary energy from its food reserves, then trees as we know them would never have been able to evolve.

PROBLEMS OF AIR IN THE SYSTEM

Sucking water to the top of the tree through the xylem will only work if there is no air in the tubes. If you crank an old village pump, usually nothing happens except a lot of clanking. But if you remove the top of the pump and prime it by pouring water in to completely fill the tube, then the pump can pull water up from a well tens of metres deep. This is because the column of water has great cohesion but not if it is broken by an air bubble. If there is an air blockage, the column of water below the bubble becomes so heavy that extra suction applied at the top simply acts to pull apart the air in the bubble, creating a partial vacuum, and water will not rise more than 9 m up the pump. This same principle applies to trees; the xylem cells begin their useful life full of water, and as long as they stay that way, water can be pulled up over 100 m. The problem is that air bubbles can and do appear. Air in the system can come from too much tension, damage to the tubes and freezing; we will look at each of these in turn.

In hot, dry conditions, leaves can lose water faster than the soil can supply it and the tension in the xylem steadily increases. The first line of defence is usually to close the stomata to reduce evaporation. But even with this, if the tension keeps on getting greater, eventually the cohesion between the water molecules fails and the water column breaks, or *cavitates*, with an audible snap, producing a bubble of water vapour supplemented by air pulled in from outside the tube. This creates a blockage, or *embolism*, which means the tube can no longer conduct water. In official parlance this is called a loss in *hydraulic conductivity*. If this happens in too many tubes, there will be *hydraulic failure* with dire consequences for the crown of the tree. Fortunately, many trees can cope with huge losses of conductivity due to embolisms and still live. Saplings of Loblolly Pine *Pinus taeda* have been found capable of losing 80 per cent of their xylem conductivity and still have a 50 per cent chance of survival. Most of the recovery after this drought was by growing new xylem full of water (Fig. 77) rather than the ability to refill embolised tubes (Hammond *et al.* 2019). Similar losses have been found in broadleaf trees (Urli *et al.* 2013).

Embolisms can also be caused by damage, either physical damage that breaks open the xylem tubes, such as a branch snapping off or the trunk being hit by a rockfall, or by biotic damage, such as fungi or an insect burrowing into the xylem. Most important of all, winter can also bring its share of embolisms. Although it is sometimes said that xylem drains in winter ('the sap going down in the autumn and rising in the spring'), with the exception of some vines, the xylem of trees stays full of water year-round, which is the only way of ensuring that the tubes

xylem grown
after drought

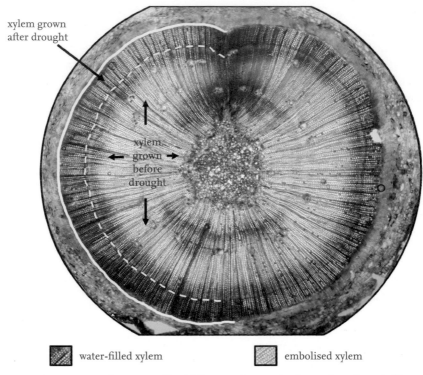

xylem
grown
before
drought

water-filled xylem   embolised xylem

FIG 77. Cross-section of two-year-old Loblolly Pine *Pinus taeda* taken 85 days after it had been exposed to drought that caused an embolism in 96 per cent of xylem tracheids. The red stain shows tracheids that are water-filled and capable of conducting water. It shows that much of the restored water movement is in xylem grown after the drought (the area between the dotted yellow line which indicates the edge of the xylem at the time of drought and the solid yellow line which is the edge of the xylem 85 days later). From Hammond *et al.* (2019).

are always water-filled. As the water in the xylem cools and freezes in winter, it can hold less of the gas that is dissolved in it (ice holds 1,000 times less gas than liquid water) and so gas bubbles form. If these bubbles do not redissolve in early spring before tension develops, then the tubes will remain embolised and useless for water movement.

Fortunately, there are various mechanisms for reducing the impact of embolisms. The first aims to prevent cavitation by good stomatal control that closes the stomata before tensions get too great. Another mechanism is to have smaller-diameter tubes, since there is proportionately more friction from the tube wall, which helps to hold the column together by increasing the cohesion of the water column. An equally important mechanism is to keep the embolisms small and local when they occur. When a tree is under great tension and an embolism forms, it is important to stop that air bubble spreading and being pulled up through all the tubes, effectively blocking them all. This is done by valves within the wood – pits and perforation plates – with variable success in conifers and different types of broadleaf trees.

Conifers

In conifers, the individual tracheids are very narrow, typically ranging from 0.025 mm in Pacific Yew, up to 0.080 mm in Coastal Redwood – small enough to be difficult to see with the naked eye on a cross-section of wood. The narrow size, particularly when less than 0.03 mm in diameter, means that cavitation is very unlikely even during hot summer spells because friction from the cell walls helps to hold the narrow column of water together. The vessels are also very short, typically from 1.5–5 mm long, reaching up to 11 mm in Monkey Puzzle. So small are the tracheids that a cubic centimetre of wood from Douglas Fir can contain 180,000 tracheids! Where the pointed ends of the tracheids touch each other (Fig. 75), they are joined by a variety of 'pits'. The commonest type is a *bordered pit* where there is a raised doughnut around each side of a central hole. Inside, a thickened lump (the *torus*) is suspended by a web of cellulose strands (the *margo*) just like a trampoline suspended by a web of elastic cords. These pits act as valves. Water can flow freely through the strands of the margo, but if damage occurs to a tracheid, allowing air to enter, the pressure difference across the pit pulls the torus over and blocks the hole, producing an *aspirated pit*. The value of these pits acting as valves is that it holds the embolism within one tracheid and stops it spreading. The importance of this to a tree is underlined by considering that in a 100 m tall Douglas Fir, water may have to pass through more than 20,000 pits on its journey from the roots to the leaves. Each one adds a little bit of resistance to the water movement such that more than half of the hydraulic resistance in the xylem of a conifer comes from the pits (Choat *et al.* 2008). This drastically slows water flow through the xylem but this disadvantage is greatly outweighed by the advantage of stopping gas bubbles spreading through the tree.

These pits are also very useful in coping with winter gas bubbles caused by freezing. The small-diameter tracheids means that gas bubbles are going to be small, and the frequent pits stop the bubbles rising up and joining into larger, longer-lived bubbles. As the spring warms up, these small bubbles will redissolve within a few minutes or at most in a few days. This means that by the time the leaves are needing water, the tracheids will be bubble-free and able to conduct water. It also means that individual tracheids can keep functioning for many years, since they are at little risk of embolism and so remain water-filled. It is not uncommon to find that water can be conducted up the trunk of a conifer over 30–40 years' worth of rings (Berdanier *et al.* 2016). Water flow is usually fastest close to the bark where the tubes are newest, with something like 60 per cent of the volume flowing in the outer 1.5 cm and 80 per cent in the outer 2.5 cm of the trunk – but older tubes deeper in the tree are still useful in conducting water. These features allow conifers to grow in a wide range of habitats, but first of all, how do broadleaf trees cope with bubbles?

Broadleaf trees: Diffuse-porous wood

Things get more interesting in broadleaf trees depending on whether the tree is diffuse-porous or ring-porous, which is to do with the size and arrangement of the vessels, or pores as they are called when viewed on the cross-section of a piece of wood (Fig. 78). In temperate areas, a new ring of wood is grown each year and can be divided into *earlywood* that is grown in the early spring and *latewood* that is grown in the summer. At the end of each year there is a boundary between rings, representing winter in temperate trees, which is visible as a thin dark line made up of small thick-walled cells.

In *diffuse-porous* trees such as birch, maple, beech, poplar, lime, mahogany and most eucalypts, the vessels, or pores, are fairly uniform in size and are evenly distributed throughout the earlywood and latewood of the growth ring – so they are *diffused* across the wood. These vessels are quite narrow, typically 0.050–0.075 mm in diameter and so similar in diameter to the tracheids of conifers (0.025–0.080 mm, as noted above). Since the vessels are so narrow, they are hard to see, and the most distinctive thing in the wood is usually the boundary between rings. As with conifers, the narrow tubes and many perforation plates will slow the water flow.

Individual vessels are made up of several cells joined together (called *vessel elements*) like sections of drainpipe slotted together, creating vessels that are typically up to 5 cm long, so more than 10 times the length of conifer tracheids. Pits, similar to those in conifers, can sometimes be found joining vessel walls, but vessels are primarily joined by *perforation plates*. Looking again at Figure 75, it can be seen that these perforation plates resemble drain-hole covers with bars. Some perforation plates have just one hole, or perforation, but most are more complex, having bars (as shown), a net-like pattern or a series of round holes like a pepper pot. Whatever the shape, these perforation plates all act in the same way as valves to trap gas bubbles, just like pits, but with no moving parts. Although the holes through the perforation plates may look quite large, the bigger bars in

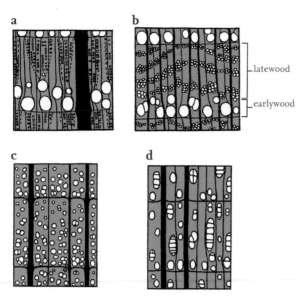

FIG 78. Ring-porous woods (a) oak and (b) elm; and diffuse-porous woods (c) beech and (d) alder. The centre of the trees is towards the bottom of the drawings; horizontal lines mark the edge of a year's growth (i.e. winter in temperate trees); vertical lines are rays running from the centre of the tree out to the bark. Earlywood is the part of the annual ring grown in the early spring and latewood in the summer. From Thomas (2014).

Figure 75 are bridged by small cellulose strands, leaving holes less than 0.02 μm (20 hundred thousandths of a millimetre) wide. These act very effectively at holding back bubbles and so will effectively stop gas bubbles moving into the next vessel no matter how dry the conditions. Perhaps not surprisingly, the percentage of diffuse-porous species with complex perforation plates, which act as better valves, increases with altitude and latitude where it is colder in winter. The narrow, comparatively short vessels keep bubbles formed in the winter small, which quickly redissolve before the buds open and the leaves need water. This obviously works because Italian Alder *Alnus cordata* has been measured as losing more than 80 per cent of hydraulic conductivity during the winter by gas bubble formation, but by early spring the loss was down to less than 20 per cent (i.e. the tubes had lost their bubbles) and remained less than 30 per cent during the summer. The reason that not all tubes were bubble-free after winter is that even when the vessels are refilled with water, there can be some residual damage from winter freezing, called *frost fatigue*. This is where the cell walls and cellulose fibres at the junctions between tubes become deformed (Christensen-Dalsgaard & Tyree 2013), allowing bubbles to merge and making future embolisms due to drought more likely. However, those species that can produce positive xylem pressure (Chapter 3) have been found to be much more resistant to frost fatigue (Yin *et al.* 2018).

From this it can be seen that the wood structure of conifers and diffuse-porous broadleaves has a lot in common. Both have fairly narrow xylem tubes (0.025–0.080 mm diameter in conifer tracheids, 0.050–0.075 mm in broadleaf vessels) which are quite short (1.5–5 mm long in conifer tracheids, up to 5 cm long in diffuse-porous broadleaf trees). As explained above, the narrow tubes of both wood types act to reduce embolisms in dry weather by helping to hold the water column together. Moreover, both conifers and broadleaves have valves (bordered pits and perforation plates) to keep gas bubbles small and stop them joining up. The valves slow water flow, but both conifers and diffuse-porous broadleaves have lots of tubes available to conduct water that work for many years, each taking a comparatively small amount of water at a slow speed. These trees are the Volvos of the tree world – safety first. They do well in the harsher environments of the world such as hot dry habitats where embolisms are most likely, and cold dry habitats at high latitudes where winter freezing is common. Given that conifers are at the extreme end of smallness of tubes, they tend to occupy the driest and coldest habitats.

Broadleaf trees: Ring-porous wood

In contrast to conifers and diffuse-porous broadleaves, trees with ring-porous wood are the Ferraris of the tree world – fast but a little risky. In ring-porous trees such as ash, elm, oak, false-acacia, hickory, catalpa and teak (see Table 9), the earlywood is dominated by large-diameter vessels (0.2–0.8 mm) – see Figure 78. These vessels are 6–10 times wider than the smaller ones of the latewood which are more similar in diameter to those in diffuse-porous wood (Fig. 78). Earlywood vessels are also very long, typically several centimetres to many metres, so the perforation plates are well spaced out.

Looking at the cleaned cross-section of one of these trees, these large holes, or pores, are big enough to be seen with the naked eye and can be seen to follow each ring around the tree; hence *ring-porous*. Nature, of course, is rarely so clear-cut, and in a ring-porous tree, the rings at the centre of the tree (i.e. the first rings grown), or ones grown while the tree is under great stress, can look quite diffuse-porous. Plus, some trees, such as walnuts and other *Juglans* species, are in the middle and described as semi-ring-porous because the vessels in the earlywood are 3–5 times larger in diameter than those in the latewood. To add more spice, trees in the same genus, and sometimes even within the same species, may be ring-porous in wetter areas and diffuse-porous in drier places, or diffuse-porous for a year or two before showing their ring-porous colours.

The very wide and long earlywood vessels are exceptionally good at conducting water up to the canopy in spring, helping to support early growth and the transpiration of the new leaves. The volume of water flowing in a tube increases by the fourth power of the diameter (the Hagen-Poiseuille Law), so making a vessel four times wider increases flow volume by 4^4, or 256 times, up to an upper limit of around 0.5 mm diameter. This is because the water flowing near the side of the tube is slowed due to friction from the wall, but in a wider tube this friction affects a proportionately smaller amount of water, so flow increases. Water can flow through these super-tubes of the earlywood at rates of 1–4 m per hour compared to usually less than 30 cm per hour in conifers and diffuse-porous trees. Flow rates at night are, of course, much slower, typically up to around 15 cm per hour in all types of trees. Indeed, so good are these ring-porous tubes at moving water that the majority of water needed by the crown of the tree can be met using just the newest, outermost ring.

But this is risky! The reduced friction that helps speed water flow also means that the water columns have less help from the walls in holding the column together and are more likely to cavitate when transpiration is high on hot, windy, dry days. Moreover, the vessels that remain water-filled through to the autumn are likely to develop large gas bubbles during winter, and the sparsity of perforation plates allows these bubbles to join together, forming very large bubbles that are unlikely to redissolve in spring. It has been shown that in North American Red Oak, around 20 per cent of vessels can be embolised by August, rising to 90 per cent after the first hard frost. These big bubbles can take weeks to disappear in spring and in many cases are highly likely to remain as permanent embolisms. Fortunately, ring-porous trees have another solution: they depend on growing a new ring of earlywood (which starts ready-filled with water) before the leaves appear and start demanding water. And, as noted above, the tree may rely just on the earlywood of that one new ring, just a millimetre or so wide, to supply all the water that the crown needs during the summer.

Due to the need to grow earlywood over the whole tree before anything else in spring, ring-porous trees tend to leaf-out later than diffuse-porous trees but their trunks start growing earlier in the spring than their diffuse-porous neighbours. This is risky because damage to these new earlywood vessels can spell disaster for the tree. Fortunately, so good are these wide earlywood vessels that in Pedunculate Oak it has been calculated that only about 2 per cent of all these new vessels are needed to supply water, even during periods of rapid transpiration. They also have an insurance policy. As embolisms accumulate in the wide vessels through the summer, water conduction is taken over by the smaller vessels of the latewood as growth slows (see Fig. 78). These are less efficient at water conduction but ensure a safe water supply, albeit limited. If there is a very bad drought in summer that causes all the wide tubes to embolise, the tree will usually survive

TABLE 9. The wide range of families and genera that contain prominent trees with ring-porous wood, although not every species of a genus will be ring-porous. There are several other small genera that could be included but are omitted for brevity. Also given, where appropriate, is a typical species found in Europe or an important commercial species.

Family	Genus	Notes
Anacardiaceae	Cotinus	Smoke-trees of warm temperate areas, includes European Smoke-tree C. coggygria
	Pistacia	Cashews of subtropical areas, includes the Mastic Tree P. lentiscus
	Rhus	Sumachs of temperate and subtropical parts, includes Stag's-horn Sumach R. typhina
Annonaceae	Asimina	North American Pawpaw A. triloba
Bignoniaceae	Catalpa	Bean trees spread across North America to Asia, includes the Indian Bean C. bignonioides
Cannabaceae	Celtis	Hackberries found across northern temperate areas, includes C. occidentalis
Elaeagnaceae	Elaeagnus	Temperate areas, particularly Asia but includes North American Silverberry E. commutata
	Hippophae	Sea buckthorn of Europe, includes H. rhamnoides
Fabaceae	Acacia	Wattles and acacias of Africa and Australasia
	Castanopsis	Chinquapins of eastern Asia
	Chrysolepis	Chinquapins of North America sometimes placed in Castanopsis, includes the Golden Chinquapin C. chrysophylla
	Dalbergia	Tropical regions around the world, includes Brazilian Rosewood D. nigra
	Gleditsia	Honey locusts of North America and Asia, includes G. triacanthos
	Gymnocladus	Coffee trees of North America and Asia, includes Kentucky Coffee Tree G. dioica
	Laburnum	Europe, includes Common Laburnum L. anagyroides
	Robinia	Locust trees of North America, includes the False-acacia or Black Locust R. pseudoacacia
	Senegalia	Previously included in Acacia, includes the Gum Arabic Tree S. senegal
	Styphnolobium	Includes the Japanese Pagoda Tree S. japonicum
	Vachellia	Previously included in Acacia
Fagaceae	Castanea	Chestnuts of temperate northern hemisphere, includes European Sweet Chestnut C. sativa

Family	Genus	Notes
	Quercus	Oaks of the northern hemisphere, includes Pedunculate Oak *Q. robur*
Juglandaceae	*Carya*	Hickories of North America and Asia, includes White Hickory *C. tomentosa*
	Platycarya	Asian trees
Lamiaceae	*Tectona*	Teak *T. grandis* of Asia; very variable and sometimes described as semi-ring-porous or diffuse-porous
Lauraceae	*Sassafras*	North America and Asia, includes Common Sassafras *S. albidum* from eastern North America
Lythraceae	*Lagerstroemia*	Crepe myrtles of Asia and Australasia
Moraceae	*Maclura*	Americas, includes Osage Orange *M. pomifera*
	Morus	Mulberries across temperate areas, includes Black Mulberry *M. nigra*
Oleaceae	*Fraxinus*	Ashes of temperate and subtropical areas, includes European Ash *F. excelsior*
Paulowniaceae	*Paulownia*	Asian trees, including the Foxglove Tree *P. tomentosa*
Rosaceae	*Prunus*	A familiar genus of northern temperate areas, many not ring-porous but some are, including Black Cherry *P. serotina* from North America
Rutaceae	*Phellodendron*	Cork trees of Asia, includes the Amur Cork Tree *P. amurense* from western Russia
Sapindaceae	*Koelreuteria*	Asian trees, includes Golden Rain Tree *K. paniculata*
Simaroubaceae	*Ailanthus*	Asia and Australasia, includes Tree of Heaven *A. altissima*
Ulmaceae	*Ulmus*	Elms of the northern hemisphere, includes Wych Elm *U. glabra*
	Zelkova	Europe and Asia, includes the Caucasian Elm *Z. carpinifolia*

Based on information from the InsideWood database (https://insidewood.lib.ncsu.edu/) and Wheeler *et al.* (2020).

on its smaller latewood vessels, although growth will be constrained by lack of water in the crown. Dependence upon a single ring, however, has important consequences for elms and Dutch elm disease (Chapter 15).

The fast but risky approach of ring-porous trees means that they can grow very fast and efficiently but they are more prone to summer cavitation and irreversible freezing damage. Ring-porous trees thus tend to be native to temperate areas where the winters are mild and the growing season is long, allowing them time to grow a new ring before the leaves emerge. In temperate Europe, around 35 per cent of species are ring-porous, declining in abundance until they form just 1 per cent of species in the tropics.

Tropical trees

Tropical trees are interesting in that most are classified as diffuse-porous, primarily because they have no annual rings and so there is no reason for the vessels to be clumped in wood grown at any particular time of year. The vessels are therefore scattered through the wood, just like in temperate diffuse-porous wood. However, the vessel diameter of tropical trees can vary from as small as that found in temperate diffuse-porous wood through to the largest vessels in ring-porous wood (McCulloh *et al.* 2010). There is obviously no problem with winter gas bubbles in tropical trees, but summer cavitation can create embolisms, especially for trees in open conditions exposed to wind and hot sun. Individual tropical species can be placed along the same Volvo–Ferrari spectrum as temperate trees, but this is expressed in vessel diameter rather than the arrangement of vessels within an annual ring.

Tyloses and gels

Pits and perforation plates are good at stopping gas bubbles from being pulled into the whole tree. However, many trees also have a longer-term method of sealing xylem tubes that are embolised. In the tracheids of conifers this is done using resins, and in broadleaf trees vessels can be permanently blocked by gels (often called gums) or by tyloses (singular *tylosis*) (De Micco *et al.* 2016). Gels are slowly secreted into the vessels from living cells surrounding them and are found mainly in trees with narrow vessels less than 0.08 mm wide. Tyloses are balloon-like outgrowths, also from surrounding living cells, including those in the rays, that explode into the vessel, just like airbags in a car (Fig. 79). These stick together to form an effective physical barrier against air bubbles and often contain phenols that make them a chemical barrier against fungal rot. Tyloses are found in around 17 per cent of the world's trees and tend to be in trees with wide vessels, but as is usual in botany, there are many exceptions. Nevertheless, whether by gels or tyloses, the vessels are permanently blocked to gas and water movement.

Not all broadleaves have these extra defences. Red Oak, for example, does not produce tyloses or gels even though it is ring-porous and has very wide vessels. Presumably the perforation plates have proved adequate in the forests of eastern North America, and the extra cost of tyloses has been dispensed with. This does, however, affect the use of the wood. As mentioned in Chapter 2, whisky barrels

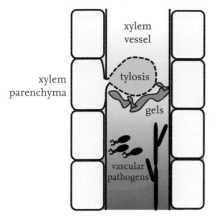

xylem vessel

xylem parenchyma

tylosis

gels

vascular pathogens

tyloses

FIG 79. To permanently resist the movement of gas bubbles and to resist pathogens moving up through the xylem vessels, trees can produce either tyloses or gels. Both tyloses and gels are produced by the living cells (parenchyma) that lie alongside the vessels, and both act to block the tubes irreversibly. The pathogens drawn are fungi with long hyphae and Oomycetes, fungi-like microorganisms, including *Phytophthora* species, that have motile spores. From Kashyap *et al.* (2021).

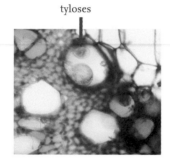

made from Red Oak would be useless, since they would leak, and certainly much of the alcohol would evaporate through the wide vessels that lack tyloses and are therefore still open conduits for liquid and alcohol fumes!

HYDRAULIC ARCHITECTURE

A tree has to ensure that water reaches every part of the crown. In a broadleaf tree, the vessels make up around 8 per cent of the cross-sectional area of the xylem (the rest being fibres), while in conifers the tracheids occupy 41 per cent of the cross-section. Part of this difference is explained by the tracheids doing both jobs of conducting water and giving strength to the xylem, but it also underlines how efficient the wider vessels of a broadleaf tree are in conducting water and why conifers generally need to use a greater number of rings over which to conduct water up the tree.

It is tempting to think that the very short tracheids in conifers means that the numerous bordered pits are more of a drag on water flow compared to the streamlined vessels with perforation plates. Since tracheids are around 10 times shorter than vessels, water moving up through a conifer will pass through around 10 times as many 'valves' as it would in a broadleaf tree. However, for tubes of the same diameter, the total hydraulic resistance (the reduction in water flow through the xylem) is 56 per cent in broadleaves and only a 64 per cent loss of conductivity in conifers. Bordered pits offer almost 60 times less resistance to water flow than perforation plates, which is down to the superb design of the torus-margo system described above. Perforation plates, with no moving parts, are very simple but offer a high resistance to the flow of water. So efficient are bordered pits that it has been calculated that if conifers had perforation plates, there would be over 95 per cent reduction in water transport. The reason why conifers need so many tubes through which to conduct water is not due to the valves slowing things down but to the high friction of the water against the sides of the narrow tubes; narrow to help resist cavitation and keep air bubbles small. These efficient valves help conifers compete with broadleaves, particularly in colder areas that favour diffuse-porous broadleaf trees.

Vessels and tracheids tend to become narrower with height in the tree. This is due to the suction against gravity becoming greater with height and the higher evaporation likely from leaves in full sun at the top of the crown. The tubes become narrower with height, giving greater cohesive strength to the water column due to increased friction with the cell walls. Conversely, vessels and tracheids below ground tend to be up to 2–3 times wider than in the trunk, especially in the deepest roots. This helps to reduce hydraulic resistance to water movement from deep in the soil. Also, roots do not need to support themselves in the same way as a trunk and so can have wider tubes and lower overall wood density. Roots are normally largely immune to freezing, at least in temperate areas, so there is less need for small tubes to keep winter gas bubbles small.

Narrower vessels and tracheids near the top of the tree help hold the water column together, but the extra friction also slows the rate of water flow to the top leaves. Since these are the ones that will usually be in full sunlight, they will usually need more water than those lower down and yet they should get less. Conversely, leaves lower down in the crown have less far to pull up the water and so will get more. An analogy I've regularly used is to think of two people with straws in a thick milkshake. One has a 20 cm long straw and the other 2 m; if they both start sucking at the same time, the one with the shortest straw will get more than their share, since there is overall less friction to overcome. So how can leaves at the top of a tree get their share of water when competing with those nearer the roots? The simple answer is that a tree has constrictions in the xylem tubes at junctions such as where branches leave the trunk, reducing the conductivity to less than half that of the branch itself. This is like putting a crimp in the short straw to even up the share of milkshake. These constrictions have two advantages. The first is that they help even out the water supply to all parts of the canopy. Secondly, these constrictions can help in preventing air embolisms that may develop in branches from getting back into the main stem and even the roots. This helps protect the main trunk, which has a lot of resources invested in it, in times of water shortage, while low-investment parts such as leaves can be sacrificed and regrown later. Palms, which have only one growing point and one set of xylem tubes through their life, have considerable hydraulic resistance at their leaf bases to ensure that any water stress is felt by the leaves and not the trunk and its growing point.

Many broadleaf trees have a *hydraulic network* through the trunk and branches that allows water to spread out sideways around the tree as it moves upwards. These are described as being *highly integrated*. Inside the xylem, the vessels intermingle like spaghetti hanging from a fork and are abundantly connected together, allowing water and nutrients to fan out around a growth ring and even between adjacent rings. This can be helped in broadleaf trees by conifer-like tracheids forming bridges between vessels, allowing water even more freedom to move sideways (Pan & Tyree 2019). This is a great advantage, since it ensures that water is shared evenly regardless of any damage to roots, especially for trees in environments where water is unevenly spread through the soil.

However, ring-porous trees such as oaks and elms, and conifers such as Eastern White Cedar *Thuja occidentalis* of eastern North America, tend to be at the other end of the spectrum and are more *highly sectored*, where water from one part of the roots moves up to one portion, or sector, of the tree above (Ellmore *et al.* 2006). This happens because individual vessels are much more isolated from each other and act as straight drainpipes, linking one part of the roots to one part of the crown. This would seem to be a distinct disadvantage, since water is not shared around the crown, but these trees may actually do better on soils where water availability is variable through the year. Their advantage comes because although they cannot share water around the crown, it also means that drought-induced embolisms are not shared around the tree either, allowing adjacent vessels to keep functioning with no risk of air bubbles passing between vessels (Zanne *et al.* 2006). It also limits the spreading of pathogens around the tree. And, it allows old trees to isolate living tissue more easily from areas that are dead or dying (Larson *et al.* 1994). The connections between the wide vessels of the earlywood and the smaller latewood vessels are also very few, so again this protects the smaller vessels from galloping embolisms that form in the wide vessels during dry periods.

HOW MUCH SOIL DO TREES NEED TO PROVIDE ENOUGH WATER?

If a tree is capable of using 20–400 litres of water per day during the growing season, it raises the question of how much soil a tree will need if planted in a pot or in an urban area where the root spread may be very restricted. There are obviously many factors that will affect this, including the species planted (how good it is at controlling water loss), tree size, what the surrounding conditions are like that affect evaporation of water from the leaves, and the type of soil, particularly how much water it can hold (sandy soils obviously holding less than organic-rich soil). Certainly, if given regular water and nutrients, a tree can survive in a ridiculously small amount of soil (Fig. 80). However, as a general estimate in temperate areas, 0.6 m³ of soil is recommended for every square metre of crown projection – the area of shadow under the tree if the sun is overhead (Lindsey & Bassuk 1991). This obviously works best for a standard spreading tree and will be an underestimate for upright, fastigiate trees. To put this into perspective, the birch and oak in Figure 81 would need soil volumes of about 1.4 m³ and 29 m³, respectively. For the mature oak, the raised planter is far too small to provide enough water. Fortunately, roots are designed to explore widely and will readily escape from their planting pit if conditions allow. This can be seen when paving slabs are lifted in streets and thick lines of fine roots can be found to follow the cracks between slabs, ready to absorb rain seeping down. Or if there is a nearby lawn, roots can usually be found to pass beneath paths and even roads to explore the lawn. The large oak in Figure 81 must have access to other sources of water beyond what is available in the raised planter. Behind the parked cars there is a grassed-over churchyard, so many of the roots providing water and minerals for the tree will be under the tarmac and wall and into the churchyard.

The soil volume required by a large tree contrasts with the standard 1 m³ planting hole often specified in urban tree-planting schemes. Small holes can be fine at first but will become more limiting as the trees grow in size, and made worse by periods of drought. This explains why urban trees can be fine for a few years and then suddenly show problems of drought, by which time the planting contractor is long gone. Shortage of water is not just a matter of

FIG 80. A specimen of Elm *Ulmus* 'Wingham' that is 3.2 m tall and yet growing in a 3-litre pot! This demonstrates how little soil is needed providing water and nutrients are regularly given. Nevertheless, it is not to be recommended, since any delay in watering will quickly be fatal, and the roots are tightly spiralled around the inside of the pot, which will cause problems when the tree is planted out. No wonder the garden centre was selling it cheaply!

survival but can also dramatically slow down growth; doubling the pot or hole size of a planted tree can increase the weight of the tree by an average of 43 per cent (Poorter *et al.* 2012).

a

b

FIG 81. (a) The young Szechuan Birch *Betula szechuanica* has an average crown diameter (from the centre of the trunk to the end of the branches) of 0.85 m, which gives a crown projection (the size of the shadow if the sun was directly overhead) of around 2.3 m². (b) The more mature Pedunculate Oak *Quercus robur* has an average crown diameter of 3.9 m, giving a crown projection of about 48 m². Using the general rule of thumb that a planted tree needs 0.6 m³ of soil for every square metre of crown projection, these two trees would need a soil volume of 1.4 m³ and 29 m³ of soil, respectively, to provide them with enough water throughout the year. For the Birch, this is just about met by the planting pit, which is roughly the size of the plastic grid in the ground, but for the Oak, the amount of soil in the 2 m diameter raised planter is obviously vastly inadequate, so roots will be passing under the tarmac and into the churchyard in the background.

Trees can, of course, be watered to prevent the soil drying out, and this is especially important when the tree is first planted until the roots can grow into the available soil to maximise water uptake. Watering can be by a tube installed deep into the planting pit when the tree was planted to ensure the water gets below the surface (Fig. 82a) or by using a watering bag or similar (Fig. 83) around a young tree that slowly releases water over a number of hours,

FIG 82. (a) The base of the Szechuan Birch *Betula szechuanica* shown in Figure 81, with a plastic grid over the planting pit, designed to reduce compaction of the soil surface. It also shows the top of the watering tube which is coiled around the root ball to ensure that water poured into the tube reaches the roots. (b) A less substantial plastic matting which reduces the impact of people walking on it but still allows some compaction of the soil. (c) A much more substantial metal grid at a high-footfall crossing, which protects the soil. (a) Keele University, UK, (b) Hokkaido, Japan, and (c) Reims, France.

FIG 83. Watering of planted trees. (a) A watering bag that is zipped up around a newly planted tree and filled with water which seeps out of holes at the base over 5–9 hours, giving the tree a constant supply of water that soaks into the soil rather than running off the surface. (b) A similar home-made system using felt (foreground) and plastic sheets (background) to retain irrigation water on a slope in a private garden. (a) Hyde Hall, Essex, and (b) northern Honshu, Japan.

helping to increase infiltration into the soil and reduce runoff that is water lost to the tree. De-compacting soil around the base of a tree that has been repeatedly trodden on or run over by vehicles will also help the tree by encouraging water to infiltrate into the soil rather than just running off to the nearest drain. Preventing soil compaction in the first place is obviously even better, and the surface can be protected by using a grid or plate (Fig. 82) or by using permeable asphalt.

If grids are not feasible, the whole planting pit can be given some protection by planting the tree in a *structured soil*. This is composed of very sandy or stone-rich soil (up to 80–90 per cent stone, with pieces up to 15 cm diameter) such that the sand or large stones bear the weight of the street surface and its users, protecting the matrix of finer soil between and thus allowing roots to grow

FIG 84. Underground structures to aid tree roots. (a) A fairly simple plastic honeycomb that will prevent the paving slabs from compacting the soil, allowing roots to grow widely and absorb water. (b) A more involved crate system being demonstrated in a car park. It consists of a large upper crate containing the tree (the trunk can be seen) but allowing roots to spread sideways from below the main crate into further crates which include a water pipe below the planting crate to allow water to spread between crates. Forcing roots to go more deeply into surrounding crates will save disruption to the paving surface as the roots become thicker. (a) Central London and (b) Myerscough College, Preston.

unhindered. The heavier the street surface, the bigger the aggregate used. Trees certainly grow better in this than in standard soil that is compacted (Smiley *et al.* 2006) and can grow very well. Swamp White Oak *Quercus bicolor* and Willow Oak *Q. phellos* in a New York suburb in a structural soil grew as well over a 17-year period as they did when planted in an adjacent lawn area (Grabosky & Bassuk 2016). The main problem with structured soils is that there is only 20 per cent or less volume of soil that can hold water and nutrients, so they are even more prone to drying, and watering and fertilising become even more important and are often needed for much longer than is usual.

A stage further is to use a *crate system* or *suspended pavement* (Fig. 84), where structural cells made of plastic or concrete bear the weight of the street, preventing the soil from compacting. These have the benefit that 90–95 per cent of the planting pit is made up of uncompacted soil and this can spread below the pavement or road, providing a much larger volume of soil for the growing tree. These also have the advantage of holding significant quantities of water and can help prevent or reduce local flooding of streets. But, of course, they are more expensive to buy and install.

CHAPTER 7

The Growing Tree

Approaching the middle of the summer, when growth is at its maximum, raises a number of questions about just how fast trees grow and how they coordinate the size of different parts of the tree that rely on each other. If you want to know more about how big they can become and how long they live, skip to Chapter 12.

INCREASE IN SIZE THROUGH A TREE'S LIFE

There is some dispute over the identity of the biblical mustard seed but it is generally considered to be Black Mustard (*Brassica nigra*, formerly *Sinapis nigra*) which 'Though it is the smallest of all seeds, yet when it grows, it is the largest of garden plants and becomes a tree, so that the birds come and perch in its branches' (Matthew 13:32, NIV). The plant gets its name from the black seeds which are about 1 mm in diameter. These produce a plant that is usually up to 1.5 m tall but can ostensibly reach 2.4 m and so, although a herbaceous (non-woody) plant, could resemble a tree. Like faith, a small seed can blossom into something large and firmly rooted. But for many trees the difference in size between seed and mature plant is even more extreme. For example, a seed of Giant Sequoia *Sequoiadendron giganteum* is in the order of 0.005 g yet the mature tree can be over 1,250 tonnes, which is roughly 250,000 million times larger than the seed. This is an extreme example, and while trees do not have to be large (think of an alpine willow just a few centimetres tall – Chapter 1), they often do grow into the largest living things in the landscape. This fact often takes people by surprise, when a tree planted in their garden grows quicker and larger than expected! Professionals are not immune to this, as trees in gardens and arboreta are often too close to each other to show their true shape. I cannot claim the moral high ground though: I am just as guilty of this when planting cherry trees as part of a National Collection at Keele University, and I battle with the desire to give each tree enough room yet wanting to squeeze in as many as possible.

RATES OF GROWTH

Tropical vines and lianas can grow almost half a metre in length per day under the right conditions. Trees that grow their own trunk do not grow as quickly, but a number of tropical species can add 8–9 m to their height in a year. A tropical tree named Batay *Paraserianthes falcataria* (previously *Albizia falcata*) is an important forestry tree in Southeast Asia because it can grow very rapidly, reaching a height of 7 m in one year, 16 m in three years and 33 m in nine years (Krisnawati *et al.* 2011). In the cooler conditions and shorter growing season of temperate areas, trees grow more slowly. Vigorous young trees in Britain might be expected to grow around 15–50 cm in height per year, equivalent to 1–2 mm per day, although this can reach 1.5 m, particularly in trees that have been pruned and so have plenty of stored food to fund new growth. This rate of growth is not unusual in cut hedges, particularly in vigorous pioneer species such as Elder *Sambucus nigra* (Fig. 85), and can be even greater in coppiced trees which can grow 3 m in the first year after cutting.

Pioneer species that invade gaps in woodland and open ground tend to grow faster than those growing in dense woodland. Tropical forest pioneers typically grow at 1.5–4.0 m in height per year while species growing in shade may only average 0.5–1.2 m per year. This is partly due to the lower light levels, but even if grown together in the open, the pioneer species will have inherently faster growth. The woodland tree is better adapted at using low light levels, and so may perform less well when there is an abundance of light in the open. Pioneer species also tend to invest more in speed rather than longevity, and so grow faster in height. A birch will produce a tall, thin trunk of less durable wood and quickly produce seeds before it is outcompeted by slower-growing, shade-tolerant trees. Long-lived trees such as oak make a greater investment in dense, durable, highly defended wood (Chapter 8), with a comparatively wide trunk, but this goes with slower height growth.

Height growth is also affected by age. As trees get taller, the height increase inevitably slows down before finally stopping once the tree reaches its maximum height, depending upon its genetic makeup and the growing conditions – this is explored further in Chapter 12. While a tree may approach a maximum height,

FIG 85. Vigorous growth of Elder *Sambucus nigra* in a hedge otherwise dominated by Hawthorn *Crataegus monogyna*. Once the hedge is cut, Elder grows quickly and can overshadow the Hawthorn either side, eventually killing it, which creates gaps. Since Elder is short-lived, once it dies it creates an even wider gap in the hedge. When hedges were maintained by hand, Elder was ruthlessly dug out to prevent this happening. Now, however, due to the creation of gaps in this way, many of our hedgerows are disappearing metre by metre.

the trunk will normally keep getting wider by the formation of new wood throughout its life to ensure water transport from roots to crown (Chapter 6).

WHICH IS STRONGEST: SLOW- OR FAST-GROWN WOOD?

Whatever you think the answer to this question is, even if you say it makes no difference, you would be correct! This is because conifers and different sorts of broadleaves grow in different ways, making the answer very simple but at the same time dependent upon what type of wood is being considered.

Conifers

The conifer tree ring in Figure 86 shows that each annual ring is made up of a pale portion followed by a darker portion. The pale portion is referred to as *earlywood*, grown early in spring, and is composed of tubes that are wide in diameter and have a very thin wall. This is because in spring when growth begins, the tree is in need of plenty of water at the top and so water transport is given preference. By contrast, *latewood*, grown later in the spring and summer, is denser and darker and gives the wood its strength. The higher density is the result of the xylem tubes having a very narrow internal diameter and thick walls to give strength. Annual rings will usually vary in width. This is because when growing conditions are good (which usually means warm and moist), the rings are wider than in years when the spring and summer are poorer. Similarly, trees grown under clement conditions will have wider rings than trees grown in harsher places. Although the ring width varies, the width of the latewood stays fairly constant – shown in Figure 86b. This means that in wider rings the extra width is created by growing more earlywood.

The result is that conifers grown in good conditions (such as in rich soils or in warm, moist climates), where the annual rings are wide, will have a greater

OPPOSITE: **FIG 86.** (a) The details of one year's growth of a pine tree, showing the transition from pale earlywood to darker latewood. The colour change is caused by the wood cells becoming smaller with thicker walls, resulting in denser latewood. The outermost ring is shown in detail. At the right there is the cambium and the inner bark, or phloem. The coloured lines show the change in tracheid internal diameter, wall thickness and wood density across the ring. (b) Relative contribution of earlywood and latewood width to total ring width in Sitka Spruce *Picea sitchensis*. As the ring gets wider, the width of latewood stays constant and the lower-density earlywood expands. (a) Based on Cuny *et al.* (2014) & (b) from Moore (2011), reproduced under the Open Government Licence.

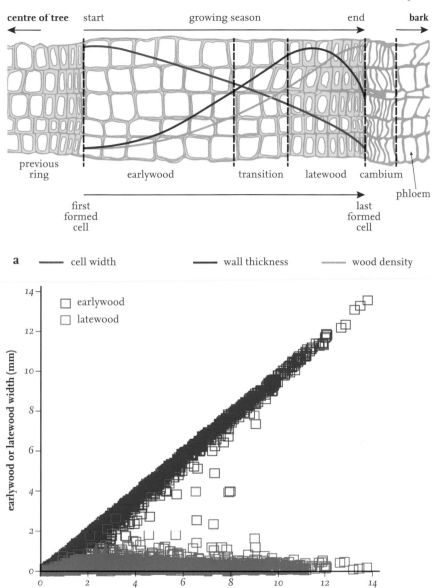

centre of tree · start · growing season · end · bark

previous ring · earlywood · transition · latewood · cambium · phloem

first formed cell · last formed cell

a —— cell width　——— wall thickness　——— wood density

b

ring width (mm)

earlywood or latewood width (mm)

☐ earlywood
☐ latewood

proportion of earlywood in a cube of wood. The wood will thus be overall of lower density, and therefore a plank of wood cut from the tree will be less strong than a plank from a slow-grown tree. In Britain, where our climate is ameliorated by the Gulf Stream, conifers grow very rapidly, and the resulting timber is of lower quality than wood grown in northern Europe, such as Sweden, where the climate is colder and the trees grow more slowly. If you want to build a house from pine, better to get the wood from Sweden rather than Scotland. Similarly, the best violins are made using sounding boards of very slow-grown Norway Spruce – the quality of sound of a Stradivarius violin comes from Spruce grown particularly slowly during the Little Ice Age between 1645 and 1750.

There is a caveat to this, since the strength of a piece of wood used in building depends on more than just wood density. Knots will tend to weaken a piece of wood, and where it came from within the tree will also make a difference, as density tends to increase from the centre of a tree out towards the bark. Wood density is thus not always a perfect predictor of wood strength... but I would still buy the pine for a new house from Scandinavia.

Conifers that grow in particularly warm areas, including members of the families Podocarpaceae and Cupressaceae, do not produce visible growth rings, and there is a smaller correlation between ring width and wood density.

Broadleaf trees

Looking at a piece of broadleaf wood, such as an oak shown in Figure 87, it can be seen that the earlywood is made up of large-diameter tubes. This type of wood is referred to as *ring-porous* (described in more detail in Chapter 6), since the pores, or vessels, can be followed around a ring with the naked eye. This allows very fast and abundant water movement. It also means that the earlywood in oak, just like that in the conifer, is fairly low density, while the latewood with smaller-diameter tubes has a higher density. However, unlike the conifer, it is the earlywood that stays roughly the same width regardless of the width of the growth ring, and in wide rings it is the latewood that increases to make up the extra width. This means that if oak is growing slowly, the majority of the ring is made up of low-density earlywood, and the wood is weak. By contrast, if the wood is grown quickly, the rings of wood each have proportionally more high-density latewood. The net result is that the strongest oak is found in trees that are grown fast, unlike conifers, where slow-grown wood is strongest. British oak is therefore much sought after, since our fast-grown oak is structurally stronger than that grown under a cooler continental climate.

This difference has been well known by woodworkers for generations. As an example, Dava Sobel in her 1995 book, *Longitude*, describes the quest for an

FIG 87. Cross-section of wood from Pedunculate Oak *Quercus robur*, showing at the bottom the large vessels of the earlywood and smaller vessels in the wider latewood above. The broad line of cells to the left is a ray, and the almost horizontal lines mark the winter breaks in growth. From Hirons & Thomas (2018). See also Figure 78 in Chapter 6 for another cross-section of oak.

accurate marine clock necessary for finding the longitude of a ship. This clock was finally provided by John Harrison, a self-taught clockmaker. But before this, Harrison built a clock made almost entirely of wood, including the cogs, into a tower above the stables for Brocklesby Park in Lincolnshire. Harrison used oak from fast-growing trees for the cogs because it was strongest. Elsewhere in the clock where he wanted lighter-weight material and strength was not as crucial, he used slow-grown oak. It is also a clock that runs without the need for oiling, since those parts that would normally need lubrication – the bushes through which brass rods pass – were made from Lignum Vitae (from *Guaiacum officinale* or *G. sanctum*), a very dense hardwood traditionally used for police truncheons, which also has a high oil content that in effect lubricates itself. The clock was completed in 1722 and is still running. John Harrison, a carpenter by profession, knew about the growth of wood!

What about other broadleaf wood? Tropical and temperate broadleaves that have the pores scattered more or less uniformly across the ring (referred to as diffuse-porous woods), including beech, maples and birches, show a much weaker relationship or no relationship at all between ring width and wood density. The speed of growth is more to do with how much wood can be grown per year rather than its strength.

WHAT CONTROLS TREE GROWTH?

It is perhaps obvious that there are many external constraints on tree growth imposed by the climate and soil conditions. A tree needs optimum supplies of light, water, carbon dioxide, nutrients and an equitable temperature range for optimum growth. How they affect growth is considered below. A number of other factors also affect growth and survival, including such things as wind (Chapter 11) as well as insects and pathogens (Chapter 15). There are also a number of internal controls on growth that allow a tree to overcome some of the external constraints and optimise how it grows; these are also considered below.

Light

Light is often the first limiting factor affecting growth that we tend to think of. A lack of light in a woodland can lead to very rapid vertical growth (etiolation), particularly if the light is coming from a small gap in the canopy far above. This leads to a taller, thinner tree with shorter branches in order to try and reach better light conditions. This is compounded by *crown shyness*. Trees tend to avoid growing through each other's crown (Fig. 88), which constrains a younger tree to quickly grow tall and thin to fit within the gap between older trees without touching them. Crown shyness is caused partly due to shade slowing growth as two crowns meet. But a tree also reacts to the shade of a neighbour by growing fewer buds and producing shorter branches where two crowns get close. This shyness may also be helped by abrasion of buds, leaves and twigs where the crowns of neighbouring trees rub together in the wind, particularly in tall, thin trees that may sway more in the wind than a stouter neighbour. White Spruce *Picea glauca* growing up through Aspen *Populus tremuloides* in Canada can suffer by being whipped by swaying Aspen branches, and the same is true in many other groups of trees. The result of crown shyness is that a woodland may be dominated by wide trees that started growing in the open and had room to

FIG 88. Crown shyness. (a) Looking up at the canopy of a Beech *Fagus sylvatica* woodland in Transylvania, Romania, gaps can be seen between the individual tree crowns. This is not always clear from the ground, since different layers of branches may overlap at different heights and so obscure the gaps. (b) The clear gaps between neighbouring crowns can be seen between the conifers, creating a mosaic, captured by a drone over a Norway Spruce *Picea abies* plantation at Keele University, Staffordshire.

spread, mixed with much narrower trees that had to fit into the gaps left by the original trees.

When light is plentiful, it can be cheaper to produce low spreading branches than a tall trunk. This is under genetic control so that an oak and a number of other trees will often be broad and spreading while a lime *Tilia* species tends to always have a fairly tall, narrow crown (Fig. 89) even when it is free to grow sideways. Other trees such as the Cypress Oak *Quercus robur* 'Fastigiata Koster' (Fig. 89c) have been bred to be upright, or fastigiate, and although they might spread a little when older, they maintain this shape for life.

Also under genetic control is the ability to cope with shade. Some trees are classified as *shade-intolerant* or *light-demanding* because they are poor at coping with low light levels, including species of aspen, birch, larch, many pines and False-acacia *Robinia pseudoacacia*. Others are *shade-tolerant*, and their seedlings are usually found naturally inside dense woodlands although they may eventually reach the top of the canopy to dominate. These include beeches, many maples, including Sugar Maple, hemlocks, firs and Western Red-cedar.

Despite being adapted to shade, growth can be very slow for heavily shaded seedlings and saplings. They may indeed be more or less standing still, barely surviving until a gap opens in the canopy or they die. These are often described as forming a *seedling or sapling bank* with a slow turnover, adding a few seedlings in good years and losing a few saplings in dry or cold years when the extra stress added to the shade tips some over the edge. There is evidence showing that heavily shaded, shade-tolerant temperate trees make a greater investment in defences to help them avoid being compromised by pests or pathogens (Jia *et al.* 2020), allowing them to hang on for longer. Other adaptations to shade include altering the leaf makeup, leading to the sun leaves and shade leaves described in Chapter 3 – shade leaves being thinner and larger to be better able to capture low light levels.

Nutrients

Trees, like all plants, require nutrients to support rapid summer growth. Some requirements, such as carbon (in carbon dioxide) and oxygen, are obtained from the air. Carbon dioxide is rarely limiting to growth even though it forms just 0.04 per cent (410 ppm) of the atmosphere, since there are very few situations that will reduce its supply. However, the opposite is now happening with an increase in greenhouse gases. Before the Industrial Revolution in the mid-1770s, the atmosphere held around 280 ppm, and while I was at school it was a memorable 333 ppm, while now it is 410 ppm and rising every year. As discussed in Chapter 2, since this is the main ingredient in photosynthesis, we are now giving our

FIG 89. The differing natural shapes of trees. (a) Keaki *Zelkova serrata* is known for its broad spreading crown if given the room; (b) by contrast, limes (in this case Silver Lime *Tilia tomentosa*) tend always to be taller than wide when open-grown; (c) but trees have also been bred to be tall and fastigiate, making them perfect street trees, such as the cultivar Cypress Oak *Quercus robur* 'Fastigiata Koster'. Some of the trees further down the road can be seen to be keeping their dried brown leaves, typical of beech and some oaks – discussed in Chapter 10. (a) Capesthorne Hall, Cheshire, (b) Keele University and (c) Belfast, Northern Ireland.

trees extra free 'food' supplies but, alas, this is not leading to more tree growth in the long term due to other factors, particularly soil nutrients, rapidly becoming limiting.

Plant nutrients are mostly taken up by a tree from the soil and dissolved in the water absorbed by the roots. However, there is evidence that absorption of nutrients through the leaves can happen. There are currently considered to be 14 *essential nutrients*. *Essential* means that a nutrient is directly involved in the metabolism of the plant, and the plant is unable to complete its life cycle without it, nor is it replaceable by another nutrient (Arnon & Stout 1939). Essential nutrients can be divided into *macronutrients* that are required in larger quantities and *micronutrients* needed in smaller amounts (Table 10). There are also another five elements in the table that are considered *beneficial* – that is, they can help growth, by either alleviating the toxicity of another element (for example, silicon helps with manganese toxicity) or by assisting a beneficial symbiotic organism (symbiotic nitrogen-fixing bacteria need cobalt) – but the plant can still grow without them.

Nitrogen, phosphorus and potassium are the macronutrients that most frequently limit plant growth and so tend to be the focus of fertilisers. As can be seen in Table 10, nitrogen is usually in low abundance in most soils. This is perhaps paradoxical, since nitrogen forms 78 per cent of the atmosphere and so is very abundant. However, plants cannot use elemental nitrogen (N_2) and it needs to be converted into a form such as nitrate or ammonia, either in the atmosphere by lightning or in the soil by bacteria, before it can be used. Phosphorus is often in short supply, since it occurs in low concentration in many soils and is often chemically locked up and hard to extract by a plant. Of the micronutrients, iron has a comparatively high availability in soils. However, whether it is soluble, and thus able to be taken up by the root in solution, depends upon the soil pH. It is much more available in acidic soils and so only tends to be in short supply in alkaline soils. Other nutrients such as nitrogen, potassium and calcium, conversely, become more available in alkaline soils. Although a pH of 7 is considered chemically neutral, in ecological terms a 'neutral' soil has a pH of 5.0–6.5, and nutrients are most available to the tree between pH 6.2 and 6.5.

Lack of any of these nutrients leads to characteristic symptoms and reduced growth. For example, a lack of nitrogen leads to chlorosis, where the leaves become uniformly yellowed, and a deficiency of magnesium or molybdenum can cause yellowing between the veins. Many books and web pages give the typical symptoms in trees of deficiency of different nutrients.

Too much of a nutrient can be as bad as too little: high levels of calcium in alkaline soil (such as on chalk and limestone), helped by iron being in short

TABLE 10. Nutrients required by all plants, including trees, divided into macronutrients which are required in relatively high amounts, micronutrients, required in smaller amounts, and beneficial elements which help plant growth but are not essential. The typical concentration of nutrients found in plants growing healthily is given in mg per kg of plant material when dried (dry weight). Average content of nutrients and beneficial elements in soil is given as g per kg of dry soil.

Element	Typical amounts found in plants (mg/kg)	Typical amounts found in soil (g/kg)
Macronutrients		
Nitrogen (N)	15,000	2
Potassium (K)	10,000	14
Calcium (Ca)	5,000	15
Magnesium (Mg)	2,000	5
Phosphorus (P)	2,000	0.8
Sulphur (S)	1,000	0.7
Micronutrients		
Chlorine (Cl)	100	<0.1
Iron (Fe)	100	40
Manganese (Mn)	50	1
Boron (B)	20	0.02
Zinc (Zn)	20	0.09
Copper (Cu)	6	0.03
Nickel (Ni)	0.1	0.05
Molybdenum (Mo)	0.1	0.003
Beneficial elements		
Sodium (Na)		5
Silicon (Si)	330	
Cobalt (Co)		0.008
Selenium (Se)		0.0003
Aluminium (Al)		70

Data based on Larcher (2003) and Marschner (2012).

supply in these soils, can lead to lime-induced chlorosis, seen as yellowing and thinning of the foliage, common in beech, pine and larch. But it is not always that straightforward. Nitrogen is a case in point due to atmospheric pollution. The natural background level of nitrogen compounds falling from the atmosphere in a form that is usable by plants is 1 kg of nitrogen per hectare per year (1 kg N/ha/yr). In many parts of the world, the current deposition rates are 20–30 times higher than this, and up to 60 kg N/ha/year in densely populated centres (Bobbink *et al.* 2010). Nevertheless, the increased levels of nitrogen may not be enough to keep up with increased growth of vegetation due to warmer temperatures, especially in tropical areas, and it has been calculated that by the end of the century, the global growth of trees will be reduced by 19 per cent due to nitrogen limitation and by 25 per cent if both nitrogen and phosphorus are considered (Wieder *et al.* 2015). However, very high levels of nitrogen can be too much of a good thing and lead to decline and death, as seen in Figure 90.

If the level of a nutrient falls below a minimum amount, then growth will not be possible; similarly, there will be a maximum amount above which it is toxic and again growth will stop. In between these will be a concentration that leads to maximum growth. The key is not just to hit this optimum for the main nutrients like nitrogen and phosphorus, but also to ensure that the relative proportion of *all* nutrients is suitable – referred to as *stoichiometry*, the balance between different nutrients and how this balance is affected by the environment (Ågren & Weih 2020).

As gardeners, we can be conscious of the need for fertilisers and worried that our trees need feeding. However, for most trees, if leaf litter is left in place to rot down, or an organic mulch is added, and if the soil is suitable for fine-root growth (Chapter 5), then there will enough nutrients from decaying organic matter to supply all that is needed. This is true even though trees reabsorb many nutrients before the leaves are dropped (see Chapter 10) and so the litter is comparatively nutrient-poor. The problem comes in urban areas where leaves are swept up or blown away, or the amount of soil is limited and can soon run out of nutrients. In these cases we are back to gardening and the possible need for regular use of fertilisers, as you would give to a tree grown in a pot (Hodge 1991).

Water

Too little or too much water will also limit growth and survival. Drought will result in insufficient water in the leaves, leading to reduced photosynthesis due to stomatal closure. Further shortage of water will eventually disrupt the biochemical pathways of photosynthesis and cause irreversible leaf wilting and death. The speed at which a lack of rainfall affects a tree can be altered by the

Control   Low Nitrogen   High Nitrogen

FIG 90. A long-term experiment of adding nitrogen to the soil was conducted at Harvard Forest, the outpost of Harvard University in central Massachusetts. The forest was already receiving 7–8 kg N/ha/year (the normal background level is 1 kg N/ha/yr). Plots of Red Pine *Pinus resinosa* receiving this were treated as the untouched controls. Experimental treatments were set up in 1988, with Low Nitrogen plots receiving 50 kg N/ha/year (six times the background level) and High Nitrogen plots receiving 150 kg N/ha/year (18 times background). (a) By 2002, 14 years after the experiment started, 12 per cent of the control trees died, but mortality was 23 per cent in the Low Nitrogen plots and 56 per cent in the High Nitrogen plots, with the look of the plots visibly changed with a progressively more open canopy. (b) By 2015, 27 years after the experiment began, mortality in the experimental plots had dramatically increased, particularly in the High Nitrogen plots where it was almost 100 per cent. Pioneer broadleaf trees had started to invade the Low and High Nitrogen plots, including Black Birch *Betula lenta*, Red Maple *Acer rubrum* and some Black Oak *Quercus velutina* and Red Oak *Q. rubra*. Nitrogen pollution was thus leading to a complete and drastic change from conifer plantation to broadleaf woodland. There can be too much of a good thing, and the global long-term consequences of high nitrogen pollution are worrying.

ability of the soil to store water. Sandy soils with relatively few capillary pores will dry quicker than clay soils with many small pores. Since nutrients enter the plant by dissolving in soil water, a dry soil will always be infertile even if it is potentially nutrient-rich. Some of the effects of soil drying from the top down can be alleviated by hydraulic redistribution (Chapter 5), where the tree re-wets the dry surface using water taken from deeper down in the soil.

Flooding can be equally damaging to a tree, primarily by causing fine-root death. Oxygen in flooded soils is quickly used up by roots and microorganisms, and diffusion down through the water is too slow to replenish what is lost. This results in soil below the top few centimetres becoming devoid of oxygen (anoxic). The coarse woody roots can survive for a considerable time in a dormant state, but fine roots quickly die in anoxic conditions, leading to greatly reduced water and nutrient uptake and, ironically, symptoms of drought in the canopy. The lack of oxygen also leads to the production of toxic compounds from the soil (such as hydrogen sulphide) and potentially also from inside the tree. The roots of many *Prunus* species (including cherries, peach, apricot and almond but not plum) contain cyanogenic glucosides as a defence against being eaten (Chapter 8). These are released as cyanide gas when oxygen is limiting, which, if trapped in waterlogged soil, can kill roots.

There is an element of genetic adaptation to flooding, since a number of mostly woodland trees are very intolerant of flooding, such as Holly and Hazel, while others are very tolerant, including willows. In North America, Swamp Tupelo *Nyssa aquatica* and Swamp Cypress *Taxodium distichum* (Fig. 91) can survive flooding for several months or even permanently. In the Amazon basin, there are more than 100,000 km^2 of forest that is flooded by 2–3 m of water for 4–10 months every year. Saplings that are completely submerged can keep their leaves and continue growing, albeit slowly.

In temperate areas, healthy trees will usually recover from short-term flooding, especially if this occurs during the dormant season. Thus, when we see flooded winter landscapes that seem to be occurring increasingly frequently in winter with climate change, although we humans may suffer hugely, most of the trees will recover.

Temperature

After light and water, temperature is usually the other most important aspect of climate to affect tree growth, and the effects of climate change are considered in Chapter 12. Here it is worth saying that photosynthesis and other physiological processes will work at temperatures from not far above freezing to around 35 °C. The optimum temperature for growth, however, decreases on a gradient from the tropics to the poles. What is more interesting is that most temperate plants, including trees, grow faster when the nights are around 10 °C cooler than the day (Yang *et al.* 2013). Plants kept at the same constant average temperature grow more slowly. The likely reason for this is that cooler temperatures at night lead to slower respiration which is less of a drain on the plant's resources.

But this need for cooler nights is not universal. For some reason trees such as the Dwarf Mountain Pine *Pinus mugo* of European mountains grows just as fast

FIG 91. Swamp Cypress *Taxodium distichum*. Native to seasonally flooded wetlands along rivers, it can survive many months of waterlogging but can grow very well on well-drained soils, as demonstrated in many British gardens. The typical 'knees' that develop in flooded areas can be seen emerging from the water. Growing up to 4 m high, these knees do not appear to act as snorkels, allowing oxygen to reach the submerged roots, as they do in other wetland and mangrove trees. Here, they appear to brace the roots and catch mud, adding to the weight of the root system and helping to stabilise the tree. As a result, Swamp Cypresses are very resistant to high winds, even hurricanes. Gainesville, Florida.

when kept at a constant temperature (Hoch & Körner 2009). Many tropical plants are also adapted to grow better with night-time temperatures that are closer to those of the daytime. It has also been suggested that a higher daily temperature range is linked to *reduced* growth in some trees. A study of Scots Pine in central Spain found this, but in this case it was because the higher daytime temperature caused by climate change led to drying of the soil and this slowed growth rather than the diurnal temperature range itself (Büntgen *et al.* 2013). There is undoubtedly still much more to learn about this.

What is known is that trees and other plants are not dormant or 'asleep' in the cool of the night. Trees do most of their physical growing at night, perhaps because water stress is lower. Moreover, tree branches move around at night.[3] It has been found using 3D-laser scanning that these movements can be quite large – Honey Locust *Gleditsia triacanthos* moved branch tips up and down by

---

3 These are termed 'nastic' movements if they are cyclic, or reversible, as compared to 'tropic', which is directional growth such as phototropism, where shoots grow towards light.

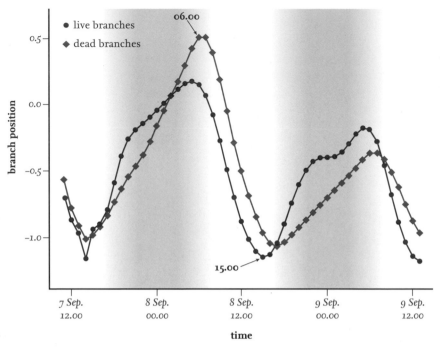

FIG 92. Movement of live and dead branches of Creosote Bush *Larrea tridentata* in New Mexico over two days, measured against a relative scale. Grey shading indicates night. From Hallmark *et al.* 2021.

up to 9 cm, although in most trees the movements were smaller and within the range of 0.5–1 cm. The patterns of movement can be quite complex. Branches of Oleander *Nerium oleander* were found to bend slightly downwards by 7:00 pm, then upwards by 9:30 pm, and down again with minor oscillations until 4:30 am. Better still, in the Southern Magnolia, there was a cycle of movement repeated three times a night, each cycle taking four hours (Zlinszky *et al.* 2017). In some trees the branches returned to their original positions but in others they did not. These movements took place at night so are not in response to light changes and are unlikely to be caused by cooler night-time temperatures. Daily movements have been found in both live and dead branches (Fig. 92). Changes in live branches are likely to be connected to changes in the water content of living cells, since the night is used to recharge water supplies within the tree, and cells with more water are more turgid and hold up the softer tissue more stiffly. The changes in dead branches are more likely in response to the wood absorbing more water from the higher humidity of the air (Hallmark *et al.* 2021). Whether there is any meaningful ecological benefit of these movements in live branches or whether they are just an inadvertent effect of water movement is, as yet, unknown.

Internal control

How well a tree grows is, as seen above, influenced by growing conditions. But trees are not just passively responding to these conditions – they are very much in control of what is happening inside in response to signals and stresses from outside.

A large part of this internal control of growth is exercised via the movement of sugar. Glucose is made by photosynthesis and is then used within the tree for respiration (the running costs of the tree). Respiration combines oxygen and sugar to release energy along with by-products of carbon dioxide and water – the reverse of photosynthesis. Sugar is also used to build *structural carbohydrates* such as cellulose and lignin, making the body of the tree. Some sugar may be given to mycorrhizas, useful fungi on the roots (Chapter 14). Sugar left over from these processes is stored in the roots and trunk for later use as *non-structural carbohydrates* which may be stored as sugars, starch or fats (lipids).

The upshot of this is that there are thousands of *sources* of sugar, mainly in the leaves but also wherever there is green chlorophyll, such as in the bark, flowers, fruits and buds. There are also many competing *sinks* where this sugar can be used. The strength of each sink will vary during the year; roots may need lots of sugar in early spring and so would be a strong sink, followed by the buds as they open and the cambium as new wood begins to form, with flowering and fruit production competing in there as well. This creates competition such that abundant fruit production can lead to branches around the tree being shorter with a smaller leaf area, less height growth in the tree and a narrower growth ring that year.

Although balancing the myriad of sources and sinks is seemingly worse than trying to organise a distribution centre for a supermarket, some of the decisions in the tree are fairly easily made, since sources close to a sink will generally meet their needs. So the first leaves to open will provide sugar to subsequent leaves on the same branch, smaller branches will fund the growth of the larger branches that they are carried on, and flowering and fruit production costs tend to be met by the branch they are on, along with any sugar they can grow themselves. However, there is also a need for long-distance transport so that, for example, the furthest roots get their share of sugar. This local and long-distance transport occurs through the phloem, or inner bark. Water and sugar flow passively through the phloem along gradients of sugar (from high to low concentrations), but these gradients are created by actively loading and unloading sugar at sources and sinks, which requires coordination within the tree to ensure that every part of the tree gets what it needs. There also needs to be coordination so that a balance is kept between the growth of different parts. Too much height growth might leave a tree with no stored food for producing next year's spring leaves; too many roots would be a drain on the tree if the tree already has enough water (more on this below). Fortunately, the tree has a highly organised internal control system of allocating resources by switching on and off different sinks at different times. Plants do not have a central nervous system like us, so the tree-wide internal control comes primarily from the production and movement of plant hormones.

Plant hormones can be *growth regulators* that promote various aspects of growth, or they can act as *inhibitors*. Growth regulators include: auxin (mainly IAA – indole acetic acid), produced primarily by shoot tips and leaves (and the basis of commercial rooting compounds); gibberellins, produced in the same place as auxins plus the root tips; and cytokinins, produced particularly by young fruit and root tips. The gas ethylene is also involved in regulating wood formation and fruit ripening (one rotten apple, producing ethylene, spoils the barrelful).

There are a number of growth inhibitors in plants but the main one is abscisic acid (ABA), produced mostly in leaves and seeds and causing bud and seed dormancy and the shedding of leaves. It is becoming increasingly apparent that other hormones are equally important, such as salicylic acid – this is famous as being a simple form of aspirin, originally derived from the bark of willows, but is a widespread compound. It was originally thought of as a defence hormone (Chapter 8) but now seems to be involved in many areas of growth and development, from seed germination to senescence before death (Rivas-San Vicente & Plasencia 2011).

Hormones are spread around the plant in the plant's plumbing – in the sap moving up through the wood (xylem) from roots to crown, and in all directions around the tree through the inner bark (phloem). This has the disadvantage of being quite slow. Sap in the xylem moves at around 30 cm/hr in conifers and diffuse-porous trees, and up to 1–4 m/hr in ring-porous wood. Phloem sap has been measured at speeds of 10 cm/hr to 125 cm/hr in European Larch *Larix decidua* and American Ash *Fraxinus americana*, respectively. Hormones are comparatively large molecules and so move slower than the sap itself, perhaps in the order of a few tens of centimetres per hour. Although a slow process compared to human nerve impulses which move more than 2 million times faster, this is fine for big seasonal changes such as flowering and breaking dormancy of buds after winter.

But it is too slow to help in an emergency, such as reacting to drought or low temperatures, or in the event of attack by pests and pathogens. Fortunately, some signalling around the plant – sensing a problem in one part and sending the message to other parts – can be quicker using a wide range of peptides (the building blocks of proteins) and other compounds (Takahashi & Shinozaki 2019) which, being very small, can travel in the plumbing at more like the speed of the moving sap. Fastest of all, sudden changes in turgor (the water pressure in a cell), along with the movement of calcium ions in and out of cells to produce nerve-like electrical impulses that move at metres per minute, allow much faster responses (Choi *et al.* 2016), such as rapid closure of the stomata to reduce water loss. This is still slow compared to human nerve impulses, which move at up to 120 m *per second*, but again, it is usually a quick enough reaction time to keep the tree safe from water loss or damage. Trees can also produce compounds that are volatile and which can disperse by air through the crown to rapidly spread messages about pathogen attacks (see *The tree's immune system* in Chapter 8).

A tree sitting in the landscape might not show that anything is going on inside, but its growth and development are controlled by a complex set of feedback loops and communication links that monitor what the external environment is doing to the tree and the internal levels of water, nutrients and sugars around the tree, and it reacts accordingly. A tree may not have a brain or a nervous system, but it is finely tuned to react to its needs and challenges, allowing it to persist for centuries or millennia.

BALANCING THE ROOTS AND SHOOTS

An illustrative example of a tree's internal control at work is in the relative balance between the weight of roots and shoots that it holds, usually referred to as the root–shoot ratio. Too many roots will be a drain on the limited sugars produced by the canopy, and too few roots will leave the crown potentially short of water. This balance is partly determined by genetics. Species invading open areas tend to have proportionately more roots, since they are exposed to full sun and need more water than those invading woodlands. There can also be genetic differences within a species so that trees growing in an arid area will have a greater proportion of roots to shoots than individuals of the same species growing in a moister climate (Ledo *et al.* 2018). If trees from the two areas are grown under the same conditions, this difference is maintained, showing that it is under genetic control.

Within this genetically determined ratio, there is room for manoeuvre depending upon conditions at any one time. This flexibility in ratio is brought about by the internal signalling described above. For example, drought leads to a higher proportion of roots, since water is essential even if the extra roots are a drain on the sugar-producing capabilities of the crown, and a very small root

FIG 93. Bonsai trees on display in Honshu, Japan.

system artificially given abundant water and nutrients can support a very large shoot (look back at the 3.2 m elm growing in a 3-litre pot in Fig. 80 in Chapter 6). In a similar way, if part of the canopy is broken off by a windstorm, some of the fine roots will die to restore the balance. If the roots have problems growing because of shallow soil or competition with other plants, the canopy will remain small. This is exploited in the use of dwarfing root stocks (Chapter 13), which are genetically limited in their root growth and will slow the growth of a tree grafted on top. Bonsai trees (Fig. 93) are kept perfect miniatures, partly by genetic selection of small-growing specimens and partly by using small pots and regular root pruning to restrict the growth of the roots, along with a low amount of nutrients to slow the growth of the crown.

TREE PLANTING AND ROOTS

The importance of the balance between roots and shoots is seen when transplanting bare-rooted trees. A tree dug from a nursery bed may lose up to 98 per cent of its roots. To help the tree, we usually move trees in the winter to allow some of these missing roots to regrow as the soil warms and before the shoots start needing water. Or we move them as root-balled or containerised trees (Fig. 94) so they keep more roots. But the shortage of roots is why a newly planted tree needs regular watering, often for several years, to ensure that the few roots it does have can take up sufficient water for the canopy while it grows more roots and restores the root–shoot balance. Methods of ensuring that irrigation water reaches the roots are discussed in *How much soil do trees need to provide enough water?* in Chapter 6.

Even with plenty of water, a newly planted tree may not put on any appreciable growth in the first few years. At that point we usually want to add fertiliser, since the tree has plenty of light and water and so nutrients must be the problem. This may sometimes be true, but in many cases it is not the lack of nutrients in the soil but competition for those same nutrients and water from herbaceous plants, particularly grasses, growing over the rooting zone of the tree. Applying nutrients may not improve growth, since it has been shown that when

FIG 94. To help preserve fine roots when trees are planted, they can be (a) root-balled – lifted from the nursery, keeping some soil around the roots. The whole ball of hessian and fine wire can be planted and both disintegrate quickly. Even here some 80 per cent of the original fine feeding roots can be lost. (b) An alternative is to containerise trees by transplanting them into pots for up to a year, even big pots – in this case 250 litres, though much bigger ones are possible. This allows time for more fine roots to grow while being irrigated through the black tube seen above the pot. A problem with growing trees in pots is that when roots meet the edge of the pot, they can start circling around the inside, which will continue when planted as *girdling roots* and impede proper root development. Here, the white pot keeps the roots cool and encourages the roots to grow down away from the light rather than around. Barcham Trees, Ely.

nitrogen is applied around trees less than 20 years old, the grass benefits far more than the trees (Hodge 1991). This gives a salutatory lesson that the roots of a newly planted tree should be kept free from competition. This is best done by creating a bare area at least 1 m in diameter under the tree. If this is covered in mulch, so much the better, since this will supply some nutrients and it also dissuades people from walking over the rooting area and compacting the soil (Fig. 95). Older and weed-free trees can, however, respond readily to nitrogen.

Removing competition and planting in a moist and nutrient-rich soil allow us to cultivate a range of trees in places they would not be found in nature. For example, in Britain, Ash is found mainly on nutrient-rich soil, including moist river valley bottoms and dry limestone, birches and pines on nutrient-poor sands and peats, and Yew on dry chalk and limestone. If we protect these trees from competition, we can grow all of them side by side on a wide range of soils.

The factors limiting seed germination are often different from those limiting subsequent growth. As an example, Alders are widely planted on very dry land reclamation sites, since they leak nitrogen into the soil from root nodules (Chapter 14) and help the growth of other plants. But they have to be planted as saplings, since they could not grow there from seed; the seed and young seedlings need waterlogged soil to do well. Once past that stage, however, they can be planted almost anywhere.

FIG 95. Protecting roots from compaction. (a) An area of mulch underneath the crown directly benefits the tree by holding water and providing nutrients, and subtly dissuades people from walking near the trunk, reducing soil compaction. (b) A more effective reduction in foot traffic can be produced by stacking up broken branches and tree prunings, but at the loss of the benefits of mulching. (a) Chinese Tulip Tree *Liriodendron chinense* at Killerton, Devon, and (b) Large-leaved Lime *Tilia platyphyllos* at Charlecote Park, Warwickshire.

CHAPTER 8

Defending the Growing Tree

BARK, THE OUTERMOST DEFENCE

Bark is often useful in helping to identify an unknown tree, and we often plant trees because their bark is so varied and aesthetically pleasing (Fig. 96). Despite the relatively small amount of bark on a tree, it is critical to its survival. Bark evolved as the outermost defence of the tree against pathogens and environmental stresses. On average, bark in Eurasian trees makes up just 14 per cent of the trunk volume – more (around 25 per cent) in small branches less than 2 cm in diameter and less (around 12 per cent) in thick trunks up to 1 m in diameter (Schepaschenko *et al.* 2017). Bark is often thicker in conifers than in broadleaves.

Bark is made up of two portions: the inner bark, or *phloem*, which conducts sugary sap around the tree, and the outer bark, which acts as the waterproof skin of the trunk, keeps out pathogens and protects the vulnerable living tissue from extremes of temperature, including fire. As described in Chapter 3, the inner bark is comparatively thin and functions in conducting sap for normally just one or two years (exceptionally up to 5–10 years), making a layer usually less than 1 mm thick. The older, non-functioning phloem on the outside is crushed as the expanding tree meets the constraining outer bark and is rapidly assimilated into new outer bark.

The outer bark is analogous to our skin, forming the barrier between the internal body and the external world. In a young stem, the skin is formed by a layer of cells (the *epidermis*) covered by a waxy cuticle. As the twig grows, it needs a new tissue that will be able to grow sideways to cope with the twig getting fatter. Underneath the epidermis a new growing zone is produced – the *cork cambium*, or *phellogen*. This is similar to the vascular cambium between the xylem and the

FIG 96. Trees admired for the beauty of their bark. This is often in the eye of the beholder, but trees planted for their bark include (a) the variety of Chinese Red Birch *Betula albosinensis* 'Septentrionalis', (b) Deciduous Camellia *Stewartia pseudocamellia* from Japan and Korea, and (c) Tibetan Cherry *Prunus serrula* native to west China.

phloem that grows new wood (xylem) on the inside and new inner bark (phloem) on the outside, except the cork cambium is formed further out, just below the epidermis (Fig. 97). This cork cambium coats the whole tree except for the very ends of the branches and the fine roots. As it grows it divides to produce new corky cells on the outside (*phellem*) that rapidly die as they are impregnated with tannins and a waxy substance called suberin, turning them into a new waterproof and gasproof layer below the epidermis. A few cells can also be formed on the inside of the cork cambium (the *phelloderm*) that remain alive and can be capable of photosynthesis (see Chapter 3). The whole of this new corky bark is collectively called the *periderm* (Fig. 97). As the periderm forms it cuts the epidermal cells off from the water and nutrient supply and so these die and scale away, leaving the phellem as the bark you touch on the outside of the tree. The periderm being gasproof also cuts off the living cells inside the stem from oxygen from outside. Some oxygen is dissolved in the water absorbed by the roots, but this is supplemented by having holes in the bark, called *lenticels*. In these the cork layer is puffed up into a loose mound of cells, with air spaces between, allowing gas exchange. These lenticels typically make up 2–3 per cent of the bark's surface and are seen on different trees as small pimples or the horizontal raised lines on birches and cherries (Figs 96c & 98d).

In some trees such as Beech and Cork Oak, the original periderm stays alive for the life of the tree. To cope with the expansion of the tree, the periderm is able to grow sideways to keep the tree covered. In Beech, relatively few corky cells

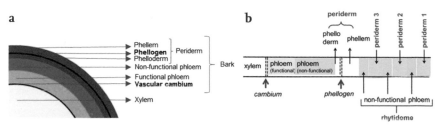

FIG 97. (a) Bark (periderm) formation starts underneath the epidermis (not shown in the Figure) and above the inner bark (phloem). (b) Some trees keep the original periderm throughout their lives; other trees produce new periderms, successively deeper in the old non-functioning phloem. In this case, the oldest is periderm 1 and the current fourth periderm is now growing inside periderm 3. Each new periderm cuts off the old ones which die, remaining as dead bark. These dead layers are called the *rhytidome*. Strictly speaking, trees like beech that keep the original periderm and do not form successive periderms deeper in the old phloem do not have a rhytidome. From Leite & Pereira (2017).

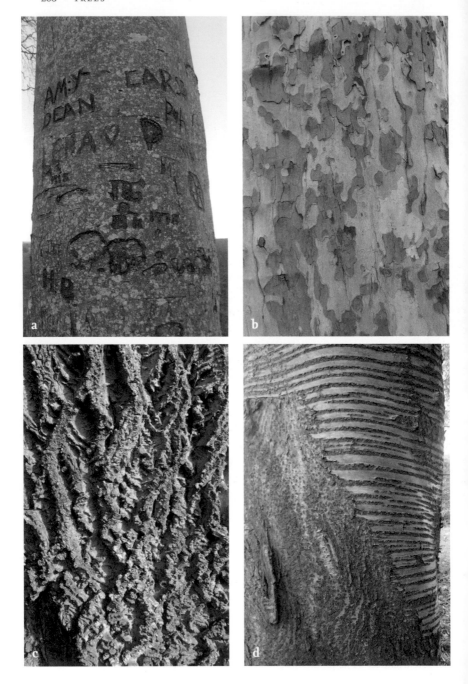

(phellem) are produced so the bark remains smooth and typically around just 6 mm thick in a 30 cm diameter Beech, with the oldest corky cells on the outside breaking away as powdery fragments. The persistence of the original periderm is why initials carved on a Beech remain for decades if not longer (Fig. 98a). These carvings become even more prominent because the tree reacts to the wounds made by the penknife by producing a *wound* or *traumatic periderm* under the wound that joins up with the original periderm. It effectively seals the wound, but the wound periderm is rougher than the normal bark, making the letters even more prominent. A similar story can be seen in Cork Oak, but here the periderm can produce many centimetres of corky bark on the outside over 20–30 years, making a wonderful fireproof blanket which protects the living cells inside the trunk from the frequent natural fires in its native Mediterranean habitat. We humans also find this corky bark useful, and the almost pure cork (made up of 45 per cent suberin) can be split from the tree by making a vertical slit in the bark and peeling off the cylinder of oak along the cork cambium. This works best in spring when the cork cambium is actively growing and so easily split. The cork tree reacts to this by producing a brand-new cork cambium (a *traumatic phellogen*) deeper into the old inner bark, which produces new cork. Stripping of the oak can be done every 8–10 years, producing increasingly better-quality cork until by the third to sixth stripping it is good enough for wine corks. It is important to note that when bark is stripped from most trees, they will die because it separates from the tree along the *vascular* cambium and so all the inner bark is lost. In Cork Oak the bark splits off through the *cork* cambium, leaving the inner bark in place and free to carry on conducting sugary sap.

Most trees cope with their growing diameter by producing new cork cambia successively deeper in the old non-functioning phloem. In Figure 97b, the oldest periderm is on the outside (periderm 1), and in this case the fourth one has now started. Each periderm may live for several years before being cut off and killed by new cork cambia below. These periderms may be continuous and coat the whole tree, but often they are smaller and shell-shaped with their edges pointing outwards, and so producing plates of bark (Fig 98b) that will flake off with time or longer lines down the bark (Fig. 98c) creating ridges formed from the old bark

OPPOSITE: **FIG 98.** Bark of (a) Beech *Fagus sylvatica* which keeps the original bark, or periderm, for life, with very little new bark production, so initials carved in Beech remain for years, just being stretched sideways as the tree expands in diameter. (b) London Plane *Platanus × acerifolia* with plates of bark. (c) Hybrid Walnut *Juglans × intermedia* which, like ashes and oaks, has ridges of bark. (d) An old Wild Cherry *Prunus avium* losing the smooth bark with horizontal lenticels of its youth for a less familiar bark of old age.

that clings to the outside of the tree. Birch bark, which can be peeled off in thin papery strips (Fig. 96a), has a large number of thin, tightly packed periderm layers each consisting of alternating layers of thick-walled and thin-walled cells that tear through the thin-celled layers. In most trees as they increase in girth and the old bark accumulates, its appearance changes from the smoothness of youth to the rugged pattern of maturity, a process that typically varies from 10 years in Scots Pine to 20 years in Alder, 30 years in oaks and around 50–60 years in cherries (Fig. 98d).

As noted above, the bark on Beech may be just a few millimetres thick and make up just 7 per cent of the trunk volume in a tree 30 cm in diameter – half that of most trees. In other temperate trees the successive periderms stick together so that in oak the bark can be several centimetres thick. This is still remarkably thin compared to some: in the Coastal Redwood the bark can be up to 15 cm thick, in a 50 m tall Douglas Fir around 25 cm thick, and in a Giant Sequoia the bark can be more than 80 cm thick, representing centuries of old periderms, creating a soft, spongy fireproof blanket around the tree. Trees growing in the shade tend to have thinner bark which can be damaged if suddenly exposed to the sun, referred to as being 'scorched', especially those trees which tend to have thinner bark anyway, such as beeches, hornbeams, maples, spruces and Silver Fir. Transplanted trees of these species are sometimes protected by being wrapped in paper, raffia or sacking when planted in the open (Fig. 99). Thicker bark will develop, and the protection is needed just for the first year or two.

When outermost layers of bark are shed this tends to happen towards the end of the growing season, especially after hot weather when desiccation helps dislodge loosening pieces. Bark shedding would appear to be wasteful but can be seen as an advantage in that it helps dislodge any parasites (such as mistletoe) or epiphytes growing on the outside that add extra weight and wind drag and can reduce a tree's wind firmness (Chapter 11). Moreover, London Plane *Platanus × acerifolia* (previously *Platanus × hispanica*) has survived so well in polluted cities because it regularly sheds bark, taking with it lenticels blocked by soot and exposing fresh open ends to the lenticels.

A little digression is called for into the role of corks and tainted wine. When a newly opened bottle of wine has a musty taste, like licking a cellar wall, the wine is said to be 'corked'. This cork taint is primarily due to TCA (2,4,6 trichloroanisole), which is derived from chlorophenols by the action of moulds and some bacteria. Chlorophenols can originate from a number of pesticides or wood preservatives used in a cork forest or a winery but can also come from the reaction of lignin in the cork with common bleach (sodium hypochlorite) used for sterilising. TCA can be detected by many people at levels of 1–5 ppt (parts per

FIG 99. Planted trees wrapped to prevent bark-scorching. (a) A flowering cherry, a variety of *Prunus*, in Hokkaido, Japan, and (b) an American Sycamore *Platanus occidentalis* on Mount Sugarloaf, Massachusetts. The unwrapped bark can be seen at the top of each picture.

trillion), which is the equivalent of a quarter of a teaspoon in an Olympic-sized swimming pool. Fortunately, most people also rapidly adapt to the taste and it becomes less bothersome after the first mouthfuls. Corked wine was one of the main drivers towards the use of plastic corks and screw caps for wine bottles. However, better husbandry of the cork trees and changing hygiene regimes during wine production seem to be solving the problem and it is to be hoped that the use of real corks will expand again. The biodiversity loss from a decline in cork oak use, which is already under threat from climate change and a resulting increase in pest problems, is potentially devastating (Bugalho *et al.* 2011, Tiberi *et al.* 2016, Kim *et al.* 2017).

TREE DEFENCES

Repelling boarders: First line of defence

Bark is a very good first line of defence, creating a physical barrier of dead cells impregnated with lignin and suberin, often backed up by tannins and phenols, that is waterproof and very difficult material to eat or digest by pests and pathogens. In a number of species, the bark also contains crystals of calcium oxalate, making it physically harder to chew through, like putting gravel in a sponge cake.

Defence of the bark – and the more easily eaten leaves, flowers and fruit – against larger herbivores can be backed up by an array of other physical defences such as spines and thorns (Fig. 100). These are formed either from modified leaves (as in *Berberis*) or stems – either the whole stem, as in *Pyracantha* and hawthorns *Crataegus*, or just making the end of an otherwise normal stem pointed, as in crab apples *Malus* and Blackthorn. Holly is unusual in having leaf prickles, in this case designed to dissuade browsing deer as described in Chapter 3. Thorns are also possible on the trunks of trees (Fig. 101).

Most European trees and shrubs do not have these physical defences, which raises the questions of why not and whether there is any pattern to their occurrence. The answer to the first question is that spines and thorns are expensive things to produce and will only be grown where they are needed. Peter Grubb of Cambridge University proposed a neat solution many years ago in the *scarcity-accessibility hypothesis* (Grubb 1992). First of all, physical defences are found in habitats where nutritious growth is *scarce*, such as in harsh environments like deserts or heathlands (for example, gorse *Ulex*). They are also found where food has limited *accessibility*. This explains why trees that invade gaps in woodlands, such as hawthorns, apples, Honey Locust and False-acacia, have thorns – they are the only food accessible to deer and other large mammals, and so the

FIG 100. The aptly named White-thorn Acacia *Vachellia constricta* (formerly *Acacia constricta*) with 5 cm long thorns that deter large herbivores. In the desert grassland where it is native, food is scarce for large herbivores, so the thorns are a necessary investment. Sonoran Desert, Nogales, Mexico.

FIG 101. Thorns and spines on trunks. (a) Honey Locust *Gleditsia triacanthos*, native to central North America, has thorns up to 20 cm long, frequently branched, and on older trees can join to form an impenetrable mass around the trunk. They probably evolved to keep the now-extinct Pleistocene megafauna which came to feed on the fruits from browsing the tree. Thornless cultivars are readily available and normally planted in gardens and parks. (b) Nibung Palm *Oncosperma tigillarium* is widespread in Southeast Asia and is renowned for the long black spines on the young trunk and leaves. Being a palm with just one growing point, the energy invested in the spines is well spent. (a) Transylvania, Romania, and (b) Singapore Botanic Gardens.

cost of growing physical protection has been worthwhile. The same principle applies to European Holly. As an evergreen tree in an understorey dominated by deciduous trees and shrubs, without its leaf prickles it would be a sitting target for deer browsing in winter. These prickles in Holly make up around 1 per cent of shoot weight; not a huge amount but still a tax on the tree's resources. Holly demonstrates a beautiful piece of evolutionary adaptation in not producing leaf spines above around 2 m – the maximum height that deer can reach – and so not wasting resources (Fig. 102).

Physical defences against insects are inevitably smaller in scale and include hairs on vulnerable young growth that can be a physical deterrent. The hairs may also contain chemical deterrents such as in the nettle-like hairs of the Australian

FIG 102. Leaves of Holly *Ilex aquifolium* have few leaf prickles above around 2 m from the ground. The leaves at the bottom of the picture have a full set of prickles but the leaves around the fruits have few, while those even higher, out of the top of the picture, have none. Cañon de Añisclo, northern Spain.

rainforest stinging trees, *Dendrocnide* spp., described in Chapter 3. Physical defences can, surprisingly perhaps, also make use of mimicry. The leaves of passion flower vines, *Passiflora* species, are eaten by caterpillars of Heliconid butterflies. The butterflies tend to avoid leaves that already carry a number of their yellow eggs, so the vine grows imitation yellow eggs on its leaves! Better still, the Chameleon Vine *Boquila trifoliolata* (Fig. 103), found in temperate rainforests of South America, takes mimicry to an even higher level. This twining woody-climber mimics the size, shape, colour, orientation, petiole length and vein conspicuousness (amongst other things) of the host tree over which it climbs, even going so far as to grow a needle-like leaf tip if that is what its host has. The same vine even changes its leaf size and shape as it grows over a number of different hosts. This mimicry reduces herbivory on the vine leaves, since they are more easily overlooked (Gianoli & Carrasco-Urra 2014). This begs the question of how the plant can possibly know what its host looks like. Ken Thompson (2018) in his book *Darwin's Most Wonderful Plants* gives a few suggestions. Perhaps the vine is detecting volatile chemicals from the host, but how would these tell the vine what its host looks like? Or if it is done by gene transfer from host to vine, when and how would that happen? The mystery continues!

FIG 103. Leaf mimicry in the climbing plant *Boquila trifoliata* in the temperate rainforest of southern Chile. (a) The normal appearance of the vine's leaf when not in contact with a host. Each row includes pictures of the same individual of the vine in contact with different host trees or shrubs. A red arrow points to the vine, and a blue arrow points to the host species. (b) The colour, veins and toughness of leaves from *Rhaphithamnus spinosus* are mimicked (c) and change to mimic the size and shape of leaves from *Aextoxicon punctatum* (and are around 10 times larger than those shown in b). (d) The vine's leaves resemble the thin and light-green leaves of *Fuchsia magellanica* and then (e) the thick and dark-green leaves of *Myrceugenia planipes*. From Gianoli & Carrasco-Urra (2014).

Physical defence can also involve the use of other organisms. This includes bacteria and fungi, particularly mycorrhizal fungi. These may make the tree more sensitive to attack by pathogens or pests, producing a quicker and stronger reaction, or they may be directly antagonistic to the attacker. Around the world, there are many examples of trees that use ants as their main defence, and many temperate trees use mites and small insects to provide a similar service. These are all described in Chapter 14.

Chemical defences

Chemical defences are more common in trees than physical defences and can be found in all parts of a tree. These chemicals are called *secondary metabolites*, since they are not otherwise involved in plant growth, development or reproduction. Some are very poisonous while others simply make the tree parts unpalatable or distasteful. These chemicals are diverse and numerous, with the following being the most abundant.

- *Terpenes*: The active ingredient in resins and effective deterrents to a wide range of herbivores, insects, pathogenic fungi and bacteria. Terpenes include essential oils such as those in cinnamon bark *Cinnamomum* spp. and Bay leaf *Laurus nobilis*. They tend to work in low concentrations; for example, leaf terpene content ranges from 0.5–8 per cent of dry weight in eucalyptus trees which are considered rich in essential oil. There are many thousands of different terpenes, of which a few hundred are universal in nearly all plants, but the vast majority are found only in a particular genus or even just one species. Some of these may be left over from old defences that have been overcome by herbivores but there is evidence that some can be repurposed once the original function is lost, and that suites of terpenes are a more effective defence than one lone type (Pichersky & Raguso 2018).
- *Phenolics*: Another large group that includes flavonoids, anthocyanins, lignin and tannins. Lignin is very hard to digest, and tannins bind with the protein in the leaf when the leaf is eaten to make the protein indigestible, hence starving the attacking herbivore and reducing further eating.
- *Alkaloids*: Made from amino acids and include compounds such as caffeine, strychnine (Fig. 104), quinine and nicotine that are highly physiologically reactive.
- *Cyanogenic glycosides*: When mixed with enzymes they produce hydrogen cyanide (hydrocyanic acid), a particularly lethal chemical that stops respiration in aerobic organisms. Found mostly in leaves and fruits, the glycosides and enzymes are stored in separate compartments within cells and only mix to produce the cyanide when the tissues are crushed, as they are when being eaten. The commonest example in Britain is Cherry Laurel; if a leaf is crushed and held in a warm hand for a minute, a distinct smell of almonds – cyanide – can be smelt unless you are one of the 5 per cent of females or 20 per cent of males who have an inherited inability to smell it (Kirk & Stenhouse 1953).

Chemical defences can be difficult to store within the tree because they are so toxic. The component parts can be stored separately, like the cyanogenic glycosides and their enzymes, or the finished product can be stored between cells or in specialised storage units such as resin canals (common in conifers) or inside hairs (referred to as *glandular trichomes*).

Which defensive chemicals are found in a particular species of tree is partly dependent upon its natural habitat. On nutrient-poor soils, where nitrogen is in short supply but the tree is still able to produce sugar by photosynthesis, defences tend to be carbohydrate-based such as terpenes and phenolics. In more shaded

a

b

FIG 104. Strychnine Tree *Strychnos nux-vomica* native to India and Southeast Asia is distinctive (a) with its rounded, glossy, three-veined leaves. (b) The seeds and bark are the source of the alkaloid strychnine. When ingested it causes the muscles to go into spasm, causing convulsions, and the poor victim can be left with only the back of the head and heels touching the ground. In extreme cases the contractions can break the spine, resulting in the head being forced back to the buttocks. Strychnine was formerly used as a rat poison and thus widely available for skulduggery. The compound evolved to stop pests eating the tree, since the muscle spasms would kill or at least prevent further feeding. Pictures from Angkor, Cambodia, where locals apparently nibble the seeds as a stimulant!

areas, photosynthesis is limited, so as long as nutrients are readily available, trees tend to have nitrogen-based defence compounds such as alkaloids and cyanide. Either way, these toxic compounds are costly for the tree to produce and trees that have high concentrations tend to be slower-growing than chemically undefended plants.

A solution is to produce the chemicals *where* they are most needed. This helps explain why many of our spices come from the tropics, since 7–8 per cent of the annual growth is eaten by herbivores. This is double that in temperate areas so it has been cost-effective in the tropics to evolve a wide range of interesting chemicals, many of which we happen to find pleasant in small doses. But even within a tree species there are different growth-defence trade-offs. For example, trees in the Amazon rainforest on nutrient-poor soil, where it is harder to regrow leaves, invest more in defences and have physically tougher leaves (Fine *et al.* 2006). In a similar way, the North American Aspen will produce abundant tannins in its leaves when growing on nutrient-poor soil, since it is cheaper to defend the leaves than produce new ones. However, if the tree is fertilised, smaller quantities of tannins are produced in the leaves grown thereafter, since it is now cheaper to replace leaves than defend them.

Another solution is to produce chemical defences *when* they are needed. For example, an oak tree can put up to 15 per cent of its yearly sugar production into chemical defences, which is quite a hefty tax on the tree. So an oak will have a background level of tannins within its leaves, enough to keep normal amounts of defoliation down to an acceptable level, but when a tree is attacked by a plague of insects, the whole tree will produce larger quantities of tannin. Trees may not have brains but they are constantly weighing up costs and benefits and altering their way of growing accordingly.

Trees can also be genetically engineered to produce more of the chemicals that we might want commercially (Peter 2018) – see Chapter 13.

The tree's immune system
How does a tree prepare itself for attack from pests and pathogens? First of all, plants have to be able to detect pathogens. This is done using *pattern recognition receptors* (PRRs) that recognise molecular patterns of microbes or pathogens (Delaux & Schornack 2021). For example, a receptor might be able to recognise chitin which makes up the cell wall of fungi, and if detected, triggers the tree to start defending itself against fungal attack before the fungus does any damage. Within trees and other plants there are a number of *defence hormones* which spread the message around the tree that attack is imminent. Most prominent of these is salicylic acid, mentioned as a plant hormone in Chapter 7. Salicylic acid is readily converted to methyl salicylate which is volatile and along with other volatile compounds such as jasmonates can escape from leaves to spread rapidly through the tree's crown, quickly relaying the message that the tree is being attacked or about to be attacked by animals or pathogens. The chemical signals may also spread to neighbours. An oak being attacked uses volatile chemicals to signal to the whole crown to produce more tannins. Surrounding oaks may also pick up the signal that a neighbour is being attacked and start

producing extra chemical defences before they are actually attacked. This has been interpreted by some as trees talking to each other and even being altruistic by deliberately helping neighbours, when it is very unlikely that there is any intent in helping others. This is discussed further in Chapter 14 under *Wood wide web*.

These chemical signals spreading through the air can be general in their message while others are very specific. For example, Moreira *et al.* (2018) found that a Californian shrub *Bacharis salcifolia* produces specific signals for different aphid species, and these each induce resistance in neighbouring plants to the respective aphid species only.

All of these receptors and responses are part of a tree's *systemic acquired resistance* (SAR) – the tree's immune system – where local attack or infections lead to the tree *acquiring* a defence that is *systemically* spread through the whole tree either internally through the xylem and phloem or externally through the air around the crown. This immune system involves rapidly produced mobile signals, such as methyl salicylate, to quickly spread the message. These then trigger genes to locally produce defence compounds (Kachroo & Kachroo 2020) that prime them against infection by pathogens or attack by pests. Importantly, the SAR has a biochemical memory so that, should the pest or pathogen attack again, the response will be stronger, hence *acquired resistance* (Turgut-Kara *et al.* 2020). This immune memory works via *epigenetic changes* within the tree. This means that the genetic code of the tree remains unchanged, but there is an accumulation of compounds that attach to the DNA, altering gene expression and so modifiying what proteins and how much of them the genes produce or even *silencing* them by stopping them producing anything at all (Fu & Dong *et al.* 2013). There is evidence that this biochemical memory can be passed on to offspring so that they react more strongly to stresses that the parent was exposed to (Alonso *et al.* 2019).

DEFENCES ONCE THE BARK IS BREACHED

If insects or fungi penetrate the bark, the next line of defence is to seal over the wound to stop them getting any further. This is done by resins, gums and latexes, full of defensive chemicals. These are stored under a slight positive pressure in special structures such as *resin canals* and *lactifers* for latex. If the bark is breached, these compounds ooze out and, as well as being toxic, they physically trap or wash out invaders. After this they set hard by the evaporation of lighter oils or water to form a solid plug blocking the wound. Insects and even larger animals found trapped in amber (fossilised resin) show that this battle has been going on for at least 100 million years.

Resins are oil-based, which is why conifer resin is so difficult to remove if you're rash enough to handle a resin-laden pine cone (Fig. 105). In most conifers, the resin ducts occur in both the bark and wood, running from the roots to the needles. But in some, such as hemlocks, cedars and firs, resin is found only in the bark, although all conifers are capable of producing *traumatic resin canals* in the wood in response to injury or infection. Some resins have proved extremely valuable, such as that from giant Kauris *Agathis australis* in New Zealand (Fig. 106) – often called 'kauri gum' but it is an oil-based resin. Great deposits have been mined from below ground in the past, and it would long ago have dissolved if it was a water-based gum. Not all conifers produce resins (e.g. yews), and resins can be found in a range of broadleaves, notably in the family Burseraceae. This includes frankincense *Boswellia carteri* and myrrh *Commiphora* species, both used in incense and perfumes.

FIG 105. Sticky resin being produced from young cones of Bristlecone Pine *Pinus longaeva*. Handle at your own risk! The White Mountains of California.

FIG 106. (a) A display from the Kauri Museum on the North Island of New Zealand showing a Kauri *Agathis australis* being cut into to cause high-quality 'gum' – really a resin – to ooze out, which would set on the trunk and be collected. This was used as an ingredient in many household products from varnish and paint to candles and cigar holders, and it was even used in making dentures! Maoris used the gum as chewing gum, as a tattooing pigment when burnt to powder and as a fire starter. (b) A Kauri in Waipoua Forest, New Zealand, showing the gum oozing from the tree before setting solid to seal the wounds.

Gums are similar to resins but are water-based. They are found in a wide range of tropical trees, notably in the family Anacardiaceae, including the Varnish Tree *Rhus verniciflua*, giving a gum which is the basis of Chinese lacquer. Plenty of temperate trees also produce gums, from cherries and plums to eucalypts (Fig. 107). Like resins, these are carried in ducts in the bark and often the wood, and traumatic gum canals can be formed in the wood of some broadleaves, including Sweet Gum and cherries.

Latex is a milky mixture of resins, oils, gums and proteins that is particularly common in tropical rainforests where around one-third of trees have latex. Included here is the Rubber Tree *Ficus elastica* (in the Moraceae family), which is used as a house plant and was the original commercial source of rubber. Most rubber now comes from the Pará Rubber Tree *Hevea brasiliensis* (Euphorbiaceae),

FIG 107. A dark-red gum called kino is found in a number of trees but particularly eucalypts, in this case one of the Scribbly Gums *Eucalyptus haemastoma*. Kino has been used in medicines and for tanning and as a dye. Scribbly Gums are named for the 'scribbles' on their bark made by the larvae of the Scribbly Gum Moth *Ogmograptis scribula* which burrows between layers of bark. The insect is well adapted to cope with the tree's defences, since it appears to be seldom killed or flushed out of the bark by the gum. Wallumatta Nature Reserve, Sydney, Australia.

native to the Amazon rainforest. The world's best rubber was collected from wild trees in the Amazon under difficult circumstances. But in 1876, Sir Henry Wickham brought 70,000 seeds of Pará Rubber to Kew Gardens and that was the start of rubber plantations around the old empire. The story is told that Wickham was a great adventurer who daringly stole the seeds, called a 'celebrated seed snatch' by Joe Jackson (2008) in his wonderful book *The Thief at the End of the World*. However, it is likely that some of this was embellishment on the part of Wickham to make it all seem more exciting. He asked permission to collect the seeds and his most daring deed was apparently to forgo obtaining an export licence from Brazil and instead smuggle the seeds out. So, a daredevil or someone cutting red tape? Either way, it became a big industry that is still economically important in Southeast Asia (Fox & Castella 2013).

COVERING A WOUND: WOUND PERIDERM AND CALLUS

A small wound in the bark initially sealed by resins, gums or latex is followed by the formation of new bark (a wound or traumatic periderm) that effectively restores the integrity of the outer bark. The resins are ready for use whenever the injury occurs, even in winter, but the new bark can only form during the growing season, backing up the resin once spring comes.

A larger wound where a significant area of bark is removed needs a different mechanism. In broadleaves (but not usually conifers), if the exposed wood remains moist, there are enough living cells in the sapwood (parenchyma cells) that can survive long enough to start to grow new corky cells, creating small buttons or ridges of new bark (Fig. 108a). Within these, a new vascular cambium

FIG 108. (a) Wound callus tissue growing in from the edge of a large wound caused by Grey Squirrels *Sciurus carolinensis* on a Sycamore *Acer pseudoplatanus*. The living cells of the rays across the wound are also forming corky periderm nodules which can in theory amalgamate to create new bark, but it is normally a slow process. (b) Thick ridges of callus on a lime in Poznań, Poland, grown over a number of years. As they grow wider, they will eventually cover the whole of the wound.

can develop, enabling the buttons and ridges to spread sideways over the wound. This is unlikely to cover the whole wound unless it is quite small. Large wounds can usually only be effectively covered by *wound callus* growing from the edge of the wound (Fig 108b). The vascular cambium at the edges of the wound produces new wood and bark that expand sideways every year and so may eventually cover the wound.

It is important to note that wounds in a tree never heal, they are just covered over. If I cut my hand, the wound will heal and, unless it leaves a scar, there will be no sign of it in the future. But a tree cannot heal a wound in the wood; it can only hide it by covering it over with callus tissue. Fungal rot will invade long before the wound can be covered over with wound callus and will remain there even after it is covered over. So why bother? The first reason is that the wood in an open wound rapidly dries, allowing oxygen to permeate into the gas-filled tubes, creating excellent growth conditions for fungal decay. When the wound is covered by callus, much of the oxygen supply to the rot is cut off, the wood becomes wetter and this combines to slow the growth of the fungal rot. Secondly, as described in Chapter 11, the callus growth strengthens the tree, reducing the chance of it snapping at that point as it sways in the wind. Both of these reasons make callus tissue a good long-term investment.

SAPWOOD AND HEARTWOOD

As shown in Chapter 3 (look back to Fig. 33), the woody centre of the tree is divided into a layer of sapwood overlying a central core of heartwood. Water is conducted through the sapwood and contains living cells, whereas the central heartwood is completely dead.

The water-conducting tubes running vertically through the wood (vessels in broadleaves, tracheids in conifers: see Chapter 6) make good corridors for fungal growth once they are no longer filled with water. However, the pits between tracheids in conifers can be impregnated with resins, and the vessels of broadleaves can contain gels and tyloses (Chapter 6) which have many of the chemical defences listed above. These act to slow the growth of microorganisms – fungi and bacteria – along the tubes. Sideways spread of decay is also impeded by the highly indigestible lignin of the woody cell walls. Equally, the dense latewood at the end of each ring is high in lignin and helps slow decay moving along the radius of the tree. Branch junctions also tend to have denser wood (part of the hydraulic architecture of the tree – see Chapter 6) which slows the spread of rot from branches back into the main trunk.

Sapwood

The living cells in the sapwood not only produce gels and tyloses but can also mount an active defence against decay microorganisms. The living cells that run vertically through the wood (axial parenchyma) and those in the rays (ray parenchyma) form a living three-dimensional network through the sapwood, making up 5–40 per cent of the sapwood by volume. When damaged, this network forms a wall of dark-coloured *wound wood* around the area, highly impregnated with defensive chemicals, with all tubes and parenchyma cells filled with insoluble deposits, similar in nature to heartwood (described below). This is called the *reaction zone* (Fig. 109), since the wound wood is a reaction to the wound. How good a defence it is in containing any rot depends upon the size of network of living cells and the defensive chemicals they produce, both of which vary between different tree species.

The process involves the tree's immune system (the systemic acquired resistance – SAR – described above). A wound causes the release of chemical signals transmitted to the intact wood to trigger wound wood formation. The signals are also sent to the vascular cambium. The vascular cambium near the wound receives these signals, and when it grows new wood after the wound (which will be next spring if the wound occurs in winter), genes involved in latewood growth are triggered (Chano *et al.* 2017), and the new wood is dense, chemically rich and sometimes with more rays and parenchyma cells. This creates a *barrier zone* that helps isolate new wood grown after wounding from fungal rot that may have invaded the wound. In small trees, the barrier zone may continue around the whole trunk or branch. In larger trees this zone may be more like a plate under the bark, typically extending around the wound at a 30° angle from the centre of the tree. The cambium usually produces reaction zone wood for a few years and then reverts to normal, leaving the barrier buried within the wood. If the original barrier zone is breached, new reaction and barrier zones are produced in an ongoing battle of attack and defence.

These zones are part of the concept of CODIT (Compartmentalisation of Decay/Dysfunction in Trees), first put forward by Alex Shigo in the 1970s, which states that trees are highly compartmentalised, walling decay into discrete compartments. This visualises the inside of the wood as a series of walls that slow the spread of decay. The first wall restricts the vertical spread of decay by the plugging of tracheids and vessels; the second wall corresponding to the growth rings slows rot radially across the stem; and the third wall corresponds to the rays that prevent tangential decay around the stem. These can be preformed in the heartwood or formed anew as in the reaction zone. A fourth wall, grown after damage, represents the barrier zone. The concept has moved on, since

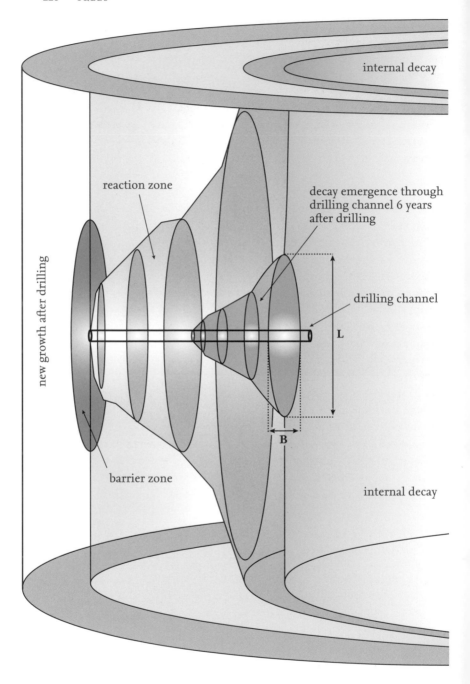

internal decay

reaction zone

decay emergence through
drilling channel 6 years
after drilling

drilling channel

L

new growth after drilling

barrier zone

B

internal decay

it is not just the walls that restrict the spread of rot; it is also being slowed by lack of oxygen in sapwood and may only spread quickly once the trunk has been damaged, allowing oxygen levels in the wood to rise. Also, disruption to water movement in working sapwood may be more important than decay, so it is suggested that 'decay' should be 'dysfunction'. The concept of CODIT is not perfect, but it gives a mental framework for thinking about how trees cope with internal decay using chemical defences, especially in sapwood. Morris *et al.* (2020) give an excellent review of the four walls inherent in CODIT and how these link with current thoughts on tree defence.

Heartwood

The network of living cells in the sapwood, helping defend the wood and storing food in the rays, is expensive for the tree to maintain. So it makes good economic sense to maintain just the amount of sapwood necessary to keep the tree functioning, and as new rings of sapwood are added to the outside, to kill off the oldest sapwood cells furthest from the bark. This is the process of heartwood formation (Fig. 110), a 'shedding' of internal wood that is no longer required for daily life, just like leaves and branches are shed when they use more sugar than they produce (Hirons & Thomas 2018). This is not just a gradual withering away of old cells but a deliberate transition of programmed cell death, usually happening in the late summer and towards the autumn. With no living cells, the heartwood cannot be defended in the same ways as sapwood, but this is compensated for by a chemical reinforcement. Some of these chemicals put into heartwood can be waste products of the tree, but the vast majority are specifically and expensively produced by the parenchyma cells before they die, usually filling the cell lumen and being deposited in the walls of neighbouring cells. These chemicals are similar to the defensive compounds used elsewhere in the tree, including lignin, phenolics and terpenes, collectively known as *heartwood extractives*. These give the heartwood its distinctive colour, usually darker than the

OPPOSITE: **FIG 109.** A Black Poplar *Populus nigra* with fungal rot in the centre of the tree (green). Six years previously a core of wood had been removed (see Fig. 175 in Chapter 12) to investigate the rot, as indicated by the drilling channel. The central rot and the drilling were detected by the tree as wounds, and the sapwood produced a *reaction zone* of wound wood, shown in brown. Once the tree resumed growth after drilling, it laid down a further *barrier zone* in the vicinity of the drilling wound, shown in blue, which helped isolate the wood grown after drilling from the old wound wood. Despite the reaction zone, the rot spread along the oxygen-rich drilling channel, particularly vertically (L), following the xylem vessels, but much less so tangentially around the tree (B). Based on Weber & Mattheck (2006).

FIG 110. Darker heartwood inside the narrow band of lighter sapwood in a Huntingdon Elm *Ulmus* × *vegeta*. The tree originally had two trunks that fused, hence the two centres and the piece of enclosed bark. The heartwood also contains dark areas of decay, producing variations in colour across the heartwood.

sapwood, and often smell – think of sandalwood from various *Santalum* species, which smells for decades after cutting.

The extractives may make up less than 1 per cent of the weight of dry wood but they greatly increase the wood's resistance to rot. As an example, woods with few extractives (such as Alder, Ash, Beech, limes and willows) may rot in less than five years when in contact with the ground, whereas extractive-rich woods (such as oaks, Sweet Chestnut, Yew and Western Red-cedar) will persist rot-free for more than 25 years. Having said that, if the wood is maintained in a waterlogged condition, and so with low oxygen levels, fungal rot is very slow and even non-durable wood will persist. For example, elm rots quickly when used as a fence post, but used underwater as a paddle in lock gates it may persist for centuries. In earlier times, the Romans used hollow elm logs as water pipes because they are very resistant to splitting. A number of more recent elm pipes were unearthed in London and were still sound after more than 300 years (look back at Figure 15a in Chapter 2). The Italian city of Venice is supported on posts of Alder submerged under water that were put in place up to 500 years ago.

The amount of sapwood a tree has at any one time varies greatly between species and the growing conditions of each tree. Conifers need many rings over which to conduct water (Chapter 6), and so the sapwood can be up to a hundred or more rings wide, as has also been seen in some broadleaves, notably the North

FIG 111. A Douglas Fir *Pseudotsuga menziesii* that began to rot only after it had died. After death, the sapwood is the part with least chemical defences and is most easily invaded by decay fungi. Trees that begin to rot while alive tend to lose the central heartwood first. Although this is chemically rich, these defences are more easily overcome than the active defences in the living sapwood. The Cascades, Washington State, USA.

American Tupelo *Nyssa sylvatica*. At the other extreme, the Northern Catalpa *Catalpa speciosa* may have just one or two rings of sapwood. Similarly, in oaks – which are ring-porous and rely on just the outermost ring for water transport (see Chapter 6) – the sapwood is usually very narrow and made up of just 5–6 rings at most, such that it is mostly cut off with the bark when planked in a sawmill. For many trees, the sapwood becomes wider over the first decade or so; in Norway Spruce grown in the open, it reaches around 8 cm wide in less than 20 years and then stays almost constant over the next century. However, in Norway Spruce growing in shade, sapwood reaches only 2 cm wide over the first 50 years, because the trees are growing more slowly and have less need of water, since most of the leaves are in shade and not losing excessive amounts of water (Sellin 1994).

Some broadleaves, including maples, birches, hornbeams, beeches, ashes and Olive, have *irregular heartwood* (sometimes referred to as *ripewood*) where the heartwood does not follow the growth rings but has an irregular shape and varies in colour across the tree. In some, such as in Beech, the ripewood is as light in colour or lighter than the sapwood and has few chemical additions, making it prone to decay. In other ripewood species, denser heartwood is produced in response to wounding or harsh growing conditions and they are referred to as *delayed heartwood* species.

It is often possible to tell whether a tree started to rot before or after it died. If it starts rotting while still alive, as happens in most trees, the sapwood with living tissue is more tightly defended and so the heartwood tends to rot first. After the tree has died, however, the heartwood, with its high levels of extractives, is more resistant to rot than the now lifeless sapwood, and this will rot first (Fig. 111).

CHAPTER 9

The Annual Bounty: Seeds and Fruit

SEED DISPERSAL, CONES AND FRUIT

As described in Chapter 1, a fundamental difference between broadleaves (angiosperms) and conifers (gymnosperms) is that the former have fruit that encloses the seed, while conifers do not have fruit and the seeds are contained in cones. So how do each of these spread their seeds?

Conifers

A pine cone (Fig. 112a) can be bent open, without breaking anything, and the naked seeds can be seen sitting on the woody scales. The base of each scale reacts to humidity such that the scale bends in when damp, closing the cone, and out when dry, causing the cone to open and allowing the seeds to fall out by gravity. A cone will usually open wider each time it goes through the wetting/drying cycle, so seeds are progressively lost, and 70–90 per cent of seeds will usually fall in the first two months of autumn, with the rest falling over winter. Cones of cypresses, giant sequoia and others in the family Cupressaceae have globular fruits (Fig. 112b) with abutting plates that shrink when mature, creating gaps through which the seeds fall. Cones are usually upright when young to help pollination (Chapter 4), and in most conifers they bend to hang downwards as they mature, allowing the seeds to slide out.

In some conifers, notably firs, cedars and the monkey puzzle relatives *Araucaria* (making up around 95 species), the cones remain pointing upwards (Fig. 112c), and the seeds are released by the whole cone breaking apart, leaving the central axis on the branch pointing up like a candle. If you want one of these cones intact, it

FIG 112. Conifer cones. (a) Pinyon Pine *Pinus monophylla* which, although the cones are small, has large seeds that are spread by birds; (b) Giant Sequoia *Sequoiadendron giganteum* where abutting scales shrink to allow the seeds to fall out; (c) Delavay's Fir *Abies delavayi* with upright cones that fall apart to release the seeds. (a) The Grand Canyon, Arizona, (b) Sierra Nevada Mountains, California, and (c) Exeter, Devon.

needs to be collected from the tree before maturity, otherwise it will just break up on your shelf as it dries. This may involve climbing the tree, or if you were a Victorian gentleman, you could simply shoot the top out of the tree to collect your cones. Health and safety regulations might have something to say about that today!

In several cases a fruit-like structure is produced in conifers but these are still based on a cone or the seed. A fleshy cone is produced in junipers (Fig 113), becoming blue and berry-like when mature. These do not open, and the seeds are spread primarily by the cones being eaten by birds and so act much like the fruits of an angiosperm (Thomas *et al.* 2007). The red 'berry' of the Yew is a fleshy outgrowth from the base of the naked seed (an aril), and in the Maidenhair Tree, the seed coat itself has become fleshy and appears like a fruit but it is still just a naked seed.

Pine cones come in a very wide range of sizes (Fig. 114), from cones just a few centimetres long to those from the Coulter Pine *Pinus coulteri* that can reach 25–30 cm in length and weigh up to 2.3 kg each when fresh. Fortunately, there is a discernible ecological pattern in these differences. The so-called 'hard pines' with needles in twos or threes produce hard cones with very robust scales that are hard to force open. Two-needle pine cones (Fig. 114e and 114f) tend to be fairly small (golf- to tennis-ball size), which has the benefit of spreading the seeds across many smaller cones, so if one is attacked, there are many others left to shed their seeds. The three-needle pine cones (Fig. 114b, 114c and 114d) are also hard but are much larger. These

FIG 113. Spanish Juniper *Juniperus thurifera* with fleshy cones. The Iberian Mountains, Spain.

FIG 114. Conifer cones of different sizes. (a) Sugar Pine *Pinus lambertiana*, (b) Coulter or Big-cone Pine *P. coulteri*, (c) Grey or Digger Pine *P. sabiniana*, (d) Monterey Pine *P. radiata*, all from California, (e) Dwarf Mountain Pine *P. mugo* from Europe, and (f) Jack Pine *P. banksiana* from Canada. The number of needles bundled together varies from five (a), to three (b, c and d) and two (e and f). From Thomas & Packham (2007).

tend to be native to dry habitats with sparse vegetation (the Coulter Pine grows in the very arid San Gabriel Mountains of California where food for herbivores is in short supply). Under this intense herbivore pressure, it is a worthwhile investment to produce large, strong and expensive cones to protect the seeds. By contrast, pines with five needles in a bundle, often called the soft pines, tend to have cylinder-shaped cones up to 15–50 cm long and even 60 cm in Sugar Pine *Pinus lambertiana* (Fig. 114a), with relatively soft, flexible scales. This is perhaps an adaptation for habitats with lower herbivore pressure and allows seeds to be concentrated at the upper and outer edges of the canopy, giving them maximum access to wind.

Broadleaves

As the seeds develop in broadleaf trees, the ovary becomes the fruit, keeping the seeds hidden inside. This leaves the seeds completely enclosed by the fruit and not visible without breaking into the fruit, whether this is a dry nut or a fleshy peach. Fruits come in many shapes, sizes and degrees of fleshiness and have accrued a terminology all of their own – Box 1 outlines the main types of fruit found in trees.

BOX 1 TYPES OF FRUIT FOUND IN BROADLEAF TREES

The fruit (botanically called the *pericarp*) is made up of three layers. In dry fruits they appear as one, but in fleshy fruits the outer skin (*epicarp*) encloses the fleshy layer (*mesocarp*), and the inner layer (*endocarp*) can take a variety of forms, being hard and woody in drupes to juicy in a berry.

DEHISCENT DRY FRUITS (OPEN ON THE TREE TO SHED SEEDS)
Legume The usual fruit of the pea family; it splits along both sides into two halves, revealing the seeds (e.g. laburnum, gorse and acacia). Some tropical species, such as the Sea-bean *Entada gigas*, break crossways between seeds into small segments that float. The seeds can be found washed up on British beaches.
Follicle Like a legume but splitting along just one side (e.g. grevilleas *Grevillea*).
Capsule Fruit that usually splits along a number of weak lines to let the seeds out (e.g. willows, spindles, horse-chestnuts and eucalypts). A few, like that of the Brazil Nut Tree *Bertholletia excelsa* (Fig. 116), have to be opened by animals.

INDEHISCENT DRY FRUITS (THE FRUIT STAYS AROUND THE SEEDS WITHOUT OPENING)
Achene Single seed with a dry fruit but not as woody as a nut (e.g. Cashew *Anacardium occidentale*). In the Cashew, the achene grows beneath a fleshy 'apple' (Fig. 115b).
Nut (or nutlet if small) A common fruit of trees, composed of a single seed and a dry fruit that may be hard as in hazel or comparatively soft and easily peeled off as in the acorn of oak. A nut may have a bract or bracts (modified leaves) loosely attached as a wing (e.g. lime and hornbeam) or more highly modified and enclosing as in hazel, the cup of an acorn and the hard spiny covering (cupule) of a beech and Sweet Chestnut *Castanea sativa*. Note that these spiny cases do not completely enclose the nut, since the stigma and style of the ovary poke out of the end; this separates a nut from the superficially similar *capsule* of a Horse-chestnut where the outer spiny case is the fruit itself.
Samara The fruit is elongated into a dry papery wing, as in ashes, elms, maples and tulip trees *Liriodendron*. Note that the double wing of a maple fruit is two samaras joined together.

SUCCULENT FRUITS
Berry A fleshy fruit enclosing the seed or seeds (e.g. gooseberry, blackcurrant, orange, bilberry and the Date *Phoenix dactylifera* – where the 'stone' is the seed). Bananas are also berries. In a citrus fruit (botanically called a hesperidium), such as a lemon, the outer two layers form the peel (epicarp) and pith (mesocarp), while the inner layer (endocarp) forms fluid-filled hairs, or juice sacs.

continued overleaf

BOX 1 TYPES OF FRUIT FOUND IN BROADLEAF TREES *continued*

Drupe A fleshy fruit with the innermost layer (endocarp) forming a hard 'stone' around the seed (e.g. plum, cherry, peach, almond, walnut, Elder *Sambucus nigra* and Olive *Olea europaea*). Sometimes we eat the fleshy part and throw away the stone containing the seed (e.g. plum and peach), or we throw away the fleshy part, crack open the stone and eat the seed (e.g. walnut and almond) – Fig. 115a). Some drupes contain more than one stone (e.g. hawthorn, Rowan *Sorbus aucuparia*, Holly *Ilex aquifolium* and Medlar *Mespilus germanica* – some consider this now to be *Crataegus germanica*). The commercial Coconut *Cocos nucifera* is just the stone (containing solid and liquid food for the germinating seed), and the coir (removed to make coconut matting and peat-free compost, etc.) is the equivalent of the flesh and skin of a plum.

FALSE FRUITS
Some fruits are supplemented with other structures. In apples and pears (pomes), the base of the flower (the 'receptacle' where all the bits like stamens and petals are joined on) grows up and around the true fruit: the core is the real fruit and the part we eat is the 'fleshy receptacle'. This is similar in rose hips (containing achenes) and figs (containing drupes) where the receptacle from one (rose) or more flowers (fig) grows up and encases a mass of individual small fruits.

Several flowers can grow together as the fruits develop to produce what appears to be one fruit (e.g. plane, mulberry, Osage Orange *Maclura pomifera* and the cones of alders and banksia *Banksia*: these are **multiple fruits**). So, strictly speaking, the false fruits of fig and the double samara of maples could be called multiple fruits.

Adapted from Thomas (2014).

This terminology is sometimes at odds with our everyday use of English. For example, we eat many 'nuts' but botanically only some are true nuts, such as hazel nuts. Others, such as almonds and walnuts (Fig. 115a), are only the hard inner part of the fruit, technically called a drupe stone, which we crack open to reveal the seed. Cashew 'nuts' are really the seed from inside a dry achene (Fig. 115b). Other stones involve the seed coat becoming woody, such as the date stone, or pit. Similarly, Brazil nuts are seeds with a hard woody seed coat. A Brazil 'nut' which contains the seeds is shown in Figure 116. Just to add icing to the cake, alders and she-oaks *Casuarina* species from the southern hemisphere seem to produce cones like a conifer (Fig. 117) but in reality they are a woody catkin with many flowers welded together to form a multiple fruit (see Box 1). When the 'seeds' fall out, they are really complete fruits – small nuts (nutlets) in the case of alder and small winged samaras in she-oaks.

FIG 115. Nuts that are not nuts. (a) An almond showing the drupe stone (what we call the nut) and the removed fleshy part of the drupe to the top and right. (b) Cashew nut *Anacardium occidentale* where the 'nut' we eat is the seed from inside a dry achene, the kidney-shaped fruit at the end of the larger cashew apple (which is really an expansion of the flower stalk). (a) Blanes, Spain, and (b) San Pedro Sula, Honduras.

FIG 116. The fruit of a Brazil Nut Tree *Bertholletia excelsa* as it falls from the tree. It is technically a capsule (see Box 1) containing 8–24 hard-shelled seeds that we call 'Brazil nuts'. The hole at the top, where it was attached to the tree, is used by animals to open the fruit – see below. The capsule can be up to 15 cm in diameter, almost twice the size of the one shown (found at a jumble sale!), and weigh 2 kg when fresh.

FIG 117. The 2 cm long cone-like multiple fruits of the She-oak *Casuarina equisetifolia*. Each one is made up of many woody flower bases pressed together in lines to give the appearance of a cone. Each woody pouch opens to release a small fruit, in this case a winged samara, seen at the top of the picture. Brisbane, Australia.

FIG 118. The three nuts of Sweet Chestnut *Castanea sativa* were surrounded by a green and spiky cupule made from modified leaves on the mother plant, which protects the fruits until they are ripe.

WHY BOTHER WITH FRUIT?

Fruits give two main advantages to the seeds. The first is that they offer protection to the seeds while they mature, just as the cone scales do in conifers. Protecting the under-ripe seeds leads to many fruits being spiny, such as in horse-chestnuts, and tough or poisonous or unappetising, such as a sour apple. In some cases part of the mother plant will take on this role, as seen in Sweet Chestnut (Fig. 118). The second advantage, which most conifers do not have, is that the fruits can help disperse the seeds when they are mature. It can help them fly through the air, float in water or be attractive to animals. Animals will eat the fruit and eject the seeds which will by then, hopefully, have been moved away from the parent tree.

Which method of dispersal is used is often linked to habitat. Trees that grow in the open where wind speeds tend to be high will frequently use wind

FIG 119. The dry, yellow fruits of Oriental Bittersweet *Celastrus orbiculatus* open to reveal the bright red, fleshy arils which are taken by birds along with the seeds. The aril is consumed and the seed is dropped, hopefully at some distance from the parent plant. This plant, native to China, grows as a large vine up host trees and is considered an invasive plant in eastern North America. Harvard Forest, Massachusetts, USA.

to disperse fruits and seeds. Those in more wooded habitats where wind speed is lower, and which also tend to need heavier seeds to help the seedling become established (Chapter 4), tend to be dependent upon mammals and birds to move fruits and seeds around. This is also the case where animals are more abundant. In animal-rich tropical forests, around 50–75 per cent of trees use animals as dispersers, and even in the tallest emergent trees that experience more wind, only 18–30 per cent of species are dispersed by wind.

But the fruit is not always involved in dispersal; the dry nut of an oak does little except offer the seed some extra protection, and other fruits may stay attached to the tree and just open *in situ* (*dehisce*, technically), letting the seeds fall out, such as in the pea-like pods of Laburnum *Laburnum anagyroides*. In some cases where the fruit is not involved in dispersal, the seeds do the job themselves by growing an aril from the base of the seed (Fig. 119). In the Yew this aril is the bright red 'berry' around the seed, and in broadleaves arils are seen as the red mace around the seed of the Nutmeg and the orange aril of the Spindle Tree contrasting with the pink fruit as a visual signal to birds.

WIND DISPERSAL OF FRUIT AND SEEDS

As noted above, wind dispersal is most common in species that grow in open conditions, such as birches, ashes and many pines, because wind speeds are highest in the open and so this is the most effective way of moving seeds and fruits. This has the benefit that seeds will be spread over a large area, so increasing the chance of finding a new open spot to invade. Wind dispersal works best if the fruits or seeds have mechanisms to reduce their vertical fall, allowing the wind to blow them further before they reach the ground. Some trees, such as eucalypts and rhododendrons, rely on fine, dust-like fruit. Larger fruits delay their vertical fall with a plume of long hairs, as in willows, poplars and bean trees *Catalpa* (Fig. 120), or by extending the dry fruit into the samara wing of maples, ashes and tulip trees *Liriodendron* (Fig. 121a). In contrast to these single-bladed seeds, the Dipterocarps of Southeast Asia have two or three wings per seed to help disperse the heavy seeds (Fig. 121b). In trees that produce small nuts, such as hornbeams and limes, a bract (modified leaf) has been added to help them fly. In the case of conifers, the seed itself has been given a wing.

Hairs act simply as parachutes to slow seed fall. However, wings that spin in the air (autorotate), such as in maples, generate lift to hold seeds up in the air for longer (Lentink *et al.* 2009). Ash fruits are straighter and more symmetrical than maple and also generate lift by spinning around their long axes as well as

FIG 120. Plumed fruits of the Mountain Mahogany *Cercocarpus montanus* where the fuzzy tail can be up to 8 cm long. Perfect for wind dispersal. Pine woodland above Lone Pine, California.

FIG 121. Samaras (a) with single wings: Sycamore *Acer pseudoplatanus* on the left and Ash *Fraxinus excelsior* on the right; and (b) with double wings on species of the aptly named dipterocarps *Dipterocarpus* spp. (from the Greek *di*, meaning 'two', *pteron*, meaning 'wing', and *karpos*, meaning 'fruit') from Southeast Asia. (a) Keele, Staffordshire, and (b) Royal Belum National Park, Malaysia.

autorotating (der Weduwen & Ruxton 2019). Table 11 lists a number of North American trees in order of vertical falling speed. It can be seen that different fruit structures can work equally as well as others. On the whole, heavier seeds do tend to fall faster but with notable exceptions! Cherry Birch *Betula lenta* is a common tree in eastern North America that invades open areas. Openings are fairly common in these woodlands, so the seeds do not have to rely on travelling

TABLE 11. Speed of seeds and fruits of some North American trees falling through still air, from slowest, at the top, to fastest. Speed is in metres per second; to convert to mph, multiply by 2.237. Seed weight (i.e. with the fruit removed) is in milligrams (0.001 g).

Common name	Scientific name	Fruit type	Seed weight (mg)	Fruit/seed falling speed (m/s)
Paper-bark Birch	Betula papyrifera	Winged nutlet	3.0	0.55
White Spruce	Picea glauca	Winged seed	2.1	0.62
Red Maple	Acer rubrum	Samara	20	0.67
Loblolly Pine	Pinus taeda	Winged seed	25	0.70
Silver Maple	Acer saccharinum	Samara	324	0.87
Box Elder	Acer negundo	Samara	38	0.92
Weymouth Pine	Pinus strobus	Winged seed	17	0.93
American Hornbeam	Carpinus caroliniana	Nut with bract	15	0.98
Sugar Maple	Acer saccharum	Samara	74	1.00
Sweet Gum	Liquidambar styraciflua	Capsule	5.6	1.05
White Ash	Fraxinus americana	Samara	45	1.40
Tulip Tree	Liriodendron tulipifera	Samara	32	1.48
Cherry Birch	Betula lenta	Winged nutlet	0.7	1.60
Red Ash	Fraxinus pennsylvanica	Samara	38	1.60
American Lime	Tilia americana	Nut with bract	82	2.92

Data from Hewitt (1998), Burns & Honkala (1990) and Thompson & Katul (2013).

far; consequently, the very small seeds have a small wing and fall comparatively fast. By contrast, the Silver Maple from the same region is fairly shade tolerant and tends to grow up below other trees, especially on nutrient-rich sites. To help it cope with shade, it has the largest seed of any North American maple, but the seed has an effective autorotation that slows down the fall, allowing this maple to disperse widely within the forest.

Wind-dispersed seeds and fruit are often released when the tree is leafless so that the leaves do not get in the way and wind speed is at its highest. Even then, they are not shed at random. The abscission layer, similar to that at the base of leaves, breaks most readily at low humidity which, in temperate areas, is typically in the early afternoon, when wind speeds are highest (Greene & Johnson 1992). Moreover, the fruit is held tenaciously by the last of the abscission zone so that the fruit is finally torn off by a strong gust of wind, often at twice the average wind speed, causing the seeds to travel at least twice the distance to that predicted from average wind speeds (Greene & Johnson 1992, Maurer *et al.* 2013). Strong gusts of wind can also lift the seeds above the canopy where wind speeds are higher still, further increasing the distance they are carried (Horn *et al.* 2001).

ANIMAL DISPERSAL

Tree seeds are generally too big for most insects to move, so unlike pollination, insects play a minor role in seed dispersal. Larger animals are needed, and although a few seeds can hitch rides on feathers or fur, most seed dispersal by animals involves supplying food in the form of fruit and sometimes seeds.

Once the seeds are mature, many fruits become attractive to animals by the flesh softening, starch and oils changing into sugars, and astringent compounds decreasing, all acting to making them more palatable. This change is often advertised by a change in colour, such as a hard, sour, green apple becoming larger, sweeter and coloured. Some fruits, however, will retain some of their poisonous or unpleasant characteristics, which reduces the chance of being eaten by the wrong animal. In all these cases the objective is for the fruit to be taken so that the seeds reach the ground some distance away from the mother tree.

Fruit shape, size and appearance are adapted to attract specific groups of animals. Bird fruits tend to ripen in the morning and be fairly small and so easily carried away. They have little smell but are highly coloured, since birds have good colour vision, and this helps the fruit to stand out against the foliage. Bird fruits are often red in the summer and on evergreen shrubs (including Wild Cherry, Holly and Yew) and black in the autumn (such as on brambles *Rubus* species, Elder, Bird Cherry *Prunus padus* and Ivy).

By contrast, fruits that attract mammals tend to be larger and less brightly coloured but with a more prominent scent, since this is the main attractant. Notable is the Durian fruit of Southeast Asia (Fig. 122), renowned for its distinctive smell. Some fruit is aimed at general feeders and offers a diet of carbohydrate, which needs supplementing elsewhere. Others, such as the Avocado, native to Mexico and Central America, offer a much more complete package of nutrition. This wins them the undivided services of a more restricted but effective set of seed distributors, including the Resplendent Quetzal *Pharomachrus mocinno*, a bird that swallows the fruit whole and regurgitates the seed. It is thought that the evolution of the Avocado and that of other large-

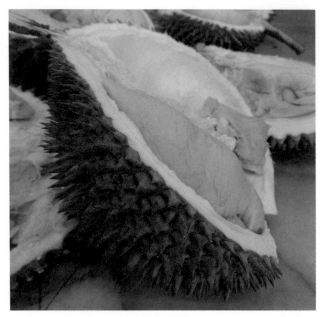

FIG 122. Durian fruit *Durio zibethinus* cut to reveal the custard-like flesh surrounding the seeds. It is normally forbidden to carry the fruit in taxis or planes or to take it into buildings due to the very strong, distinctive smell. In Southeast Asia where the Durian is native, this smell attracts orangutans from long distances to feed when the fruit are ripe. The taste of the fruit is also interesting to us humans. Alfred Russel Wallace described the taste of Durian as 'A rich butter-like custard highly flavoured with almonds... but intermingled with it comes wafts of... cream-cheese, onion-sauce, brown sherry, and other incongruities' (Wallace 1869). Personally, I think it tastes of spicy rotting garlic but yet is strangely moreish. Penang, Malaysia.

fruited species was driven by the Pleistocene megafauna, such as mammoths, horses and ground sloths (Janzen & Martin 1982). They would eat the fruit whole, and the seed would survive the journey through the gut. When these animals became extinct 13,000 years ago, other dispersers moved in to do the job. Think of mammoths when you next eat an avocado!

The seeds inside fleshy fruits are usually hard and often toxic (especially in the tropics) so that, as the fruit is eaten, the seed is either spat out or regurgitated. Others will be eaten with the fruit. They then run the gauntlet of being crushed by teeth or gizzard before being swallowed and exposed to stomach acids and passing through the gut to appear in the droppings. The seed needs protection, which can be in the form of the hard inner shell of a drupe or just a very hard seed case as in Wild Service-tree *Sorbus torminalis* (Thomas 2017). Some trees have turned this to their advantage, since the passage through the gut is not just endured but used to stimulate germination. For example, abrasion of the seed coat of Holly by passing through the grinding gizzard of birds improves germination. Seeds can also gain some protection against the hostile conditions by speeding the journey through the gut. It is no coincidence that senna pods from various *Senna* species, or figs from the large number of *Ficus* species, are superb laxatives. They hasten the passage of the seeds and ensure that they are dispersed widely and messily in all directions.

In seeds with dry fruits, such as nuts, the hard woody shell does not help the seed to be moved, it simply protects it. Nuts that are carried away are likely to be eaten, but this can still work to the advantage of the tree. Some may be dropped as they are being carried away, as happens with beech nuts being carried by small birds, and so survive to germinate. Other systems are more beneficial to the trees and involve animals storing caches of the seeds for later, usually in the ground. Some will be forgotten or not needed, leaving the seed ready-buried to germinate. Caching is found in various pines around the world where the large wingless seeds are moved by corvid birds. This includes the Chinese and Japanese White Pines (*Pinus armandii* and *P. parviflora*), as well as the Macedonian Pine *P. peuce*, which hedges its bets and is dispersed by animals and also by wind. In western USA, the Clark's Nutcracker *Nucifraga columbiana* collects seeds of various pines, including the Pinyon Pine *Pinus edulis*. Each bird will carry an average of 55 seeds (up to a maximum of 95) per trip, taking them up to 22 km away. A flock of 150 birds investigated in 1969 stored an estimated 3.3–5 million seeds, weighing between 650 and 1,000 kg, with individual birds capable of stashing over 33,000 seeds each (Vander Wall & Balda 1977). A very effective system for moving and planting seeds.

Nearer to home in Europe, jays are the major disperser of acorns in wooded areas. A pair of jays may store several thousand acorns in an autumn, which are buried at numerous locations and re-found in winter by means of the jay's superb visual memory (see Chapter 4 for how jays are responsible for the size of acorns). Acorns rapidly die if dehydrated (termed 'recalcitrant' – Chapter 4) and so burial by jays is important to their survival. Judging by the number of oak seedlings that appear around the edges of my lawn every spring, jays regularly err on the side of caution and bury far more acorns than they need, even after a long, cold winter. In a bare field in 1968, Mellanby found up to 5,000 oak seedlings per hectare arising from acorns carried by birds from the adjacent woodland. In open agricultural land where the jay is scarce, magpies can take on the role of acorn caching, moving acorns from isolated trees far out into abandoned fields. In central Spain where there are no jays, magpies were found to be capable of caching 41–56 per cent of the acorn production of Holm Oaks. Individual birds could cache up to 1,372 acorns in six weeks, which resulted in 56–439 seedling Oaks per hectare the next year (Martínez-Baroja *et al.* 2019).

In other cases, it is not clear how the tree benefits from seed removal. Acorn Woodpeckers *Melanerpes formicivorus* in California are cooperative, social birds that work in groups to collect acorns from various *Quercus, Lithocarpus/ Notholithocarpus* species between September and December, which they store communally in a granary tree (Fig. 123) or sometimes a wooden post or telegraph pole! Insects, buds and sap form a large part of their diet but the thousands of acorns stored get them through the winter. Each acorn is wedged into a hole in the bark, and as it dries and shrinks, it is moved into a better-fitting hole. Of course, such a large food supply also needs to be robustly defended by the group.

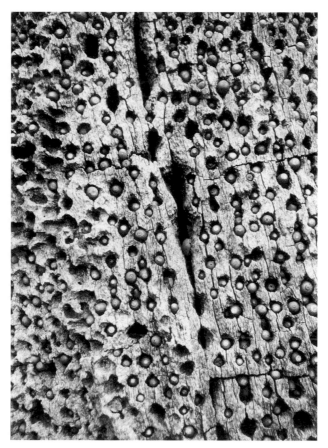

FIG 123. A communal Acorn Woodpecker *Melanerpes formicivorus* granary tree with hundreds of acorns wedged into the wood. Photograph by Walt Koenig.

Acorns have been found to be typically collected from 1–8 trees over a distance of 150–550 m from the granary trees (Pons & Pausas 2007b, Scofield *et al.* 2010). It is difficult to see how this is of any benefit for the trees except that the occasional acorn might be dropped away from the mother tree. In fact, it is best to think of the Acorn Woodpecker as a seed predator rather than a disperser. In that case, how do the trees persist? Woodpecker groups have been recorded as storing up to 50,000 acorns, with each bird adding on average 325 acorns per year (Koenig & Mumme 1988), but this is only a very small proportion of the acorns produced in most years, and the remainder will be moved by other animals, particularly scrub-jays that cache the acorns in the ground. While scrub-jays are important seed predators due to their high numbers, they are also a very important seed disperser, which allows enough acorns to be forgotten to ensure the oak trees have enough offspring for the species to persist.

Mammals can also cache seeds. In South America, the very tough, woody, cannonball-like fruits of the Brazil Nut Tree (Fig. 116) can only be opened by agoutis, large rodents related to the guinea pig, which then cache the seeds in the ground. In Europe, the introduced Grey Squirrel *Sciurus carolinensis* will cache acorns, but the squirrels frequently bite off the pointed end of the acorn that contains the embryo and thus prevent many of the acorns from germinating.

Masting

To help overcome the effect of seed predators, a small but significant number of tree species go in for *masting*, where groups of individuals synchronously produce a large number of seeds every few years, with only small numbers in between (Kelly 2020). Masting trees such as Beech and oaks *Quercus* produce abundant seed years every 2–3 and 3–4 years, respectively. In the case of oak, a good seed year might yield 50,000 acorns per tree, while in between there might be none or just a few hundred (Fig. 124).

The value of masting is usually described in terms of *predator satiation*; predators of seeds will be swamped in mast years, and so some seeds will be left uneaten and survive to produce a new tree. This works less well in trees like oaks that rely on seeds being cached, since many seeds will be left below the parent tree, but still works if the jays bury more than they need in mast years. The whole argument is in reality a little more subtle (Kelly 2020), since if the predators are very mobile, such as birds and other vertebrates, compared to relatively immobile invertebrates, they can move to areas of abundant seed. The selection pressure may therefore be for high variability in seed numbers between years but high synchronicity of mast years between groups of trees over large areas to avoid

FIG 124. A mast year for Pedunculate Oak *Quercus robur* when a superabundance of seeds overwhelms seed eaters, allowing some to germinate.

attracting birds from far away. Beech showed high synchronicity of mast years over large parts of England in the previous century; the reasons why this is less so in recent years are explored in Chapter 4.

DISPERSAL BY WATER

Walking through a European woodland, you might not think about water when it comes to moving fruit and seeds. But it turns out that while many of our trees are primarily dispersed by wind, they also have fruits that float and can easily move along streams and rivers. For example, Säumel and Kowarik (2010) dropped fruits of Norway Maple, Box Elder and Tree of Heaven into the Spree River in Berlin and found that 95 per cent were still floating three hours later, and a quarter of the seeds had floated 1,200 m in that time. Tree of Heaven seeds can float for up to 20 days and still be viable and may even germinate better and quicker afterwards. This may also apply to smaller seeds, since the cottony hairs of willow fruits help them to float for a week or so. Whether water movement counts as successful dispersal will depend upon whether the seed is deposited in a place where it can germinate and establish.

Trees of riparian or coastal habitats, such as mangroves, have seedlings that establish in wet margins and so use water as their main means of movement.

FIG 125. Many mangrove plants are viviparous, giving birth to live young. The seed (seen at the top) germinates while still attached to the parent to produce these spear-like seedlings. If they fall into mud, they literally plant themselves like arrows hitting a target, giving them stability against the next high tide. However, those that fall in water will float and can be washed to new mudflats to start a new mangrove. These are *Rhizophora apiculata* seedlings in Penang, Malaysia.

These often have fruits or viviparous seedlings (Fig. 125) with an oily outer coat that can float in still water often for more than a year, and can be moved by running water or, if small, blown across the surface of still water considerably further than they could be blown through the air. A number of tropical legumes which grow beside rivers drop their pods into the water where they break into small one-seeded segments that float away. Many wash up close by, but if they end up in an ocean can be moved worldwide. Occasional 'sea-beans' (seeds from the pods of *Entada gigas* – Box 1) can be found on British beaches, looking like fat chocolate coins up to 4–6 cm in diameter and 1.5 cm thick, having travelled from the Caribbean on the Gulf Stream.

Over 100,000 km² of the Amazon jungle are flooded each year during the wet season just when the fruits of many trees and vines ripen. The fruits of Pará Rubber Trees explode on the tree, throwing the seeds 10–30 m out over the water. The seeds have a shell nearly as hard as that of a Brazil nut, but fish with molar-like teeth, resembling those of a horse, can crack them open. Many seeds are eaten but inevitably some survive, floating around for the next 2–4 months until the floods recede. Seeds of other species, encased in fleshy fruits (built-in buoyancy aids), are either swallowed by fish or drop to the bottom as the flesh decays; either way the seed is moved away from the mother tree.

HOW FAR DO FRUITS AND SEEDS MOVE?

Measuring how far seeds and fruits are carried by the wind can be quite difficult to quantify. It is easier if there is a lone tree or small group, since the fruits can be watched, picked off the ground or caught in traps, and you can be fairly certain where they are from. Fruits can also be marked by fluorescent dye, and large fruits can be tagged and radio-tracked. Increasingly it is also possible to match the genetic makeup of a fruit (using *microsatellite DNA markers*) to that of the trees in an area to make absolutely certain which tree a fruit came from and thus how far it has travelled (Godoy & Jordano 2001).

Most winged fruits and seeds typically drop within 20–60 m of the parent although a few are likely to go much further. For example, Sycamore samaras have been recorded to travel up to 4 km and fluffy poplar seeds up to 30 km. Small seeds such as birch *Betula*, or Heather seeds still inside their spherical capsules, can be blown many kilometres over hard-packed snow to reach more than three times the distance expected by normal wind dispersal.

In temperate woodlands, mammals and birds usually carry acorns, hazelnuts and berries in the order of 10–30 m from the tree, but some dispersers can carry nuts further. For example, radio-tracking of acorns moved by jays in Spain found that acorns of Cork and Holm Oaks were taken anywhere between 3 m and 550 m from the tree, with an average distance of 69 m (Pons & Pausas 2007b). However, as noted above, the Clark's Nutcracker can carry seeds of the Pinyon Pine up to 2 km away. But acorn movement by small mammals like mice was seen to be stopped by a two-lane road in central China, and average acorn dispersal distance near the road was reduced by half (Chen *et al.* 2019). Roads might not affect jays and squirrels in Europe but they are an example of the sometimes unexpected problems we impose on trees.

Water is able to move fruits and seeds much further than any other means of dispersal. The Coconut that we buy in shops is really a drupe stone (like a plum stone), with the outer two layers of the fruit forming the coir husk. This coir helps the coconut to float, and it is suggested with little evidence that Coconuts may be transported up to 3,000 km and still be viable. More likely, floatation is a device for moving the seeds around within island groups rather than long-range dispersal but is still likely to be successful over many kilometres. The reason for Coconuts being so big is explained in Chapter 4.

Dispersal can be more complex, using more than one means of movement. This often involves wind and water or wind and animals. Many pines are first moved by wind but then cached by small mammals in the same way that jays cache acorns (Vander Wall 1994). As a more complex example, the Australian Quinine Bush *Petalostigma pubescens* uses a three-stage dispersal process. The round fruits up to 2.5 cm in diameter are swallowed by Emus *Dromaius novaehollandiae*, which digest the flesh from the drupe stone. Having passed through the Emu, there can be over a thousand drupe stones in a single scat, which would lead to intense seedling competition. This is avoided because as the stone of the drupe dries in the sun, it explodes 2–3 days later and fires the seeds 1.5–2.5 m away. The seeds have a conspicuous fatty appendage (an eliasome) at one end and are carried off by ants into nests. The ants eat the appendage and discard the seed, leaving it underground where it is protected from mice and fires and in a favourable germination site (Clifford & Monteith 1989). Although less spectacular, a similar method is used by European broom and gorse (*Cytisus* and *Ulex* species) where the pea-like pods explosively scatter the seeds. The seeds also have an eliasome and are carried off and buried by ants in the same way.

Using the methods described above, it is possible to find out how far fruits and seeds of different trees normally spread, but it is much more difficult to work out the maximum that they can achieve. This is partly because it requires looking for a small number of fruits and seeds over a large area, often confounded by secondary dispersal as noted above. But in many cases, this secondary dispersal is more random. For example, the samaras of Ash can exceptionally be blown up to 200 m from the mother tree but can also be found to be moved at least another 150 m along roads by being swept along in the slipstream of vehicles, and over a kilometre further if they end up in an urban river (Säumel & Kowarik 2013).

The important thing is that seeds do not necessarily have to go long distances, since the likelihood of a suitable spot for germination, such as a gap in the canopy from a tree falling over, is just as great nearby as it is far away. The most crucial aspect of dispersal is for the seeds to clear the shade and root competition of the mother tree. Beyond that, finding a suitable site to germinate is a random process. Indeed, the extreme distances reached by a few seeds may not often be ecologically important, since the probability of a solitary seed establishing a new plant is small, especially as it is the smaller seeds that tend to go furthest, and these are less competitive as they have smaller food reserves. But looking at the whole landscape over a long period, the occasional tree that does establish a long way away from others may be the nucleus of a new population or add new genetic material to a remote population (Jordano 2017). Under current climate change scenarios where suitable conditions are moving north, for a small population of a rare tree this occasional long-distance movement could be the difference between extinction and survival.

CHAPTER 10

The Annual Show of Autumn Colours

PREPARING FOR WINTER

As the end of the growing season approaches, the tree needs to prepare for winter, which includes dropping leaves if it is deciduous. But how does a tree know that winter is coming? Some trees, such as North American Tulip Tree *Liriodendron tulipifera* and Honey Locust *Gleditsia triacanthos*, respond mainly to the declining temperature of autumn. This can work well in a continental climate where the temperature reliably declines and gives time for the tree to prepare for winter before prohibitively cold temperatures arrive. In a temperate oceanic climate like Britain's, however, where a sudden frost can happen after a few warm days, this can catch the tree out, so most trees react to the shortening daylength of autumn to prepare for winter before it arrives (Lüttge & Hertel 2009). In some, the change in spectral quality of light in autumn can also be used as a cue, especially in high latitudes (Brelsford *et al.* 2019). How plants measure daylength is described in *The plant clock and calendar* in Chapter 5.

That daylength is involved in many trees is most easily demonstrated by the effect of streetlights in causing leaves to be shed more slowly. Figure 126 shows that the branches of a Norway Maple immediately around a streetlight hold their leaves for much longer than the rest of the tree, since they perceive the days as still being long.

Once leaf shedding is triggered by daylength, temperature can play a role in how quickly leaf shedding occurs. Warmer temperatures extend the process, and a sudden cold spell can cause leaves to be rapidly lost. Other factors can also play a part; drought, early frosts and high winds can lead to earlier leaf drop, while abundant soil moisture and nitrogen can delay it. How sensitive a tree is to temperature once leaf shedding has been triggered can be quite variable. Horse-chestnut *Aesculus hippocastanum*, for example, is particularly unresponsive to temperature, and so warmer years do little to change leaf-fall date. In contrast, leaf fall in Pedunculate Oak *Quercus robur* in a warm year can be more than five days later than in a cold year, since it is more responsive to temperature (Chen *et al.* 2020).

While it is clear that spring bud burst is starting earlier due to global warming (Chapter 3), there is some debate over how much autumn leaf fall is being affected. This is partly because the reaction to temperature is complex. For example, Chen *et al.* (2020) have shown that for temperate trees, night-time warming leads to later leaf fall, while daytime warming (linked to water shortage

FIG 126. (a) This Norway Maple *Acer platanoides* has lost most of its leaves except for those around the streetlight. (b) (opposite) The illuminance graph shows that near to the lamp (in this case a metal halide lamp), the night can be as bright as daylight, although the upper part shows that the amount of light quickly decreases away from the lamp. Note that the illuminance axis (in lux) is logarithmic. This high light intensity around the lamp is detected by the nearby branches as the days still being long, and leaf shedding in preparation for winter is delayed. (b) From Bennie *et al.* (2016).

on warm days) leads to an earlier autumn. Nevertheless, autumn does appear to be getting later such that the length of the growing season has increased by 18–24 days per decade over up to a third of Europe (Garonna *et al.* 2014). Generally this increase is being caused by autumn being delayed rather than spring beginning earlier, although the exact changes vary around Europe. Certainly, when I moved to the English Midlands in the mid 1980s, leaf fall finished around the middle of October, but now there are often still some leaves on trees in the middle of November. Leaf loss is still being triggered at the same time each year in response to daylength, but the time taken for leaf shedding is being stretched, leading to a delayed autumn.

The extended growing season is potentially useful for us in combating climate change. Every extra growing-day in autumn fixes another 98 kg of

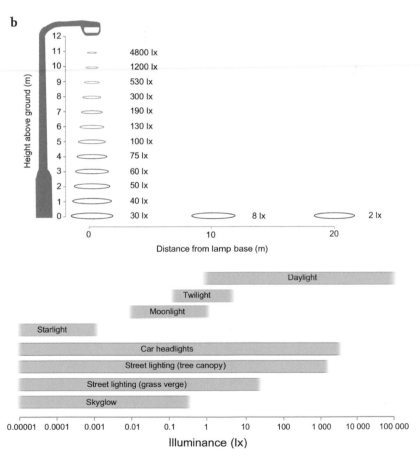

carbon per hectare in temperate forests (Richardson *et al.* 2012), although higher respiration in warmer soils may negate some of this extra photosynthesis (Zhang *et al.* 2020). However, Zani *et al.* (2020) suggest that there might be limits to how much carbon a tree can use or store in a single year, so if spring comes earlier and all the carbon needs are met earlier, then leaf fall might eventually start earlier rather than later. Indeed, after detailed modelling they concluded that in central Europe, increases in spring and summer growth would mean that by the end of the century, instead of autumn coming 2–3 weeks later, it may actually start up to a week earlier. A similar conclusion was reached in eastern USA by Keenan and Richardson (2015). If the tree is 'full' of stored food, the leaves become a costly and unnecessary expense, since the extra sugars they produce have nowhere to go. In this case, there is an incentive for the tree to shed its leaves early despite conditions being good enough for them to stay longer (Rollinson 2020).

THE TIMING OF LEAF SHEDDING

As noted above, leaf shedding in deciduous species is triggered primarily by shortening days. However, different species respond to slightly different daylengths, which is why autumn colours tend to appear in different species in more or less the same order each year (Fig. 127). In New England, Red Maple reliably shows good autumn colours before Sugar Maple. But the start of the sequence may begin earlier or later in any one year depending upon the weather, and the duration that the colours persist may also be shortened or lengthened by the weather. This explains why there are many websites in New England predicting when the colours will be at their best, aimed at 'leaf peepers' – people who travel to see the spectacular autumn colour displays.

Walk through a tropical rainforest at any time of the year and you are likely to see a number of freshly fallen dead leaves but not in the huge numbers seen in a temperate-zone autumn. This is because leaf shedding is not in response to seasonal change. Leaves eventually reach the end of their useful lives and will be shed, but this is usually in response to some internal cue and so can happen at any time of the year. Individuals of the same species may all lose their leaves together, especially when the trees are older, or as an isolated tree, but each species has its own timetable. In some species, leaves are shed from one or more branches of a tree at a time. Such loss may occur fairly regularly, or it may be irregular, and a tree or branch may grow new leaves immediately or wait for a few weeks or even months. The overall impression in a rainforest is that you are always surrounded by leaves, so even though there are trees that may be

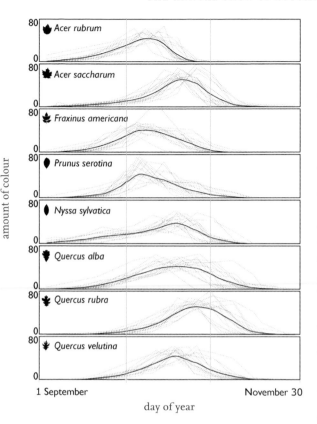

amount of colour

80 Acer rubrum
0
80 Acer saccharum
0
80 Fraxinus americana
0
80 Prunus serotina
0
80 Nyssa sylvatica
0
80 Quercus alba
0
80 Quercus rubra
0
80 Quercus velutina
0

1 September   November 30
day of year

FIG 127. The build-up of red autumn colour in trees growing at Harvard Forest in central Massachusetts between 1993 and 2010. The amount of colour was calculated by multiplying the percentage of red leaves on the tree by the percentage of leaves still on the tree. The faint dotted lines show the pattern in individual years, and the thicker lines are the average of these years. The species included are Red Maple *Acer rubrum*, Sugar Maple *A. saccharum*, White Ash *Fraxinus americana*, Black Cherry *Prunus serotina*, Black Gum *Nyssa sylvatica*, White Oak *Quercus alba*, Red Oak *Q. rubra* and Black Oak *Q. velutina*. From Archetti *et al.* (2013).

completely bare for a short time (generally called leaf-exchanging trees, since they exchange one set of leaves for another), most keep some leaves all year round and count as evergreen.

Evergreen temperate trees, where leaves and needles may live for up to several decades, will still shed their oldest leaves each year. In some evergreens, leaf shedding happens throughout the year, with some, notably *Cupressus* species, having a midsummer peak. Others, such as Holly *Ilex aquifolium*, drop their oldest leaves in one go in spring and early summer in response to competition for sugars from new leaf growth. Some conifers drop their oldest needles in the autumn. I regularly pass a plantation of Sitka Spruce *Picea sitchensis*, and in the autumn the road is covered in brown needles. When I first saw this, I was looking in the plantation for deciduous conifers (notably the larches, Dawn Redwood *Metasequoia glyptostroboides* or Swamp Cypress *Taxodium distichum*) which, of course, drop all their needles in the autumn. I eventually realised that it was the spruce shedding just its oldest needles and, being big trees, there were enough to coat the road.

LEAF AND FRUIT ABSCISSION

The death of a leaf in autumn involves more than just dying. The first step is a breaking down of the contents of the leaf and reabsorbing the useful bits for future use, including many of the macronutrients and micronutrients (Chapter 7). In a study of Aspen *Populus tremula* in Sweden by Keskitalo *et al.* (2005), the internal breaking down of the leaves took 18 days. A week after it started, chlorophyll levels had fallen by 50 per cent, and after three weeks, when the leaves began to fall from the tree, less than 5 per cent of the chlorophyll was left. Mobilisation of nutrients out of the leaf continued until several days before the leaf was shed, by which time 60 per cent of the nitrogen and phosphorus, 50 per cent of the sulphur and 20 per cent of the iron had been moved out of the leaves, and leaf weight had halved.

Across the range of trees, about 50–80 per cent of nitrogen, phosphorus and potassium is reabsorbed from the leaves (Fig. 128), while around 10–30 per cent

FIG 128. Leaves from which the chlorophyll and other useful components are being broken down and taken back into the tree for storage. The main veins are the transport routes and are the last parts to shut down and hence the last to stay green. Beech *Fagus sylvatica* from Aberdeen, Scotland.

of calcium and magnesium is reabsorbed (Aerts 1996, Prieto & Querejeta 2019). Reabsorption is at the lower end of these figures on nutrient-rich and fertilised soils where it is presumably easier to take up more nutrients from the soil than recycle them (Vergutz *et al.* 2012, Yuan & Chen 2015). Reabsorption is also likely to be further reduced where the climate is becoming warmer and drier simply because lack of water can impede the process (Prieto & Querejeta 2019). Evergreen conifers tend to reabsorb more nitrogen and phosphorus from their old leaves, which explains why leaf litter from deciduous trees is more nutrient-rich than that from conifers. The nutrients taken back out of the leaves are mostly stored in the trunk, with progressively less in smaller branches and roots, which reflects the risk of these different parts being lost or damaged over winter. As useful assets are taken out of the leaf, unwanted things such as silicon, chlorine and heavy metals are put into the leaf as a way of getting rid of them.

Once useful components have been scavenged back, the leaf can be shed. But even the cutting-free, or *abscission*, of the leaf is not just a matter of it dying and falling off. If a branch or a leaf dies suddenly due to disease or environmental stress, the leaf will wither in place and be remarkably hard to pull off. In reality, leaf shedding is an active process that takes energy, in the same way the other parts of the trees such as fruits, flowers and even branches are shed. At the base of the leaf a line of weakness forms – the *abscission zone* – consisting of layers of corky cells lacking lignin. As this zone matures, the cells become responsive to abscission signals given primarily by the plant hormone ethylene. As ethylene increases it causes the cells to loosen until the leaf tears away across the abscission zone, leaving the leaf scar covered in some of the corky cells and sealing the wound.

Many of the leaves will finally be lost to high winds or forced off by frost when the expanding ice crystals break the leaf free but hold it in place until the ice melts. One Sunday morning in New England, as the sun rose in late autumn, I heard a noise that sounded like a horde of people tap dancing on the wooden decking outside. In reality it was the walnuts (and leaves) from a large Black Walnut *Juglans nigra* raining down as the sun melted the overnight frost (Fig. 129). Hundreds of walnuts must have fallen over a quarter of an hour.

Why are apple trees increasingly holding on to their fruit into the early winter?
It has become increasingly common over the past decade to see fruit trees, particularly apples, holding their fruit long after all the leaves have dropped, leaving them looking like early Christmas trees with baubles (Fig. 130). Having looked at leaf abscission above, it is useful to consider why the apple leaves fall but the fruit are staying longer.

Apples will drop excessive or damaged fruits early in the year – 'the June drop' (see Chapter 4) – but after that, wild apple species around the world will naturally hold their apples long after the seeds are mature and the sugar content of the fruit is at its highest. This seems to be an adaptation to the seeds previously being

FIG 129. (a) Black Walnut *Juglans nigra* at Harvard Forest in rural Massachusetts, which (b) dropped much of its crop of fruit and leaves early one morning after a hard frost. The ice crystals had finally broken the abscission zone, and the walnuts and leaves fell as the sun hit the tree, melting the ice.

FIG 130. An Apple *Malus domestica* holding some of the apple crop long after the leaves have abscised and fallen.

spread by the now extinct megafauna, including mammoths, horses and ground sloths, and carried on now by bears. Being large, the megafauna could reach the apples and eat them off the tree. The seeds would then pass through these animals (and modern bears) and still be viable. By contrast, if the fruit fell to the ground, they would more likely be eaten by ruminants, and the seeds would be killed as they passed through the multiple stomachs (Spengler 2019). Hanging on to fruits is a natural feature of apples that has been further selected for in domestication, since picked apples are better quality than 'windfalls' (Li & Olsen 2016).

But in these commercial apple varieties, why is fruit drop getting later than leaf abscission? The bottom line is that while leaf fall is initiated by daylength, fruit fall is more dependent upon temperature, and by the time the temperature is just right to signal the dropping of fruit, the tree is beginning to enter dormancy, and the abscission process is interrupted or delayed. This is probably an oversimplification and needs a more detailed unravelling of the complex process of abscission.

Different genes are involved in leaf and fruit abscission (Xie et al. 2013). This is made more complex still by there being various abscission genes acting at different stages in the process. Moreover, the hormones these genes produce interact with each other in interesting ways and are affected by the weather, particularly temperature. Auxin is produced by the growing seeds to stop the fruit dropping before the seeds are ripe. It does this by making the abscission zone insensitive to ethylene. As the seeds mature, auxin production decreases and another hormone, abscisic acid (ABA), increases, which acts with the ethylene, leading to fruit fall (Sun *et al.* 2009).

The turning on and off of genes that start and control fruit abscission is, as in leaf drop, affected by temperature. Just how temperature affects gene expression is not very well known (Tisné *et al.* 2020), but keeping the apples long after the leaves have dropped is almost certainly related to the short days of autumn now being warmer.

SHEDDING BRANCHES

In the same way that leaves are shed, whole branches can also be deliberately shed. This can be used to get rid of branches that are progressively shaded as the crown grows outwards. These branches would otherwise be a drain on the tree's resources or potential invasion sites for pathogens and would also create extra drag from the wind (Chapter 11). In a large tree, the centre of the crown is made up of large branches (the *scaffold branches*), leading to fine twigs only at the edge of the canopy. In the shaded centre, the *dysphotic zone*, the small branches are redundant and are quickly shed. Shedding of branches can also be used in extreme drought, since once the leaves are shed, most water will be lost through the thin bark of small twigs. The Creosote-bush *Larrea tridentata* of western American deserts sheds progressively bigger branches during extreme heat or drought. This is a fine balancing act, since if it loses too much wood, it dies, but if it keeps too much wood, it will also die, from losing too much water. In the tropics, branch shedding is also a useful way of repelling boarders in the form of lianas by creating a long bare trunk that is difficult to climb.

Shed branches can be useful for vegetative propagation. Most branches are drained of nutrients before they are shed, but those of poplars and willows found along rivers are not, and their dropped branches are capable of rooting when washed up on muddy banks further downstream. Poplars and willows have separate male and female trees, and it is not uncommon to find that trees along a river course are predominantly either male or female, since they have vegetatively spread from one original tree.

There is some variation in how the terms are used, but generally *natural pruning* is used if a branch is passively broken or rotted off, but if the branch is deliberately shed, in the same way that leaves are in autumn, this is termed *cladoptosis* (from the Greek *klados*, meaning 'branch', and *ptosis*, meaning 'falling'). Natural pruning involves small branches being broken off by the wind or dying and falling but can also be more impressive. Elms and to a certain extent others, including oaks, Deodar *Cedrus deodara* and London Plane *Platanus* × *acerifolia*, can drop large branches up to half a metre in diameter with no warning, usually on calm, hot summer afternoons; hence the quote from Kipling, 'Ellum [elm] she hateth mankind and waiteth.' This sudden collapse appears to be due to a combination of internal water stress coupled with heat expansion affecting cracked and decayed wood.

FIG 131. A small branch of Pedunculate Oak *Quercus robur* that has been shed by the process of cladoptosis. The abscission zone forms a dome of corky tissue with a corresponding saucer-shaped hollow left on the twig.

Deliberate shedding by cladoptosis (Fig. 131) involves a similar abscission zone to that used in leaves. The wound left by the falling branch is ready-sealed over with cork which is then covered by new wood the following year. In this way, small branches up to half a metre in length and up to several centimetres in diameter are shed by a variety of trees and is particularly noticeable in maples, ashes, walnuts, poplars, willows and oaks. Shedding normally happens in the autumn, although maples shed mainly in spring and early summer. Oaks tend to shed small twigs up to the thickness of a pencil (Fig. 131), beeches may shed larger ones, and birches drop whole branches of thin, dead twigs looking like a besom broom. Conifers such as pines, cypresses *Cupressus*, Coastal Redwood *Sequoia sempervirens*, Swamp Cypress *Taxodium distichum* and Dawn Redwood *Metasequoia glyptostroboides* regularly shed small twigs complete with leaves towards the end of summer. In broadleafed trees, around 10 per cent of terminal branches are typically shed each year through cladoptosis and natural pruning. Not surprisingly, shed branches can make up a third of the weight of woodland floor litter.

AUTUMN COLOURS

Some trees, such as alders and ashes, drop their leaves while they are still green. In alders this may be due to their ability to fix nitrogen in their roots, and so there is little incentive to reabsorb nitrogen and other useful components back from the leaves. This may also be true of Ash *Fraxinus excelsior* which is native to nutrient-rich soils. Other trees produce at least a muted yellowing of leaves and the minority produce bright yellow, orange or red colours, each tree species having a typical autumn colour which can help in their identification. Archetti (2009a) surveyed 2,368 temperate tree species around the world and noted yellow coloration in 16 per cent and red in 12 per cent. Although not a huge number, they are concentrated in certain areas, notably eastern Canada and New England where there are 89 red-colouring trees (30 per cent of the total number of tree species), and East Asia where there are 152 red-coloured (21 per cent of tree species). Within these high-colouring areas, there are even more extreme concentrations, such as in central Massachusetts (Fig. 132), where 70 per cent of the 89 woody species turn red in the autumn (Lee *et al.* 2003).

Why do we have to go to New England to join the autumnal 'leaf peepers' to enjoy good colours? We can grow the same species in Europe but we just don't normally get the same intensity of colour. The reason comes down to high light intensities and low temperatures. Red coloration (from anthocyanins – see below) is at its best under high light levels, and the leaf coloration hotspots around the world are bright places in the autumn: New England receives an average daily

FIG 132. Trees with a range of autumn colours from fiery reds to lemon yellow. Harvard Forest, Massachusetts.

amount of light in September of 168 W/m² (watts per square metre), and East Asia 165 W/m², compared to only 114 W/m² in European forests, which are thus receiving 32 per cent less light in autumn than New England and so have poorer colours (Renner & Zohner 2019). An experiment growing 396 species from the New England area, East Asia and Europe in the same garden found that the lifespan of leaves was three weeks shorter for trees from New England, caused by spring bud burst being later and leaf shedding being earlier (Zohner & Renner 2017). The upshot of this is that the North American trees are shedding their leaves when the light intensity is at its highest, leading to more coloration. Low temperatures are also involved. The brightest red colours are produced with warmth and bright light during the day and cold nights that slow the transport of sugar out of the leaf, common features of New England but less so in the often cool and overcast autumns of Britain and western Europe. Under our conditions, the biochemical pathway that produces red pigments (anthocyanins) tends to be halted and produces proanthocyanidins instead, responsible for the brown autumn coloration of many European leaves (Fig. 133) and the brown colour of many fruits and woods (Rauf et al. 2019).

FIG 133. Leaves of Red Oak *Quercus rubra* which in their native New England would normally turn brilliant red in the autumn (Fig. 135). But growing in the cloudy environment of Staffordshire disrupts the production of red anthocyanin pigments, frequently resulting in the process stopping at brown.

In Japan, Koike (1990) noted that trees which produce leaves continually through the year (indeterminate growth – see Chapter 7), such as birches and larches (Fig. 134), usually begin autumn colouring in the inner part of the crown, and the colouring then moves outwards. By contrast, those that produce all their leaves in one flush in spring (determinate growth), such as maples, start their autumn coloration on the outside of the canopy. Which part of the tree colours first is also dependent on the water moving from roots to branches in 'sectors' (Chapter 6). In trees that have a sector of roots supplying water to one part of the canopy, such as maples, each sector will go into autumn at its own rate depending upon how much water stress it is under and other growing conditions, so different branches may be different colours at the same time. By contrast, in those that are more hydraulically integrated, sharing water around the tree, such as in aspen, coloration tends to be synchronised throughout the canopy.

Leaf pigments

The intense yellow colours (Fig. 135a), merging to orange, are produced by *carotenoid* pigments, including carotenes and xanthophylls. These are present in the leaf throughout the growing season, acting to protect chlorophyll from being damaged by high light intensity. The carotenoids are broken down in autumn but more slowly than chlorophyll, so they show through, particularly in those trees that started with high concentrations. A similar thing happens with the ripening of bananas and citrus fruit that turn yellow or orange. Keskitalo *et al.* (2005) found that when the leaves of Aspen in Sweden turned yellow, chlorophyll levels had decreased by 75 per cent and carotenoids had decreased by only 50 per cent.

FIG 134. A hybrid Larch *Larix* × *marschlinsii* (formerly *L.* × *eurolepis*) self-sown in a forestry plantation. It tends to lose the deciduous needles from the centre of the crown first and from the youngest outer branches last of all.

Red colours (Fig. 135b) are produced by *anthocyanin* pigments which when mixed with the yellows in different proportions produce the reds, purples and sometimes even blues. These can be present in the leaves all year, and if in high enough concentrations will produce red or 'copper' foliage, such as in Copper Beech *Faqus sylvatica* 'Purpurea' (a naturally occurring form but the best have been propagated commercially). However, the bulk of the anthocyanins are commonly produced in great quantities as the leaf heads into senescence.

Producing anthocyanins in autumn is a significant cost to the tree and so must have a function. There are two main ideas as to how they help the tree (Archetti 2009b). The first is that these pigments are involved in *photoprotection* of the chlorophyll against high light levels and free radicals (particularly types of *reactive oxygen* which are a by-product of metabolism), supplementing the work of the carotenoids. This is particularly important in the autumn as it gives

FIG 135. (a) Silver Maple *Acer saccharinum* can turn a delicious golden yellow when the chlorophyll is broken down to reveal the yellow carotenoids in the leaves. Under suitable weather conditions, leaves can also turn fiery red, (b) as has this Red Oak *Quercus rubra* opposite. New Hampshire.

longer for the photosynthetic apparatus to be dismantled and for nutrients to be reabsorbed, as described above. There is support for this idea from a range of trees. For example, red leaves in Sugar Maple *Acer saccharum* maintain better links with the twig than yellow leaves, allowing nutrient recovery to go on for longer (Schaberg *et al.* 2008). Moreover, anthocyanins are most common in plants growing on nitrogen-deficient soil where a greater recovery from leaves would be advantageous. On the other hand, it has been a puzzle just how the anthocyanins can protect photosynthesis, since they are dissolved in the liquid centre of the cell (the *vacuole*) and so are not close to the chlorophyll in the chloroplasts. Other evidence suggests that, while carotenoids are known to be involved in protection against light, the role of red pigments is less certain, mainly because most protection from light is needed in cold weather, and while the yellow leaves are more common in areas with lower minimum temperatures, there was no such relationship with red trees (Pena-Novas & Archetti 2020).

The other most plausible idea, first put forward by William Hamilton at Oxford University, is that the red coloration is a negative signal to aphids and other insects. This could be as a warning that the leaves are chemically defended and so unpalatable. However, if true, this is a subterfuge, since the red coloration is greatest in those trees with fewest chemical defences! It could also be that the leaves appear dark to aphids, since insects cannot see red light, and thus the leaves may look to be dying or dead (White 2009). Although the leaves will soon be shed, aphids are looking to lay eggs that will provide the next generation of aphids on the new spring leaves. So by repelling autumn aphids, trees are offering some advance protection for next spring's leaves. To support this idea, autumn colours have been found to be most intense in tree species that host the highest diversity of specialist aphids (Hamilton & Brown 2001). It is clear, however, that there is still a lot more to find out about the role of red colours in autumn!

LEAVES ON THE LINE – WHY LEAVES CAUSE PROBLEMS FOR TRAINS

Every year, train users are delayed by 'leaves on the line'. This has always been a bit of a joke, like having the 'wrong kind of snow', but there is good justification. Leaves that are crushed on the line form a black paste, which reduces friction between the wheels and track to something like that between ice and an ice-skater, so the wheels just spin or slide. If they slide, this quickly wears a flat spot on the wheel that is expensive to put right. The black paste is iron tannate, formed by a reaction between tannins in the leaves (and not cellulose as previously thought) and iron molecules in the track, which acts as a viscous lubricant (Watson *et al.* 2020). So leaves high in tannins such as oaks and Sycamore *Acer pseudoplatanus* are not the best trees to have near railways.

Despite better technology, the problem is apparently getting worse because older trains had brakes that bore directly on the train's metal tyres and would effectively scrape the black goo off, but modern trains have disc brakes and so this no longer happens. Plus, track-side management in the past kept woody plants further back from the tracks than happens now, a reflection of increasing costs. To counteract the problem of leaves, rail companies now use special units which spread Sandite ahead of rush hour, a sticky mixture of sand and antifreeze, aiming to temporarily stick the sand to the tracks. There are also moves to use dry-ice pellets fired onto the rails, which initially freeze the leaves and then, as the dry ice turns back to gas, explode the leaves away from the metal track, leaving no residue behind.

WHY DO SOME TREES KEEP SHRIVELLED LEAVES THROUGH THE WINTER?

Beech *Fagus sylvatica* is well known for keeping some of its withered brown leaves on the tree in the autumn (Fig. 136), but this also happens in various oaks, hornbeams, American Hop Hornbeam *Ostrya virginiana* and Sugar Maple – a process formally called *marcescence*. This normally happens on the bottom 2 m of the tree but occasionally up to 10 m. In all these cases, the dead leaves are held due to incomplete abscission. During the leaf-shedding process, the leaf blade and most of the petiole die, except for the very base around the abscission zone, which remains alive until the following spring. As the weather warms in spring, abscission is completed and the leaf finally falls.

The cause behind leaves being retained appears to be related to the difference between juvenile and adult leaves. In some trees the juvenile leaves look very different from those on the older, mature tree. This difference is referred to as *heteroblasty*,[4] from the Greek *hetero*, meaning 'another', and *blastos*, meaning 'shoot' (Zotz *et al.* 2011). Heteroblastic species include ivy, many eucalypts, and junipers that produce juvenile leaves before the mature foliage. For example,

---

4 For the purists, *heteroblasty*, change of leaf shape and size through life, driven by development with age or ontogeny, is not to be confused with *heterophylly*, which is the variation in leaf shape on a plant caused by differences in the environment, such as sun leaves and shade leaves, as described in Chapter 3.

FIG 136. A Beech *Fagus sylvatica* in early spring, keeping a few wizened brown leaves on the bottom 2 m of the tree. Even more are being held on the Beech hedge behind.

Chinese Juniper *Juniperus chinensis* has spiky juvenile needles on seedlings and young bushes and adult scale-like leaves on mature plants (Fig. 137), although some juvenile leaves may be kept on branches at the bottom of the plant. Other junipers, including the European Juniper *J. communis*, keep the spiky juvenile foliage throughout their lives. So whole trees that keep their juvenile foliage are possible, and these can be artificially produced by hard pruning or coppicing, such as with *Eucalyptus glaucescens* and other eucalypts prized for their round, silvery juvenile foliage (Fig. 138).

The majority of trees, however, are *homoblastic*, in which the physical changes between juvenile and mature leaves are negligible or gradual and so usually go unnoticed. Whether the difference between juvenile and mature leaves is visible or not, a young tree usually has a cone of juvenile leaves inside the canopy, while the branch ends and upper crown increasingly have mature-type leaves, the change starting when the tree is 5–10 years old. William Robbins (1964) suggested that it is these juvenile leaves at the bottom of the canopy that are retained in the winter in oak and beech. This would certainly explain why they are only

FIG 137. Chinese Juniper *Juniperus chinensis* shoots with needle-like juvenile leaves on the left and adult scale leaves and immature male cones on the right. Photograph by MPF from Wikimedia commons, reproduced under the Creative Commons Attribution-Share Alike 3.0 Unported.

FIG 138. Juvenile (left) and mature (right) leaves of Blue Gum *Eucalyptus globulus*. Silwood Park, Ascot.

held at the bottom of the tree and rarely at the very tips of branches. The most convincing evidence for this comes from beech hedges where all the growth remains juvenile (Fig. 136), and the hedges remain coated in leaves through the winter. However, if cuttings are taken from older, adult parts of a beech tree, and rooted, the new trees remember that they were producing mature leaves, and the resulting hedge does not keep any of its leaves in winter.

That explains how but not why the leaves are kept. It has been suggested that retained (marcescent) leaves may help keep the buds of young growth safe, as herbivores such as deer and rabbits don't like the texture of the old dry leaves or the noise they make. However, the reason is perhaps more likely to be down to the nutritional quality of the twigs versus the leaves. Claus Svendsen (2001) in Denmark found that deer browsed more on beech and Hornbeam *Carpinus betulus* if the old leaves were removed, but it made no difference to the amount eaten in Pedunculate Oak *Quercus robur*. This appeared to happen because the beech and Hornbeam twigs were higher in protein and had less indigestible lignin than their dead leaves, so the dead indigestible leaves just got in the way and reduced the amount of quality browsing. In oak, by contrast, the dead leaves had similar amounts of protein and lignin to the twigs, and so twigs were eaten with or without leaves.

So, why does oak keep the dead leaves if it is not to deter browsers? There is some evidence that oak leaves held on the tree over the winter undergo some photodegradation and so will decompose better once they hit the ground in spring (Angst *et al.* 2017). This extra pulse of nutrients in spring, just when the tree needs nutrients for new growth, may also be helped by the fact that the leaves of oaks and beech hanging on the tree are drier than those on the ground, and so they don't have valuable nutrients leached out over winter when the roots are dormant. Rather, when they fall in the spring, the roots are working again and can rapidly absorb the nutrients as they are released (Otto & Nilsson 1981). Keeping leaves only on the lower branches may improve the chance that they fall within reach of the tree's own roots.

CHAPTER 11

The Long, Cold Winter and Storm Damage

O nce trees are triggered to expect winter as a result of shortening days modified by lower temperatures, they continue their preparation. Over-wintering buds which will have formed in summer and early autumn become dormant and require chilling before they can open in spring. Cold temperatures can slow and stop photosynthesis in evergreens or inside the bark of deciduous trees (Chapter 3). Freezing can also damage cells, since ice crystals can physically puncture the cell membrane, causing the cells to leak and die, leading to the death of small twigs and branches nipped by frost. The freezing of water inside or between cells can also result in desiccation of the cell contents. Water stress on a larger scale can also be a problem, particularly for conifers on warm winter days when the foliage loses water but the ground is frozen, leading to the leaves losing too much water.

To survive all these problems, trees have developed a number of mechanisms that enable them to survive winter stress given the right 'hardening off', or preparation period, allowing *cold hardiness* to develop. In most trees this is in response to shorter days but in some this is also in response to low temperatures. These signals trigger *cold-responsive genes* which begin the development of cold hardiness to help survival during the long periods of low temperature and freezing during winter. This includes the production of dehydrin proteins which are involved in coping with dehydration. Such preparations allow twigs and evergreen foliage to survive repeat freezing and thawing, heavy frosts (Fig. 139) and other trials of winter.

The first line of defence in cold hardening is to avoid freezing of the cells by accumulating organic acids, amino acids and particularly sugars (*cryoprotective*

FIG 139. Frost on foliage of a Yew *Taxus baccata*.

compounds) inside the cell, which depress the freezing point. This works in the same way that adding salt to a road prevents it freezing. Cryoprotection prevents damage down to –1 °C or –2 °C.

The second mechanism is to allow supercooling of a cell's contents without freezing, done by adding antifreeze proteins that prevent the formation of ice crystals. This can give protection down to –38 °C but only if preceded by a hardening-off period of several days below 5 °C. This explains why an early sharp frost can kill when the same temperature later in winter does not. Supercooling is accompanied by modifications to the cell membrane to allow it to cope with the changes (Chang *et al.* 2021).

The third mechanism is needed in only the most severe of climates. Here, water is removed from the living cells and allowed to freeze between the cells and inside non-living woody cells in the xylem where it causes less damage. The upshot is that it causes *cytoplasmic vitrification* where the cell contents set like glass, and dehydrins are essential in allowing the contents to tolerate the extreme desiccation and recover come spring. This extreme mechanism is found in a wide range of trees of high latitudes, including evergreens such as firs, spruces, pines and larches as well as deciduous species such as birches, willows and poplars. Again, there is a need for a slow, continual cooling or hardening off ahead of time, but if these conditions are met, it allows dormant twigs of many northern woody plants to survive down to at least –70 °C to –80 °C and even to survive down to the temperature of liquid nitrogen (–196 °C) and liquid helium (-269 °C) without harm (Sakai 1965)!

Freezing of the water in the woody tubes of the xylem can also cause problems, with gas bubbles forming as the water cools and freezes (Chapter 6), blocking the tubes and stopping water movement. This can be a particular problem for evergreens where the leaves warm during a mild spell in winter and are using water.

FROST CRACKS

Freezing temperatures can cause external physical damage. This is seen as frost heave of young seedlings where the repeated freezing and thawing of soil can physically jack seedlings out of the ground. Freezing temperatures can also damage the trunk by causing *frost cracks* in the largely dead cells of the wood. The main constituent of wood is cellulose, which is hygroscopic and readily absorbs and loses water, causing it to swell and shrink. Wood cells tend to shrink in girth rather than length, so wood is fairly stable along the grain. Similarly, rays running from the centre of the tree act as restraining rods, reducing radial shrinkage, so most of the shrinkage of wood is tangentially around the tree. This is why a dried slice of wood from a tree nearly always has a crack running to the centre. It also explains why frost cracks form, especially common in broadleaves such as oaks, ashes, elms and maples, but also in firs.

During cold nights, as the outermost part of the wood begins to freeze, the woody cell walls shrink as the water they contain migrates inwards to join the ice forming inside the tubes. This shrinkage of the outermost rings of the trunk occurs over the still-wet and unshrinking centre, causing tangential tension (Bräuning *et al.* 2016). This is added to by further contraction of the outermost wood due to the cold, especially after sunny days when the tree has warmed up in the sun and the outside cools and contracts more quickly than the centre when night comes. The cumulative tension is eventually relieved by the wood cracking open (Fig. 140) with a sound that can be like a gunshot. Cracks on big trees in continental climates can be big enough to insert a whole hand. In the spring, the crack closes as the wood warms and becomes moister due to melting of the ice in the xylem tubes. New wood formed the following spring papers over the crack, but the break in the older wood cannot rejoin. The ring of new wood bridging the old crack may not be strong enough to resist cracking the next winter. In this way a frost crack may reopen over many years. The tree reacts to this stress by forming a ridge of thick callus tissue (Chapter 8) over the crack, creating a *frost rib*. This along with a few mild winters can build a strong enough bridge of callus tissue over the crack to prevent future opening, and all that is left to mark the previous wound is the frost rib.

FIG 140. Frost cracks in trees. (a) A crack that has opened during the winter and closed over the summer. Now in late autumn, it is likely to reopen during the following winter. (b) A crack that reopens in successive years develops extensive callus tissue that builds up over the crack to produce a frost rib which, if there are several mild winters, may be strong enough to stop the crack reopening in future years. (a) Red Oak *Quercus rubra*, Harvard Forest, Massachusetts, and (b) Mongolian Oak *Quercus mongolica*, Bichevaya, north of Vladivostok, Russia.

RESISTING WINTER STORMS

An upright tree in still air is a well-designed, safe structure. The wood itself is stiff and very resistant to compression and is in no danger of being crushed by its own weight. So gravity has little direct effect on a tree standing up, and it is possible to balance many tonnes of wood on end and do so for millennia. Branches can be a different matter. They have to spread out from the trunk in order to display the leaves to catch the most sunlight and are, in effect, cantilevered beams, attached at just one end. Branches that are fairly vertical are not much affected by gravity, but as they become more horizontal, the centre of gravity moves further from the trunk and creates a longer *lever arm* that creates a turning force at the base of the branch. This is in the same way that a longer

FIG 141. (a) A young, beautifully balanced tree with branches evenly spread through the crown to reduce the turning force on the trunk. (b) A much larger, heavier tree but still with the weight balanced around the trunk. (a) Bastard Service-tree *Sorbus × thuringiaca*, Staffordshire, and (b) London Plane *Platanus × acerifolia*, Tyntesfield, Somerset (a National Trust property).

spanner puts more turning force on a bolt. If the force becomes too great, the branch may break near or at the trunk. Fortunately, trees can detect this turning force and the branch will stop growing long before this becomes too high. The trunk itself can also detect turning forces created by the weight of the branches which, if they were all on one side, would act to bend the trunk. Trees counter this by growing branches evenly around the trunk so that their weight is balanced around the tree, resulting in no net force acting to bend the trunk and creating the perfectly balanced tree (Fig. 141).

Even trees growing close together can often keep the weight of branches balanced around the trunk (Fig. 142). But there comes a point at which this balance is not always possible. A tree growing on the edge of a block of trees or up close to another tree will produce large branches out into the light in one direction (Fig. 143), putting a lot of turning force on the trunk. In this case the need for light outweighs the need to reduce stresses on the trunk. After

FIG 142. Beech *Fagus sylvatica* growing far enough apart to keep the weight of branches balanced around the trunk. The canopies interweave but maintain crown shyness (Chapter 7). The Dark Hedges, a famous beauty spot in County Antrim, Northern Ireland.

FIG 143. An unbalanced tree. The young tree in the centre is a seedling from the 500-year-old Sweet Chestnut *Castanea sativa* 3 m to the left. Due to competition, the young tree has the mass of branches on the side away from its mother. As the tree gets older, the weight of branches on one side will become even greater. In this case the normal balance of branches around the tree is prevented by competition for light with its mother.

all, without enough light the tree is dead, and so accessing light is the highest priority; with an unbalanced canopy, it may be subjected to intolerable strain in a winter storm or may survive for centuries. Trees with the stress removed from their branches by being propped (Fig. 144) can also become very unbalanced. Trees cannot think and just respond to what they can perceive in terms of stress.

Despite the mechanical disadvantage, one tree is renowned for consistently growing with a lean – the Cook Pine *Araucaria columnaris* native to New Caledonia and widely grown in warm temperate areas up to 40 degrees north and south of the equator. Interestingly, it always leans towards the equator, with fewer than 9 per cent leaning the other way (Johns *et al.* 2017). The further away from the equator, the more they lean towards it. Why, nobody knows!

FIG 144. Trees can become very unbalanced if the weight of their branches is supported and the tree can thus not detect the stress at the base of the branch. This old Sweet Chestnut *Castanea sativa* was presumably considered to be in danger of collapse so the large branch on the left was propped, causing it to continue growing outwards because it could no longer detect any stress at the junction of the branch and trunk. Killerton, Devon (a National Trust property).

BREAKING THE TRUNK

As well as balancing the branches around the trunk, branch length is also controlled so that they will stop growing in length before they are too heavy. This creates a balanced structure. However, a tree cannot be designed to withstand everything that the environment can throw at it because the tree can only respond to what it has experienced in the past. Branches will stop growing before they become too heavy, but may still bend or break if too much snow and ice land on the branch. Trees that have evolved in snowy areas tend to have downward-hanging branches that are very flexible and able to shed the snow to prevent the branches or the trunk from snapping (Fig. 145).

FIG 145. Trees native to eastern Canada are adapted to abundant snow. The conifers have flexible branches to shed snow, and the broadleaf trees, having lost their leaves, hold less snow but are equally flexible. White Spruce *Picea glauca* and Paper-bark Birch *Betula papyrifera* in New Brunswick, Canada.

Wind is a much more important force than gravity or the weight of snow in determining whether a tree stays standing in winter. Wind speeds in temperate areas are generally highest in the winter, since there are greater differences in temperature and barometric pressure driving the winds within a winter hemisphere. The Great Storm of October 1987 in southern England broke or uprooted 19 million trees and killed 18 people. Although a particularly ferocious storm, it was by no means unique, and we are well used to seeing pictures of fallen trees in the media after a bad winter storm.

When high wind hits a tree, the whole trunk acts like a cantilevered beam, attached at the ground. The wind pushing on the crown of the tree (creating *drag*) acts to bend the tree. If the drag of the wind is sufficient, the tree could snap or uproot. In fact, snapping of sound trunks is quite rare, since trees are remarkably stiff and have a very high safety margin in resisting snapping (Vogel 2012). Taller trees, exposed to the highest wind and with longer lever arms, are stiffer than shorter trees and resist that initial bending (Jagels *et al.* 2018).

FIG 146. Poplars (probably hybrid Black Poplars *Populus* × *canadensis*) in north Kent snapped by the Great Storm of 1987. The damage to this stand was caused by strong gusts of wind hitting the tops of the trees, which were unable to flex quickly enough to avoid snapping. The highest recorded gust in Britain during this storm was 196 km/h, or 122 mph.

Most open-grown trees, where competition for light is small, tend to grow outwards rather than upwards as the cheapest way of displaying leaves. These are often only a quarter of the height at which they would start breaking when bent by wind. In forests where trees grow taller to compete for their share of light, and it is more sheltered, trees are still only up to 65–88 per cent of the height at which they would break. Trunks do, of course, still snap. When they do, it is usually due to exceptionally strong gusts of wind that slam into the top of the tree, and the stiffness of the trunk is a disadvantage, preventing it from flexing under the blow, and the top then breaks (Fig. 146).

Trunks are in more danger of breaking if they start to become hollow. This raises the obvious question of just how hollow can a trunk be and still be safe. When I was setting out in the tree world, I was repeatedly told that any loss of wood in the centre of the tree would weaken it. Fortunately, this is not true, and Figure 147 presents a fascinating story of why it is not the case. It is based on a study of over 1,200 trees, both broadleaves and conifers, of various sizes up to 80 cm radius, or 160 cm diameter, with various degrees of hollowness. The figure shows that as long as the solid wood around a central cavity is at least a third of the radius of the tree (t/R = 0.3), then it is in little danger of snapping. In other words, the middle two-thirds of the tree can be lost and the tree is still as functionally strong as one that is solid through to the centre. To put this into perspective, a tree 50 cm in diameter (i.e. 25 cm radius) needs to have solid wood under the bark at least 7.5 cm thick (0.3 x 25 cm) to ensure safety, and a tree 100 cm in diameter should have solid wood 15 cm thick. This may not seem like much wood to hold a tree up, which is gratifying if it is your favourite tree that you thought was getting too hollow to survive, but perhaps worryingly little wood if the tree is near your house. But think of a tree bending towards you in the wind: the wood nearest you is in compression, and on the windward side, the wood just under the bark is under tension, while the centre of the tree is comparatively stress-free. So it is this outer layer of sound wood that is most important in holding up the tree, not the centre.

It is also worth pointing out that a number of trees in Figure 147 have much bigger hollows, where the wall of wood is less than a tenth of the radius (t/R of 0.1 or less), and they are still standing. It may be that these trees have lost a large part of their crown and so there is less for the wind to push against, or they are in particularly sheltered areas amongst other trees or behind a wall or building. It is excellent news for many tree owners that even very hollow trees are not automatically unsafe and need to be removed. Indeed, hollowness can be an advantage. After the Great Storm of 1987, many hollow trees were left standing while their solid neighbours were snapped or uprooted. This may be partially

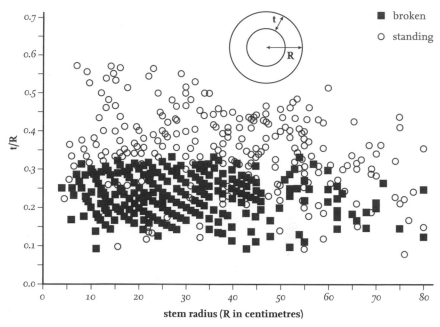

FIG 147. Using hollow broadleaves and conifers of various sizes, the thickness of the solid wood around the central cavity (t) was compared to the radius of the tree (R) and used to calculate the ratio t/R. At the bottom of the vertical axis where t/R approaches zero, the cavity fills almost the whole of the tree, and moving up the graph the trees become progressively less hollow until, if t/R = 1 is reached, the tree is solid wood all the way through. For each tree, the ratio t/R is plotted against the trunk radius. The solid squares show trees that have snapped, and the open circles are trees that are hollow but still standing. All the snapped trees, regardless of type of tree or size, have a t/R ratio of less than around 0.3, where the solid wood around the cavity is less than a third of the radius. From Mattheck & Breloer (1994).

explained by old hollow trees having a smaller crown, and so less for the wind to push against, but it also seems that hollow trees are more flexible and act like shock absorbers able to withstand the buffeting.

Hollowness may also benefit a tree by allowing it to recycle its own central heartwood. Many tropical trees with central rot but also some hollow temperate trees, such as yews and elms, can grow adventitious roots (roots that arise from non-root tissue, in this case living cells in the sapwood or cambium) down into the rotting wood (Fig. 148). Nutrients absorbed from the decayed wood, plus any from animal droppings or dead bodies accumulating in the hollow, can be used for new growth and may give the tree a competitive advantage on nutrient-poor soils. The interesting question is why more species don't have internal roots. The

FIG 148. Internal roots in rotting trees. (a) Fibrous roots spreading through the rotting heartwood of a Yew *Taxus baccata*. (b) As the wood continues to rot and disappear, the by now quite large roots can be left as relicts inside the hollow stem, as in this old Hornbeam *Carpinus betulus*. (a) Oviedo, northern Spain, and (b) Hatfield Forest, Essex.

probable answer is that since wood contains little nitrogen, the extra supply from internal roots is not usually huge.

UPROOTING AND WINDTHROW

We saw in Chapter 5 that the majority of trees have a set of shallow framework roots just below the soil surface, which spread out towards the edge of the crown. As they radiate out and branch, they graft together (Fig. 149) to make a solid *root plate* of dense wood, stiff enough to resist crushing and bending. With the weight of soil and embedded rocks, this root plate can weigh several tonnes, which can be added to the weight of the above-ground part of the tree. As the wind pushes on the tree, it acts to try and lift the tree around the edge of the root plate. What holds the tree up in wind is the weight of the tree and its root plate pushing into the ground, in the same way that a wine glass stands up (Fig. 150).

This process is shown in more detail in Figure 151. As the wind blows and the tree starts to rock, the soil begins to crack on the windward side and the roots are being pulled out or broken. This puts more downward pressure on the roots

FIG 149. Lateral roots of a young Beech *Fagus sylvatica* that have begun to graft together (indicated by the red arrows) to form the start of the rigid root plate that holds up the tree.

FIG 150. (a) As the tree is blown by wind coming from the left, what keeps it upright is its weight pushing into the ground. To fall over, this weight must be lifted over the edge of the strong woody root plate just under the soil surface. This is the same principle (b) as a wine glass which stands up when leant over, in this case by a finger. When a tree does fall, it is often because the root plate has failed, and it would be the same if we nibbled off the foot of the wine glass and tried to balance it on the stem – it would be much easier to push over. (a) Norway Maple *Acer platanoides*, Newcastle-under-Lyme, Staffordshire, and (b) at home, since experience has shown that waiters get nervous if you try this in a restaurant... and are photographing it!

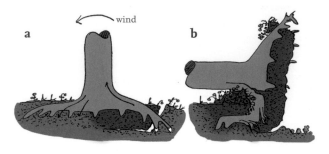

FIG 151. Windthrow of a tree with a stiff root plate. The weight of the tree and root plate must be lifted as the tree pivots over the edge of the root plate. Most trees fall when the root plate gives way, as in (b). From Mattheck & Breloer (1994), reproduced under the Open Government Licence.

to leeward side. If the tree does fail and uproot, it is normally because the roots on the leeward give way. As can be seen in Figure 151b, the leeward roots have buckled, creating a *hinge failure* – the whole root plate hinges along the leeward side and the tree falls. This is equivalent to taking the wine glass in Figure 150b and nibbling off the base on one side; it would take much less effort to push the glass over. The hinge failure usually leaves the trunk perched a little way above the ground, balanced on the hinge. So the secret to staying upright is strong, stiff roots.

Hinge failures usually occur when the roots have been weakened by mechanical damage (such as when services are installed underground using a mechanical digger under the tree crown) or when fungal rot is present (Fig. 152). Sometimes, however, a perfectly sound tree will experience wind that is far stronger than it has previously met and it will fall – a tree can only react to what it has experienced in the past.

A tree is a compromise between growing tall enough to be a good competitor and being sturdy enough to be safe. This compromise is usually arrived at by following the most frugal path of being safe enough to cope with what it has previously experienced. Due to this, a tree growing in strong prevailing winds will invest more in roots than a sheltered tree. A problem arises when a tree that has been previously sheltered is exposed to stronger winds when, for example, the surrounding trees are felled. The previously sheltered tree will react by growing new roots in response to the higher wind speeds it now experiences, but this takes time and the danger is in the first winter. This is seen in forestry plantations where areas are felled, leaving the line of trees at the new edge weak and liable to blow over in the next high winds. If they survive long enough, they

FIG 152. A Sweet Chestnut *Castanea sativa* that was uprooted. The roots on the windward side (now at the top of the picture) were broken or pulled out of the ground, and the root plate on the leeward side has broken, like that in Figure 151b, weakened by fungal rot.

will invest in more roots and become progressively more wind-firm. In a similar way, as the climate changes and a tree is experiencing increasingly high winds, it will, if it survives, grow extra roots to compensate. In the interests of frugality, an individual tree will also only grow roots in the direction they are needed. A tree will usually grow more roots in the direction of the prevailing winds and to a lesser extent on the lee side, with fewest at right angles to the wind. This makes it most stable in the face of prevailing winds but also makes it vulnerable to strong wind from an unusual direction, which can be quite devastating. Forces experienced by the tree in winter when wind speeds are highest will cause the tree to react by growing more wood and roots in the spring to cope with what it experienced in winter. Reacting in this way shows that the trees have a biomechanical memory.

The conservation of resources described above also means that as a tree becomes larger and heavier, the root plate is proportionately smaller, as the trunk grows faster than the root plate. Small trees may have a root plate that extends more than 15 times the stem radius, while in larger trees it may be no more than three times the stem radius. The extra weight pushing into the soil will hold the tree up with a smaller root plate – why waste resources where they are not needed? Different species will have their own compromise in having enough but not too many roots, which is why some species are more reliably wind-firm than others.

A number of modifications to the roots can help the tree to stand up. *Sinker roots* growing down from underneath the lateral roots act like tent pegs into the soil, which helps the root plate grip the soil and also makes it heavier still due to the enclosed soil (Fig. 153). This can create a more compact rectangular *root cage* with even more weight so the framework roots do not have to extend as far out from the trunk. Some trees have a deep tap root or descending lateral branches to produce a *heart-root* system more akin to a rounded root ball, such as in some maples. Here, the tree will tend to fall as a result of the root ball rotating in a

FIG 153. (a) A tropical tree washed into a lake, showing the spreading lateral roots but also a set of sinker roots that would have penetrated vertically down into the soil, forming a root cage rather than a flat root plate. (b) A Scots Pine *Pinus sylvestris* on dry, sandy soil, which has also formed a root cage before being blown over. (a) Royal Belum National Park, Malaysia, and (b) Cannock Chase, Staffordshire.

socket of moist soil. As the tree rotates, it presents a smaller target to the wind so there is less drag, but being off-centre, gravity takes over and the tree falls. In this case, it is the friction between the root ball and the surrounding soil that holds the tree up. Thus, on wet soils (where the water reduces friction), trees with root plates are more successful, and on drier, lighter soils, a heart-root system may work just as well. Some trees, including many pines, have a particularly long tap root when young and stand up to the wind in the same way as a fence post resists being pushed over. As the wind blows, it is the stiffness of the tap root and the resistance of the soil to sideways movement that hold it upright.

Trees with very shallow roots, such as tropical trees, may brace the junction between the trunk and roots by growing buttresses (Fig. 154a). These are often quite thin (and have been used to make the walls of houses) and so are weak in compression, but they help hold the tree up by acting more as guy ropes, resisting the push of the wind on the windward side, which strengthens the junction between trunk and root plate. Temperate trees may also show the start of buttressing, as seen in the oak in Figure 154b, but the buttresses rarely get to any great size. Since buttresses are comparatively narrow, they may be more affected by winter freezing, limiting their size outside of the tropics.

To make life more interesting, trees may use different rooting methods at different ages as they change in size and experience different amounts of wind, so many temperate trees will rely on a tap root when young and develop a root plate with age, and a tropical tree might rely on its root plate when young and develop buttress roots as it becomes taller. And, of course, individual trees may use various modifications together, by having, for example, a wide root plate and a tap root. An individual tree will modify its root structure to cope with the growing conditions. Other trees are capable of using pillar roots (Fig. 155) to add to their stability – roots that grow down from the branches and act like crutches.

FIG 154. Buttress roots. (a) A large buttressed tropical tree where the buttresses act as brackets between the vertical stem and the horizontal roots to give the tree more rigidity. (b) Buttresses are also found to a lesser extent in temperate trees as in this Pedunculate Oak *Quercus robur*. (a) Singapore Botanic Gardens and (b) Powis Castle, Welshpool (a National Trust property).

FIG 155. A fig *Ficus* species producing pillar roots down from the branches to add support (like the artificially supported Sweet Chestnut in Figure 144). The City Botanic Garden, Brisbane, Australia.

RESISTING THE WIND

Despite appearances, trees are not completely at the mercy of the wind and have a variety of mechanisms to help them cope. Flat leaves are sitting targets to be torn or ripped off by high winds, and they can add considerably to the drag on the tree. Fortunately, they are not such a liability as they might seem. As the wind blows, they at first flutter, but as the wind gets faster, they tend to curl into streamlined cones or cylinders which get tighter as the wind gets stronger (Fig. 156).

This flexibility is also used by the whole tree. As the wind blows, the thin outermost branches of the canopy will bend away from the wind. This has two benefits. Firstly, it reduces the apparent height of the tree, reducing the length of the lever arm and so reducing the turning force at the base of the tree. Secondly, it reduces the sail area of the tree so the wind has less to push against. If the wind becomes stronger still, the smallest branches will snap off, which

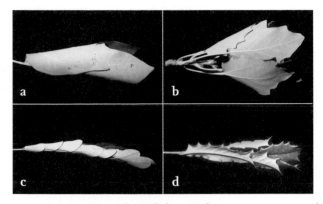

FIG 156. Leaves exposed to turbulent wind at 20 metres per second (45 miles per hour) fold up, reducing damage to the leaves and reducing drag on the tree. Leaves are: (a) Tulip Tree *Liriodendron tulipifera*; (b) White Poplar *Populus alba*; (c) False-acacia *Robinia pseudoacacia*; and (d) American Holly *Ilex opaca*. In Holly the prickles around the leaf edge appear effective in further stabilising the cylinder. From Vogel (2009).

helps save the trunk from damage by further reducing the sail area and height of the tree.

The growth of the tree may also be affected in the long term by strong winds. Generally, the stronger the prevailing winds, the shorter and more tapered is the trunk of the tree. A tree being bent by the wind will detect the turning stress this causes and will grow less in height the next year than a similar tree not experiencing the wind. Strong winds lead to trees staying shorter. In extreme cases, the tree will become wind-shaped. This shape is partly imposed by the environmental conditions such that the buds and young leaves on the windward side may be killed by desiccation or by being physically scoured by icy winter winds or salt spray if near the coast (Fig. 157). But the tree can also respond by permanently bending its branches out of the way of the wind, in effect pulling its ears back and becoming wind-shaped (Fig. 158). How it does this is described below under *reaction wood*.

A tree being intolerably bent by the wind will snap at the place that has most stress on it. These *point stresses* are monitored by the cambium and, providing the tree survives intact, the cambium will react by growing extra wood at the points of stress. This is why trees, especially in windy areas, taper. With wind pushing the crown of the tree, the turning force increases down the tree as the lever arm becomes longer. The tree reacts by growing extra wood towards the base, creating a strong taper along the trunk so that as it bends all parts are experiencing the same stress. The stronger the wind, the greater the taper. Next

FIG 157. A set of trees shaped by environmental conditions. The prevailing wind from the right kills evergreen foliage on the windward side in winter. The other trees are doing better in the shelter of the unfortunate tree at the front. Eventually the tree bearing the brunt of the wind will die, but new trees can establish in the lee and so these clumps of *Krummholz* (German for 'crooked wood') are slowly blown across the landscape. The skirt of healthy foliage around the upright trees and especially to the right in the picture – all that remains of earlier trees – is protected by snow and so escapes damage. Subalpine Fir *Abies lasiocarpa* at Ptarmigan Cirque, Canadian Rocky Mountains.

time you are in a woodland, have a look at the shapes of the trunks; those on the edge of the woodland will generally have more conical stems than the trees in the sheltered interior. The sheltered trees do not feel the full force of the wind and tend to grow more parallel-sided. The world's tallest trees, the coastal redwoods of California (Chapter 12), grow in dense forests and buffer each other against the wind and so they can afford to be narrower for their height, investing a lot of their energy in growing upwards and less in growing outwards.

Stress is particularly high around holes in the trunk. As the trunk flexes, the stress is felt most at the sides of holes and any cracking of the trunk will start here. This is why more callus tissue is added to the side of holes rather than at the top and bottom (look back at Fig. 108 in Chapter 8). In the same way, if the

FIG 158. A Manchurian Apple *Malus mandshurica* that has been shaped by strong prevailing winds but has also bent its own branches to present less of a target to the wind. Lake Khanka, north of Vladivostok, Russia.

FIG 159. (a) A dead stump that when alive had grown along the railings to reduce point stresses. (b) Similarly, a Sycamore *Acer pseudoplatanus* expanding sideways to spread the point stress when it met a metal grave marker. (a) Ambleside, Cumbria – note the coins pushed into the decaying wood – and (b) Rogalinek, near Poznań, Poland.

tree detects a heavy object leant against the trunk, or if it meets an immoveable object as it gets wider, the tree will detect this as a point stress and grow more wood there to spread the load over as large an area as possible to reduce the stress on any one point, as in Figure 159. This is like leaning against the corner of a table and putting a cushion behind you to spread the painful point stress more evenly.

Tree staking

A consequence of trees growing wood where they feel stress is that parts of a tree experiencing *less* stress will have *less* wood expended on them. This leads to a discussion of tree staking. If a newly planted tree has a stake attached high up on the trunk, as in Figure 160a, the bottom of the trunk will not bend in the wind. This reduced stress will lead to the tree not putting as much wood at the base as it would if it could move. The stake stops movement, so the cambium cannot detect stress; the tree cannot rationalise what is happening and can only respond to what it detects. The problem for the tree comes when the stake is removed. It will initially be very weak below the staking point and is liable to snap; but if it survives without breaking, the stress it now feels will lead to new wood being added rapidly, creating the tapering trunk of a well-grown tree. Sometimes, however, longer supports are needed to give extra support during storms such as tropical typhoons (Fig. 160b).

A particular weakness point in trees is the attachment of branches. If you watch a tree flexing in the wind, different branches often move independently. This creates stresses where they join onto the trunk, which can be larger than anywhere else on the tree. The branch collar (the bulge seen at the base of a branch) helps to spread this extra stress, and the wood at the base of the branch is very dense, holding it firm. However, the wood at the centre of the branch base is fairly low-density, which allows it to deform and flex in the wind, like a shock absorber. If the wind gets too strong, this is also where the branch will break, sacrificing the branch to protect the trunk from damage (Slater 2016).

REACTION WOOD AND BENDING

As the crown grows and as the conditions experienced by the tree change, the tree can produce new branches in different places to cope. But it can also reposition the woody skeleton it already has by bending it using *reaction wood*. Reaction wood is produced in response to areas of high mechanical stress detected by the cambium

FIG 160. Tree supports giving (a) support a little too high up the trunk and so stopping the tree moving and developing a stout trunk. (b) The high support here is more for protection from uprooting or snapping by the very high winds expected during the annual typhoon season in Japan, and the poles are often put in place just for the season from May to October. These are especially important in this case, since the mound-planting leaves the tree inherently unstable. Hokkaido, Japan.

under the bark. This can help reduce the biomechanical strain on the woody frame due to the turning forces described above. So, a leaning tree can be bent into a more upright position to bring the centre of gravity of the tree above the base, reducing the turning force on the trunk to a minimum. As noted above, it can also be used in wind shaping to move parts of the canopy out of the way of strong prevailing winds. Reaction wood is also permanently present in branches, where it is used to move and support branches to optimise the light reaching the leaves. This is particularly important for side branches that tend to sag as they become fatter and heavier; reaction wood can be used to move them back up into the light.

Reaction wood comes in two forms and, with a few minor exceptions, conifers push and broadleaves pull. A conifer being bent by the wind will form *compression wood* on the leeward side of the stem (the side that is being compressed), and a broadleaf will form *tension wood* on the windward side (the side being stretched and thus under tension). Similarly, in a branch parallel to the ground, a conifer will produce compression wood on the underside, and a broadleaf will produce tension wood along the top. Compression wood can also be seen in hemlocks

FIG 161. Compression wood seen as darker wood on the right of these cross-sections of Black Spruce *Picea mariana*. The trees were growing on unstable permafrost islands and began producing compression wood when they started to lean to the right 24 years before being felled (the rings can be counted), and continued producing compression wood each year to try and straighten the trees. Northwest Territories, Canada.

Tsuga species, where the topmost shoot droops as it grows but straightens itself over the year using compression wood. Stems with reaction wood are often asymmetrical because more wood is laid down on the reaction wood side (Fig. 161).

Compression wood in conifers is usually visible as a red or brown area with noticeably wider growth rings (Fig. 161). The wood is denser and the tracheids are shorter and rounder with thicker walls, with around 40 per cent extra lignin, and have fissures and spaces between them. But it is also weaker and more brittle than non-reaction wood and has exceptionally high longitudinal shrinkage (10 times higher than normal), making the wood difficult to work with.

The tension wood of broadleaves, by contrast, is not as readily seen, since the change in colour is less pronounced. Nevertheless, it is structurally different from surrounding wood, with fewer and smaller vessels, and is dominated by thick gelatinous fibres with abundant cellulose but little lignin – the opposite of compression wood. The resulting wood is prone to splitting and cellular collapse on drying and is as difficult to work with as compression wood.

If an upright tree ends up at an angle during the growing season by, for example, being partially uprooted by wind, reaction wood will start forming immediately. Tests done so far show the genes involved begin operating within half an hour of leaning (Lopez *et al.* 2021). If it happens in winter, reaction wood will start forming as soon as new growth starts. A record of when the reaction wood started to form and when it had finished its job and the tree was again upright will be recorded in a cross-section of that tree (Fig. 161). Different parts of the tree will, however, record a different picture. Recovery to the vertical is quickest at the top of the tree where it is thinnest and so most easily bent by the reaction wood. This can result in a J-shaped tree, since there is more wood to bend at the bottom of the tree and it is slower to bend (Fig. 162b). For many trees, the story ends there, but since trees continue to respond to the stress that they detect, the thicker base will keep on producing reaction wood while it detects strain. So once the top of the tree has returned to the upright, the lower part will be slowly continuing to bend, with the result that some years later the top of the tree can end up leaning in the opposite way to which it started. The top now feels strain on the opposite side and will produce more reaction wood to re-straighten itself, leaving the tree with an S-shaped stem (Fig. 162a). The problem is caused because trees cannot think; they just react (and very cleverly so) to the stresses they feel. The bending of each part of the trunk is a result of the stress at that point, with no mechanism for co-ordinating the bending throughout the tree. But the system works – a bent tree gets back up to the light, even if it's not pretty!

FIG 162. J-shaped trees.
(a) When soil on this steep slope in the Dolomites, Italy, has periodically slipped, bending the Beech trees *Fagus sylvatica* over, they have responded by producing reaction wood to straighten themselves, creating the distinctive J- and S-shapes. Since these are broadleaf trees, they produce tension wood along the top of the bend.
(b) These young Erman's Birch *Betula ermanii* in Hokkaido, Japan, have been repeatedly bent by the weight of winter snow slipping down the slope and pushing on the bottom of the trees, which are responding by straightening.

WHY DO TREES STILL FAIL?

As with all living things, the growth of a tree is always a compromise. If a tree grew so that it was sufficiently over-engineered to never fail under any weather conditions, it would be hopelessly short and shaded. And a tree that grew excessively tall and thin to win every competition battle for light would likely fail even in mild wind. So the design of trees is a compromise between competition and safety. But this is an informed compromise in that a tree will react to its environment and modify its growth in response to the stresses it detects. Trees may look fairly inert most of the time, but every tree is constantly monitoring the stress it is under (even in winter) and is fine-tuning its compromise between growth and durability.

However, a natural consequence of this compromise is that trees will sometimes break under extreme conditions. In our litigious age this leads to the interesting concept that if a tree snaps or is uprooted in high winds and damages your property, it does not necessarily mean that the tree was diseased or otherwise damaged. As a consequence, it cannot be said that the tree was unsafe and should have been spotted and remedied by the land owner. It does not mean that someone is to blame – despite current thinking, accidents do happen. The breakage or uprooting of a tree under extreme conditions is the natural price that trees must pay for achieving an energy-saving, lightweight structure (Mattheck & Breloer 1994).

From a biological point of view, it is also important to bear in mind that a tree that has been damaged by wind is unlikely to be dead. A tree left leaning by high winds will regrow any damaged roots and attempt to straighten itself using reaction wood. If we wish to intervene, a leaning tree can be winched back to the vertical and if held in place will regrow roots to hold itself upright. A tree that has lost a large part of its investment in its tall woody structure is still not beaten. A completely prostrate tree with some roots intact can produce suckers from the surviving roots or create new trunks out of branches now pointing skywards (Fig. 163). The stump of a snapped tree can usually grow new shoots from stored (epicormic) buds or completely new adventitious buds – described in the next chapter.

WHEN WILL WINTER END? CHILLING REQUIREMENTS AND WARMTH

Buds produced during the summer do not open straight away. They are not dormant but are prevented from opening by hormonal signals from surrounding leaves and branches. If the shoots are removed or damaged, or

FIG 163. An American Lime *Tilia americana* that was partially uprooted – the broken base of the trunk is seen at the bottom left. The branches left pointing upwards have taken over the role of leading shoots and are growing as new trunks. The branches that were forced against the ground have rooted and are producing lines of their own shoots, as in the front centre of the picture. The result is intense competition, with some shoots rapidly dying and others looking to win the competitive race upwards to get most light.

growing conditions are improved, for example by warm moist conditions or adding fertiliser on poor soils, the buds will be quickly released to grow new shoots. As winter approaches, however, buds acquire a self-induced dormancy, or *endodormancy* (Fig. 164), and even if given favourable conditions are incapable of responding. This ensures that the buds do not start growing during a mild midwinter spell. In Britain, this dormancy is most commonly broken by exposure to low temperatures below 5 °C for around 300 hours. Different species of trees, however, have a genetically determined amount of chilling that they need. Beech currently only just receives enough chilling in very cold British winters (needing perhaps 120 days below 5 °C), while species from colder areas, such as European Larch and Wild Cherry, have their chilling requirement met by half the number of cold days needed by Beech.

Once the chilling requirement has been met, the buds are once again able to respond to favourable conditions and are said to be *ecodormant* (Fig. 164); that is, bud burst is controlled by environmental conditions rather than internally. At this point buds are primarily governed by temperature, as described in Chapter 3, waiting for a suitable threshold temperature to be reached (Wisniewski *et al.* 2018). In European broadleaves, very cold winters (extra winter chilling) reduce the temperature needed for bud burst. It is as if the tree is convinced that winter must have passed and is willing to accept a lower temperature as evidence of spring. But it can be a little more complicated in urban areas, since it has been found that the longer amounts of artificial night-time light from street lights during the winter speed up bud burst. This varies between trees, with late-opening species

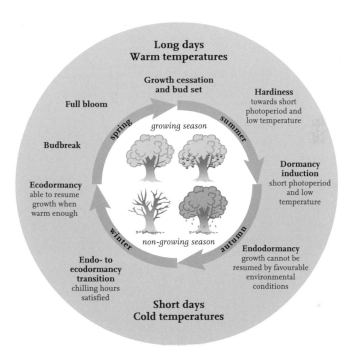

FIG 164. The yearly cycle of growth and dormancy of a temperate tree. Trees grow actively during the growing season (top of the circle) until this slows and stops in middle to late summer as the days become shorter and colder. In the autumn, trees increase their resistance to cold (hardiness) and buds enter dormancy, called endodormancy, since it is caused by internal conditions in the bud rather than imposed by the environment. This ensures that the buds will not start growing in a warm midwinter spell. Chilling temperatures during the winter periods trigger dormancy release. Once chilling requirements are met, the buds are then said to be ecodormant, since they can now respond to warm temperatures and start growing when it is warm enough. This is followed by active growth at the end of the spring and in the summer. From Falavigna *et al.* (2019).

such as oaks and Ash opening their buds in brightly lit urban areas by up to 7.5 days earlier. This appears not to be an artefact of heat islands in urban areas but a genuine response to the light (ffrench-Constant *et al.* 2016).

Trees that have not had their winter chilling met (such as Beech in a warm winter) will still move to ecodormancy but will require higher spring temperatures in order for the buds to open, which goes some way to explaining why Beech is not showing the progressively earlier bud burst in spring that many trees are under climate warming. In fact, in rapidly warming areas, Beech buds may be showing the opposite trend and opening later rather than earlier. Fortunately for Beech, when it has not had sufficient chilling, it appears to become more sensitive to daylength, adding a back-up trigger to ensure spring growth starts (Vitasse & Basler 2013).

As the spring warms, the cold hardiness of winter is progressively lost. This is why late spring frosts, or a warm spell followed by a return of winter, can kill buds that would survive much lower temperatures in midwinter. Increasingly this is becoming more of a problem because warmer temperatures are leading to spring bud burst being earlier each year. When the buds are opening, the nights are still quite long, so even though the days are warmer, there is a greater probability of frosts and of buds, young leaves and flowers being damaged or killed. However, trees are not usually killed by late frosts, since if they are healthy, they will have stores of carbohydrates from previous years (D'Andrea *et al.* 2019) that they can use to grow a new set of leaves. They may be temporarily disfigured but trees are tough and able to endure!

CHAPTER 12

The Size and Longevity of Trees

SIZE OF TREES

The height and width of trees are usually easy things to measure, and since they are the biggest living things around us, we have a fondness for listing which are the biggest. The Tree Register (www.treeregister.org) keeps a list of the biggest British trees – called *champion trees* – by species, giving the height and circumference (girth). Foresters, by contrast, think of trees in terms of diameter, which is much easier to visualise if a little less accurate. Circumference takes into account the eccentricity of a tree that is not completely round in cross-section, while diameter (usually calculated from the circumference) gives an average view which on a very misshapen tree may be hard to actually see. Although not perfect, measurements here are given as diameter.

Diameter is usually measured as *diameter at breast height*, or DBH. Officially this is 1.3 m, or 4.5 feet, above ground, but for most of us it is roughly the height of our chest, hence the name. The reason for measuring at this height is to avoid the swelling of the trunk as it approaches the ground, and the diameter at this height usually represents the diameter of a good proportion of the length of the main trunk.

The widest trees in Britain generally have a trunk in the order of 3–4 m in diameter. For example, the widest Beech *Fagus sylvatica*, growing in Windsor Great Park, Surrey, and the widest Ash *Fraxinus excelsior* at Bodmin, Cornwall, are both 3.5 m in diameter, while the Fredsville Oak in Kent, which is only 24 m tall, is 3.8 m in diameter. Try pacing that out to get a feel for just how impressively large they are. Exceptional trees can be even wider; the widest living Sessile Oak is the Marton Oak in Cheshire, with an average diameter of 4.46 m (Fig. 165).

FIG 165. The Marton Oak in the village of Marton in Cheshire. It is currently 4.46 m in diameter, making it the widest Sessile Oak *Quercus petraea* in Britain and Ireland. The scale is clearly seen by the car.

The key word here is *living*, since there have been trees up to 6 m in diameter recorded historically but with varying degrees of accuracy.

The widest living tree in Britain is the Fortingall Yew *Taxus baccata* found on the shores of Loch Tay in Perthshire, which originally had a diameter of 5.48 m. This is particularly unusual because Yews over 2.2 m in diameter (7 m in circumference) are almost unknown in Scotland (Bevan-Jones 2004). As can be seen from Figure 166a, the tree has just two remaining portions left and it could at first be thought to be the remains of two trees. The wooden posts in the picture, however, show where the trunk remains can be found. To add credence that this was all originally one tree, a woodcut by Thomas Pennant from 1769 (Fig. 166b) shows that even then the tree was tending towards two remaining parts bridged by the decaying trunk. Later, Jacob Strutt (1822) described the tree:

> *This prodigious tree was measured by the Hon. Judge Barrington, before the year 1770, and is stated by him to have been at that time fifty-two feet in circumference [5.04 m diameter]; but Pennant describes it as measuring fifty-six feet and a half*

[5.48 m diameter]. The same elegant tourist also speaks of it as having formerly been united to the height of three feet; Captain Campbell, of Glenlyon, having assured him that when a boy, he had often climbed over the connecting part. It is now however decayed to the ground, and completely divided into two distinct stems, between which the funeral processions were formerly accustomed to pass.

FIG 166. (a) The Fortingall Yew *Taxus baccata*, in the village of Fortingall, Perthshire, as it is now, a shadow of its former self with just two parts of the trunk still alive, although the remains of the original trunk circumference are traced out by the wooden posts. (b) As the tree was when Thomas Pennant toured Scotland in 1769 and wrote, 'In Fortingall churchyard are the remains of a prodigious yew-tree, whose ruins measured fifty-six feet and a half in circumference [5.48 m diameter]' (Pennant 1771).

Professor De Candolle wrote an article in the Edinburgh *New Philosophical Journal* a little later, in 1833, where he quotes a Mr Neill, the secretary of the Caledonian Horticultural Society:

> *From him we learn that considerable spoliations have evidently been committed on the tree since 1779; large arms have been removed, and masses of the trunk itself carried off by the country people, with the view of forming quechs or drinking cups and other relics, which visitors were in the habit of demanding.*

So that explains where a lot of the main trunk went to, leaving the two current portions! If this still leaves you sceptical that this ever was just one tree, it is worth looking at some of the fragmented Yews scattered through Britain that are nearer to still being whole, such as the one at Payhembury, Devon (Fig. 167), that is now separated into four trunks.

FIG 167. The old Yew *Taxus baccata* at Payhembury Parish Church, Devon. Once a whole tree, it has split to leave four 'stems'. The process by which this happens is described in *Ancient and Veteran Trees* below.

To put these trees into perspective, African Baobab trees can reach a diameter of 10 m and the Kauris of northwest New Zealand are probably the widest trees in the southern hemisphere (Fig. 168), while a Montezuma Bald Cypress *Taxodium mucronatum* in Mexico is considered to be the world's widest tree at 11.42 m in diameter.

Tree height, also recorded by the Tree Register, can be equally impressive. Across much of lowland Britain, urban and isolated individuals average around 20–25 m tall, but most of Britain's major forest trees are capable of reaching 30–40 m. There are several dozen individual broadleaf trees in Britain that are over 40 m (including eucalypts, beech, ash, various limes and Sessile Oak), with the current record height for a broadleaf tree in Britain standing at 49.7 m tall: a London Plane in Blandford Forum, Dorset. Given that these trees over 40 m are almost twice as tall as the trees we see every day, they do stand out as being truly magnificent.

FIG 168. The Kauri *Agathis australis*, named Yakas, which is 44.5 m tall with a diameter of 3.9 m. This is a very convenient tree to photograph, since the boardwalk goes right to its base. There are fatter ones: Te Matua Ngahere (its Maori name meaning Father of the Forest) is over 5.3 m in diameter. Waipoua Forest, North Island, New Zealand.

To put these British trees into context, around 50 species (less than 0.005 per cent of all tree species) grow to over 70 m in height (Fig. 169). The world's tallest broadleaf trees, all over 90 m, are congregated around southern Australia and Southeast Asia, particularly Borneo. The very tallest is widely considered to be a Mountain Ash *Eucalyptus regnans* in Victoria, Australia, called Centurion, which is 99.6 m tall. However, Shenkin *et al.* (2019) reported a specimen of Yellow Meranti *Shorea faguentiana* in Sabah, Malaysian Borneo, that was measured from top to bottom with a tape measure at 100.8 m, or 330 feet.

Conifers as a whole tend to be even taller than these broadleafed trees, and the west coast of North America is home to many of the tallest. The world record is currently held by a Coastal Redwood named Hyperion that was measured at 115.7 m (379 feet 7 inches) by Steve Sillett and Michael Taylor in August 2006. Uncorroborated records exist in British Columbia of a felled Douglas Fir that

FIG 169. Global distribution of the tree species known to reach 70 m in height. Most of the tallest species are either conifers from the west coast of North America (represented by blue stars for the top five species and light blue dots for the remainder) or eucalypts in Tasmania (red stars for the three tallest species and light red dots for the remainder), although one conifer from New Guinea (blue star) and one dipterocarp species from Borneo (yellow star) rank among the top 10. Other broadleaf species that can exceed 70 m (pale yellow dots) are found in Southeast Asia, especially Borneo. One tall conifer (pale blue dot) occurs in Eurasia, the Caucasian or Nordmann Fir *Abies nordmanniana*. Based on Tng *et al.* (2012).

was over 140 m tall. The tallest conifer in Europe is a Caucasian or Nordmann Fir *Abies nordmanniana*, recorded at 78 m (Fig. 169). This is a common Christmas tree in Britain, so you might want to think twice before planting your Nordmann Fir Christmas tree in a small garden! It is interesting that comparatively few of the tallest conifers are in the tropics. A relative of the Monkey Puzzle in New Guinea reaches 89 m (Fig. 169). Generally, however, the tallest tropical trees average no more than 46–55 m – shorter than the tallest trees in Britain.

Given the height of conifers, it is perhaps not surprising that the tallest recorded trees in Britain are Douglas Firs. The very tallest of these are found in Gwydyr Forest in Conwy, Wales, and Laird's Grave, Ardentinny, Argyll, Scotland. There is an ongoing competition as to which of these sites holds the record for height. Both contain trees that are 68 m tall, but as I write, the Argyll tree at 68.4 m is currently considered to be the tallest. Which is really the biggest can be a difficult decision to make because they and their neighbours grow close together, making it difficult to get a clear measurement of height from the ground. Aerial laser systems (LiDAR – Light Detection and Ranging) can be used, but the only cheap way to measure really tall trees is for some fearless soul to climb to the very top with a tape measure. Even these techniques do not necessarily solve the problem for a tree growing on a steep slope where the soil surface may be over a metre lower on the downslope side than on the upper. To even this out, measurements are made to the middle of the trunk perpendicular to the slope but it still adds a degree of uncertainty when heights are down to a few centimetres' difference. As well as these Douglas Firs, there are another 15 conifers in Britain of various species that are over 50 m tall. These include a 57 m tall Coastal Redwood on the Longleat estate in Wiltshire and a 58 m Giant Sequoia, also in Wiltshire, both the tallest of their kind in Europe.

Due to their height and width, trees are the world's largest living things by weight. Chief of these are the Giant Sequoias (giant redwoods) growing in the Sierra Nevada mountains of California. The largest known tree is General Sherman (Fig. 170) in Sequoia National Park, currently 83.8 m tall (others have reached 94.8 m) with a trunk width of 11.1 m at the base. It has been calculated that the trunk alone has a volume of 1,487 cubic metres (m^3), or 52,500 cubic feet. Given that Giant Sequoia wood fresh from the tree has a density of around 0.44 tonnes per cubic metre, the trunk is estimated to weigh around 650 tonnes. The weight of the branches is more difficult to estimate, and so General Sherman has been suggested to weigh anywhere between 1,250 and 5,500 tonnes[5], with the true figure probably being nearer to the lower end at around 1,250–1,400 tonnes. To

---

5 A metric tonne is 1,000 kg, or 0.98 of an imperial ton.

give an idea of just how big this tree is, branches broke off in 1978 and 2006 and in each case these were over 30 m long and around 2 m in diameter, bigger than any tree east of the Mississippi River. This makes General Sherman the world's biggest single organism. For comparison, blue whales, the largest animals, are in the order of 100 tonnes.

FIG 170. The largest tree in the world, General Sherman, a Giant Sequoia *Sequoiadendron giganteum*. It is 83.8 m tall with a diameter of 11.1 m – it is worth pacing this out to see just how wide this tree is! Sequoia National Park, California, USA.

A number of colonies of honey fungus, an *Armillaria* species (a harmful pathogen – see Chapter 15), have been found covering huge areas in America and China, typically over 800 hectares, and may weigh more than General Sherman. However, although each colony is made up of genetically identical material, it is doubtful whether they are connected together and work as a single individual. More likely they are clones of separate individuals. Similarly, the huge aspen clones described below, including Pando in Utah, are suckers from the same original tree, but again, is it still all one big tree or a collection of genetically identical but physically separate clones? I may be biased, but seeing and touching a giant redwood gives the feeling of immensity, and I am inclined to give the General the record.

For comparison with General Sherman, Hyperion, the tallest Coastal Redwood, has been estimated to contain 527 m^3 (18,600 cu ft) of wood in the trunk. In the southern hemisphere, the Kauris *Agathis australis* of New Zealand may not break records for height but are still immense trees. The largest living specimen, Tāne Mahuta (Lord of the Forest), is 51.2 m tall with a diameter of 4.91 m, giving an estimated trunk volume of 245 m^3, or 517 m^3 if the branches are included. The largest broadleaf trees are the magnificent African Baobabs *Adansonia digitata*, the largest of which have an estimated volume of 300–500 m^3 (Patrut *et al.* 2018).

To put these giants into perspective, a 100-year-old oak in Britain will normally contain 1–2 m^3 of wood in the trunk and 2–4 m^3 in the whole tree. When in the tree, oak wood has a density of almost 1 tonne/m^3, and when air-dried, the density of seasoned oak wood falls to around 720 kg/m^3, so the trunk of the 100-year-old tree will contain around 0.75–1.5 tonnes of dry wood.

As well as being impressive, large trees play a hugely important role in forests. Using data from 48 forests around the world, Lutz *et al.* (2018) found that on average the largest 1 per cent of trees (in diameter) contain 50 per cent of the above-ground live biomass and hold an average of 23 per cent of the species richness of a forest. They thus play a disproportionately high role in providing habitats for biodiversity and in carbon storage, and the list of other benefits goes on. Moreover, these largest trees are usually amongst the oldest in a forest and so impossible to replace in any meaningful human timescale (Lindenmayer *et al.* 2012). Worryingly, large trees are also the ones most likely to die from drought, wind or lightning storms brought about by climate change (Gora & Esquivel-Muelbert 2021).

WHAT LIMITS THE HEIGHT OF A TREE?

Following a good deal of debate, it has been recognised that what ultimately stops a tree growing taller is a mixture of genetics, site conditions and mechanics. Ultimately, tree height is controlled by genetics via natural selection for the size of tree that is most successful in its natural habitat – not so short as to be regularly outcompeted for light by its neighbours, but not so tall that it is mechanically unsafe. This genetic control can be illustrated by planting an oak and a Giant Sequoia next to each other; no matter how good the growing conditions, the oak will never be as tall as the Sequoia.

Within this genetic constraint, the maximum height of a particular individual of a species is predictable from the quality of the site it is growing on. The better the growing conditions, the taller the tree will grow. Nutrient availability and temperature have their greatest effect on rate of tree growth, while water availability is more likely to limit ultimate height. As seen in Chapter 6, water is pulled up the tree against gravity, and the taller the tree, the greater the weight of water hanging in the water conduction tubes in the wood (the xylem), and the more difficult it is for the foliage to pull water from the xylem. The leaves at the top of the tree can be frugal in their use of water by controlling water loss and can adapt to using less water, but eventually stress will slow and then stop upward growth. The drier the environment, the harder it is to get water to the top of the tree, and so the sooner upward growth will be stopped. As an example, Coastal Redwoods produce the tallest trees in the world, over 115 m, on the deep, moist alluvial soils where they are watered by the frequent coastal fogs, but they reach only 30 m on the drier inner edge of the fog belt.

Mechanics is also important since trees can detect stress through the woody frame and react to it by changes in growth. This is discussed in Chapter 11 in relation to standing up in storms, but here it is worth noting that persistently high winds will stunt the growth of a tree. Some of this may be due to poor growth or death of the uppermost parts, but there is also a deliberate response by the tree to stop or reduce its upward growth, based on the stress detected in the trunk when the wind is blowing. A tree can respond by putting more wood into the trunk and large coarse roots, but for a tree in the open, a doubling in height requires the diameter at the base to increase by between three and eight times for the tree to remain reasonably safe from breaking when bent by the wind. This is a huge investment and eventually it becomes uneconomical to carry on growing taller. As noted in Chapter 11, some species invest more in strength than others, such that birches tend to be tall with a narrow trunk and so more at danger from high winds.

As in most things biological, these different factors controlling tree height tend to interact. So for example, at high altitude and towards the poles, trees are shorter, since wind speeds are higher, leading to the trees slowing and stopping height growth sooner. In addition, certainly in Norway Spruce, the lower temperatures in these areas will slow the growth of xylem at the top of the tree, creating a greater water shortage that also leads to reduced height growth (Petit *et al.* 2010).

HOW OLD ARE OUR TREES?

A body of knowledge of the 'normal' maximum age of different species has accumulated (Table 12). It is hard to find concrete evidence for these data, but they are generally correct as a rule of thumb and give an indication of how long you are likely to make your mark on the landscape if you plant one of these species. Pioneer species invading open areas tend to be short-lived at around 100–200 years old. Most forest trees have limits in the range of 200–500 years, with some species, such as yew, often living much longer. Part of the reason for this variation is because each species has a different position along a longevity–reproduction continuum. At the 'reproductive' end, pioneer trees invest less in producing dense, well-defended wood and instead put their energy into producing as many seeds as quickly as possible. This fits well with the needs of growing in an opening in a woodland and trying to get everything done before other trees shade out the pioneer. At the 'longevity' end of the scale, oaks do just the opposite, defending their dense wood, having a wide, squat shape that is safer in high winds, and producing fewer seeds per year but over a much longer time period. They are in for the duration.

Maximum lifespan

For many animals the normal range of 'old age' and the maximum age tend to be reasonably close. A person can be reasonably old at 85 and very old at 110, just a few short decades apart. This is reflected in the shape of typical survivorship curves (Fig. 171), where humans and most primates show Type I survivorship (late loss): mortality is low when young due to parental care, and death occurs mostly at the oldest normal age. For some Type II organisms (constant loss), typified by birds, mortality is fairly constant through life, at least after the first year. But for trees, marine invertebrates and most fish (Type III; early loss), mortality is very high in the first part of life (as seeds and young seedlings in trees), and the few individuals that make it beyond this precarious young stage

TABLE 12. The normal upper age for a range of trees.

Common name	Scientific name	Usual maximum age (years)
Birch	*Betula* species	80–200
Holly	*Ilex aquifolium*	200–300
Juniper	*Juniperus communis*	200–850
Beech	*Fagus sylvatica*	200–400
Ash	*Fraxinus excelsior*	200–450
Hornbeam	*Carpinus betulus*	300+
Scots Pine	*Pinus sylvestris*	400–500
Lime	*Tilia* species	600–800
Oak	*Quercus* species	700–1,000
Douglas Fir	*Pseudotsuga menziesii*	750
Phoenician Juniper	*Juniperus phoenicea*	927[1]
Maidenhair Tree	*Ginkgo biloba*	1,000
Limber Pine	*Pinus flexilis*	2,000
Coastal Redwood	*Sequoia sempervirens*	2,000+
Yew	*Taxus baccata*	2,000–3,000
Giant Sequoia	*Sequoiadendron giganteum*	3,000+
Bristlecone Pine	*Pinus longaeva*	5,000+

[1] Still alive in 2019 (Camarero & Ortega-Martínez 2019) and thought to be the oldest living Iberian shrub; aged by carbon dating.

tend to live for a long time. For a tree, therefore, the tail of very old trees beyond 'normal' or 'reasonably old' can be very long. For example, Scots Pines rarely live longer than 400–500 years in Scotland, but in places nearer the Arctic where they grow more slowly, they can easily live for 800 years – a doubling of their normal maximum age.

Although the normal lifespan of an oak may be under a thousand years, there are specimens that are undoubtedly much older, such as the Marton Oak in Cheshire (Fig. 165), which is likely to be around 1,400 years old. Yew trees appear to regularly live for longer than this and are almost certainly our oldest trees, often with interesting histories. The Ankerwycke Yew in Berkshire is a venerable tree (Fig. 172), and there are suggestions that the Magna Carta was signed by King John in 1215 below this tree. The fields of Runnymede on the south side of the River Thames are usually given this credit, but at the time, the

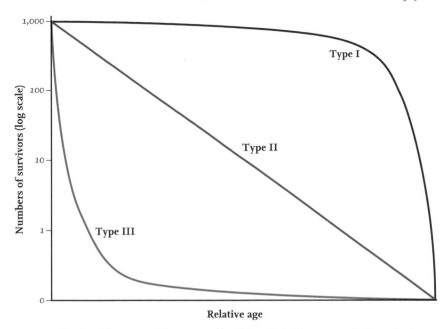

FIG 171. Survivorship curves of three types. Type I is typical of humans and other primates with high parental care that ensures high survivorship of the young; Type II is typical of birds, where mortality is fairly constant through life; and Type III is typified by trees, marine invertebrates and most fish, where mortality is very high when young but with a long tail of very old individuals.

Ankerwycke Yew stood on one of the many islands in the Thames, alongside St Mary's Priory, a nunnery founded in the twelfth century, which would have formed a perfect place where neither side in the argument could ambush the other. Given this history, the Yew tree is likely to have been a significantly sized tree 800 years ago and makes the suggested age of 2,500 years (Chetan & Brueton 1994) seem plausible.

The oldest tree in the country is almost certainly the Fortingall Yew on the north side of Loch Tay in Perthshire, shown in Figure 166. It is fondly suggested to be up to 5,000 years old but is more likely to be 2,000–3,000 (Bevan-Jones 2004, Hindson *et al.* 2019). Even so, this still makes it the oldest tree in Europe. The media regularly report on Yew trees that are said to be considerably older, often up to 6,000 years old, but these stories are largely wishful thinking.

The age of these trees is very respectable in terms of the world's oldest trees. The oldest-known tree is a Bristlecone Pine *Pinus longaeva* in the White Mountains

FIG 172. The Ankerwycke Yew *Taxus baccata* in Berkshire. Currently 2.55 m in diameter (8 m in circumference) at 0.3 m above ground and suggested to be the site of the signing of the Magna Carta in 1215.

of California (Fig. 173), which was known to be 4,900 years old when it died (Oldlist 2021), but another called Methuselah is currently only 50 years younger, at 4,850 years old. This began life back in 3000 BCE and so is older than the Egyptian pyramids and still going strong! Another Bristlecone Pine was thought to be 5,062 years old in 2010, but this cannot be confirmed because the core removed from the tree (on which the rings were counted) has unfortunately gone missing. In addition to these Bristlecone Pines, a Patagonian Cypress *Fitzroya cupressoides* in Chile and four Giant Sequoias in California are known to be over 3,000 years old, and trees older than 2,000 years are found in six conifer genera: Patagonian Cypress, Bristlecone Pine, Coastal Redwood, Giant Sequoia, Swamp Cypress and Przewalski's Juniper *Juniperus przewalskii*. The secret to longevity in trees appears to be genetic (see below), since these oldest trees can be found growing in a wide range of different areas from dry to wet (Piovesan & Biondi 2021).

Older trees that are clonal

There are regular claims of trees being much older than the 3,000–5,000 years quoted above, often many tens if not hundreds of thousands of years old. Can these really be true? Well, yes and no. Broadleaf trees that can sprout from the

FIG 173. Bristlecone Pines *Pinus longaeva* in the White Mountains of California.

base of the trunk or produce suckers from roots (as well as the small number of conifers that can do this, including Giant Sequoia, Coastal Redwood and Douglas Fir) have led to these claims. Even though individual stems may die comparatively young, new stems and roots appear, so the whole tree can keep growing almost indefinitely and so be potentially immortal.

In Britain, such clonal sprouting can greatly prolong the life of a tree. Ash normally lives for 200–450 years, but a tree at Bradfield Woods in Suffolk has a stump (called a stool in the language of coppicing) that is 5.6 m in diameter and is undoubtedly at least 1,000 years old. Similarly, in the Lake District near the northwest limit of its range, Small-leaved Lime is practically immortal. Uncut, or *maiden,* trunks can be perhaps 800 years old (Logan *et al.* 2015), but coppice shoots, forming stools more than 3 m in diameter, have an estimated age of 1,300–1,900 years and show no sign of dying (Pigott 1989). Another Small-leaved Lime at Westonbirt Arboretum is suggested to be 2,000 years old; DNA tests show that the 22 individual trunks are genetically the same (Logan 2016) and so likely to be, in effect, all from one original tree. For comparison, the world's oldest clones include a Huon Pine *Lagarostrobos franklinii* in Tasmania that is possibly 10,000 years old, and a shrub called King's Holly *Lomatia tasmanica,* also in Tasmania, that may be over 40,000 years old. It has even been suggested that a clone of Aspen in Utah, USA, called *Pando* (meaning 'I spread'), covering 43 hectares with around 47,000 stems and weighing over 6,000 tonnes if the roots are included, could be a million years old, although there is certainly no direct evidence of this (Barnes 1966). In all these cases, the tree/shrub has been growing on the site for many thousands of years but there are no stems anywhere near that old. In the case of the Huon Pine, the oldest stem is 2,000 years old, and in the *Pando* Aspen clump, the average age of the stems is 130 years.

Trees can also persist by the layering of branches (Fig. 174). Branches at the bottom of a tree's crown can droop to the ground where their weight keeps them still and moist enough to allow adventitious roots to form, and the branch can grow to form a tree of its own. In this case, although initially joined to the mother tree, if that link becomes broken, then this leaves the layered stem as a tree in its own right. Although independent above ground, the roots of all the stems are likely to be grafted together, allowing the transfer of sugars, water and chemical signals such as plant hormones, so it could be argued that it is still working as one tree, and as new stems repeatedly form in this way, the clone can reach an impressive age.

In all these cases, when the mother stem dies, leaving a circle of young offshoots, can we still say that this is an 'old' tree when there is nothing that can be touched that is more than a few hundred years old? It's like having a broom that has had numerous new handles and new heads – is that still the old broom? Perhaps the best solution is to refer to individual, long-lived trunks as *old trees* (such as our many Yews) and groups of young stems that have successively arisen from an original tree as *old clonal trees* (such as the coppiced Ash and limes). Both are an important part of our landscape, but while the clonal tree may reach very impressive ages, there is nothing quite like touching a trunk that has been growing for thousands of years.

FIG 174. A branch of this large Sweet Chestnut *Castanea sativa* has touched the ground, rooted and given rise to the tuft of new shoots in the foreground. These new shoots are capable of growing into a new tree. If this happens around the tree, a grove of trees will develop, with the mother tree in the middle, and potentially spread ever outwards like ripples from a stone dropped in a pond. Sandon Hall, Staffordshire.

PUTTING AN AGE TO TREES

Given that trees can grow very large and can be very old, how much of a link is there between size and longevity? The simple answer is often very little, since the biggest trees are not necessarily the oldest and size often says little about age. Tree height is perhaps the most misleading. Different species tend to have a genetically predetermined height, so a Dogwood *Cornus sanguinea* would never be as tall as an oak. But overlaid on top of this is a huge variation caused by the environment. Trees in windy areas may be a third or less of the height of sheltered individuals, and an individual growing in a dense woodland can be drawn up towards the light, quickly reaching a height that would never be seen in an isolated tree.

Diameter is more useful for putting an age on a tree. Most trees add new wood under the bark each year, as described in Chapter 2, and so the trunk grows predictably wider each year, which can be linked to age. However, a slow-growing tree may be small but of a great age, while a fast-growing tree, although wider, can be but a fraction of the age of the slow-growing specimen. For example, the oldest Scots Pines in the Caledonian pinewoods of Scotland are almost 600 years old, while others are half that age but of a very similar diameter (Mason *et al.* 2004).

So, how can a date be put to the trees around us? One of the most reliable methods is to use historical records where they exist. For example, Moir (1999) lists an avenue of Yew trees at Hampton Court Palace known to have been planted between 1700 and 1703 (and which were subsequently felled in 1993). Such records can go back a long way. A Sacred Fig *Ficus religiosa* in Sri Lanka is recorded as being planted in 288 BCE, making it the world's oldest living tree with a known planting date (Hall *et al.* 2011).

Trees growing in temperate areas produce one ring of wood under the bark each year, and so with access to the rings the age can be simply counted to get the exact calendar age. The problem, of course, is how to get access to the rings. Specialist equipment such as an increment corer can be used (Fig. 175), which extracts a core along which the rings can be counted (Fig. 176).

Despite the core being so narrow, the tree is still wounded, leading to discoloured or decayed wood vertically along the path of the core (Fig. 177). It is sometimes advocated that the exit hole is plugged with wax to stop fungi getting into the wound but there is little evidence that this works. In fact, leaving the hole open and letting the wood dry can be better or at least no worse. Since it is an invasive way of finding the age, it should not be used on valuable timber trees or trees of great landscape value.

More realistically, for those of us without increment corers to hand, the easiest way to get access to count tree rings is on the stumps of felled trees. A path to the centre of the stump may need to be smoothed with a sharp chisel or even sandpaper before the rings are visible below the mass of bent fibres left by the saw, but it is relatively easy to do.

Even with something as seemingly foolproof as counting the rings, there can be some problems in reaching a definitive age for a tree. Cores cannot be taken

FIG 175. (a) Removing a core of wood from a tree with an increment corer. In this case the tree is a Norway Spruce *Picea abies* in Italy, felled for timber with the bark removed. Being able to see the cut end of the trunk makes it easier to hit the pith at the centre of the tree, which is often not at the geographical centre of the tree; in this case it is indicated by the yellow mark. (b) Once the hollow corer is wound into the tree using the screw thread at the left-hand end, the 'spoon', resting top and bottom of the two corers, is used to extract the core from inside before the corer is laboriously unscrewed from the tree. Corers of two diameters are shown. The more usual size at the top removes a core 5.5 mm in diameter, while the lower corer gives a core 12.0 mm in diameter, which is more useful if chemical analyses of individual rings are wanted, simply because it supplies more wood. The holes left in the tree by the corers are approximately double the width of the core.

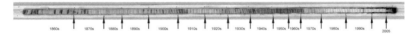

FIG 176. A core taken from a Deodar *Cedrus deodara* at Keele University in 2005, which was 40 cm in radius at the time. The core has been glued into the groove of a wooden support for protection and to keep it straight as it dries. The bark can be seen at the right-hand end, and since the tree was alive when cored, the date of the first ring is known. The individual rings arc clearly visible and can be counted back to before 1870. The centre of the tree had not been reached, since the corer was only 30 cm long.

at ground level, since there has to be room to rotate the handle of the corer, and trees are also rarely cut at ground level. In such cases some guesswork is needed in estimating how many years the tree took to grow to the point where the rings are sampled – typically between 2 and 5 years for cores. The rings in some species such as birches, Sycamore *Acer pseudoplatanus* and sometimes willows can be incredibly hard to distinguish, especially in narrow cores. To make it worse, there can also be missing or partial rings that form just partly around the tree. For example, yew is renowned for missing rings or, at least, growing very narrow and

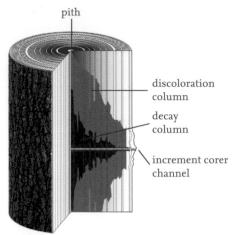

pith

discoloration
column

decay
column

increment corer
channel

FIG 177. Section of a tree following the removal of a core using an increment corer and the resultant discoloration and decay of the wood above and below the ensuing wound. How the tree copes with this wounding is described in Figure 109 in Chapter 8. Modified from Tsen *et al.* (2016).

thus easily missed rings. Lowe (1897) noted a tree in the Kew Museum that had 250 rings on one side and just 50 on the other. This is a particular problem in trees with fluted trunks, such as yews and hawthorns where rings can be squeezed to the point of invisibility in the troughs between flutes, but can be a problem in any species, particularly with individuals growing in poor conditions or very old trees that may not have enough energy to grow a new ring of wood completely around the tree.

There is, of course, the additional problem that the oldest trees are very likely to be hollow due to internal rot. Ring counts of 400–500 are not uncommon in yew (Thomas & Polwart 2003), but older trees are invariably hollow and so rings cannot be counted through to the middle. The oldest tree in Europe that has been given an age by counting the rings is a Bosnian Pine *Pinus heldreichii* (named Adonis after the Greek god of beauty), which in 2015 was 1,075 years old (Konter *et al.* 2017). This grows in the Pindos Mountains in northern Greece at 2,000 m above sea level, where the cool, dry climate reduces fungal rot so the rings could be counted through to the middle. The same is true of the Bristlecone Pines in California. Just the sort of conditions we do not often get in the oceanic climate of Britain!

Where rings can be counted, or where the planting age is known, it is possible to build up a local chronology of trunk diameter against age such as shown in Figure 178. The age of an unknown tree can then be estimated using the line, or curve, for that species. This can work very well for one site, where the tree you are wanting to age and the ones used to form the line are growing under similar conditions. It will obviously be less accurate if comparing a tree grown

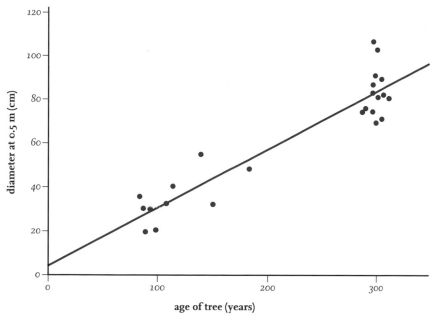

FIG 178. The relationship between the diameter of Yew trees *Taxus baccata* (measured at 0.5 m above the ground) and their age, determined by counting rings in felled trees at Hampton Court Palace, Middlesex. The line is the best fit through the data (a linear regression line with an r^2 value of 0.86 which, being fairly close to 1.00, indicates a good fit of the data with the line). Ideally, when measuring a tree of unknown age to compare to this line, it should be measured at 0.5 m above ground to be comparable. The average width of an annual ring across all these trees is just 1.32 mm. Modified from Pigott (1995).

from a different area. It should also be noted that the spread of points above and below the line is generally larger with older trees, as small differences in growth accumulate over the years into big differences in size. This does, of course, mean that age estimates tend to be less accurate for larger trees.

Fortunately, to make life easier there are various non-destructive ways of estimating age from the girth or diameter of the trunk. Inevitably these will be less accurate than ring counts, since growth rates will vary between species and sites, but they can give a surprisingly good rough estimate of age. Alan Mitchell (1978), a doyen of tree identification and tree growth, used a rule of one inch (2.54 cm) girth per year at breast height for trees in the open and half that for trees growing in the shade for most trees, whether conifers or broadleaves. In the open this equates to an average width of growth rings of 2–4 mm which is pretty close to the norm of 3–6 mm expected in trees growing in Britain, although fast-growing trees can have rings up to 10–12 mm wide, especially poplars, young oaks and many introduced conifers. Exceptionally, this can increase to 25 mm wide rings in conifers such Giant Sequoia, Coastal Redwood, Sitka Spruce, Douglas Fir and Cedar of Lebanon *Cedrus libani*. However, for most of us looking at everyday trees of less than a metre in diameter, this rule is surprisingly useful.

Estimating the age in this way is fine for trees in their prime, but in older trees, growth slows down considerably once the tree reaches maturity. As a very rough rule, for older trees over 0.5 m in diameter, if you can get your arms round the trunk, it is around 300 years old, and add another 300 years for each extra person, arms extended, it takes to circle the tree. So a 'three-person-hug' tree would be around 900 years old. With most people's arm lengths this equates to a ring width of 0.9–1.0 mm, which is fairly typical for an older tree (Williamson 1978, Moir 1999). But bear in mind that in very old trees, or trees on poor sites, growth can be even slower. In British Yews, the White Cedar *Thuja occidentalis* growing on cliffs of the Niagara escarpment in eastern North America and Bristlecone Pines *Pinus longaeva* in California, annual rings can often be as small as just 0.05 mm wide, taking 10 years to add just 1 mm to the trunk diameter (Adams 1905, Newbould 1960). A White Cedar a little more than 5 cm in diameter was found to be 422 years old, with an average ring width of just over 0.05 mm (Kelly *et al.* 1992).

More refined methods exist that take into account the change in growth rate with age. That by John White (1998) is particularly useful and gives a method for calculating the age of different species of trees using the diameter based on empirical knowledge of how annual ring width changes over time on different sites. White gives estimates of ring widths while the tree is growing to maturity, followed by a calculation to factor in how the ring widths will decrease after maturity. So for a yew tree, White's data give an average ring width for growth up to maturity of 3–4 mm per year for the first 30–60 years (reached more quickly on most-favourable sites), followed by increasingly narrower rings thereafter. Why does this work? The reasons are explained below in *Changes in trees with age*.

For very old trees it can be quite difficult to estimate their age. They are often hollow so the rings cannot be counted, nor can the oldest wood at the centre of the tree be carbon dated, since it has rotted away. The girth is useful but this becomes less accurate with age, especially if parts of the tree are missing or it has large burrs, making a sensible estimate of the girth/diameter very difficult. Putting an age to our oldest trees often comes down to guesswork and, like the fish that got away, the age can often go up with the retelling of the story!

TREE RINGS AND DENDROCHRONOLOGY

As noted above, annual growth rings generally become narrower with age. There will also be some variation in ring width through an individual tree. The rings are often wider on the side that gets more light or due to the presence of reaction wood (Chapter 11); rings also tend to be wider at the base of the crown compared to the bottom of the trunk.

On top of this, there will always be variation in ring width between years. This is where *dendrochronology* comes in: the study of how past events and environmental conditions have affected tree growth. A tree that is shaded more and more by neighbours will produce narrower rings; and if those neighbours are removed, the new rings in the surviving tree will be wider again. The date at which this happened can be found by counting the number of rings of wide growth. A heavy seed crop can also reduce ring width as available energy supplies are used to grow seeds rather than wood. This is reflected in trees such as the Yew where female trees will have narrower rings than male trees due to the burden of producing seeds (Thomas & Polwart 2003). But most variation in year-to-year ring width is caused by how the tree reacts to changes in weather. In years with better weather, trees grow wider rings than in unfavourable years.

Trees react primarily to summer temperature and precipitation. Which of these is most important partly depends upon where the tree is growing. In cold environments, ring width is often closely linked with summer temperature, whereas in arid environments, rings are, perhaps unsurprisingly, most affected by annual changes in precipitation but tempered by temperature, since the higher the summer temperature the higher the evaporation of water. In less extreme environments, different tree species have their own sensitivity to a mix of temperature and precipitation. For example, in Northern Ireland, ring widths in Beech and Ash are mainly influenced by summer rainfall and soil moisture in early summer and very little by temperature (Körner *et al.* 2016). In contrast, Scots Pine is more sensitive to summer air and soil temperature (García-Suárez *et al.* 2009). In general, ring widths across Europe and Asia are more affected by temperature than precipitation.

The responsiveness of trees to climate also varies depending upon where they are within their geographical range. While Beech in the mild climate of Northern Ireland is unresponsive to temperature, trees growing in the continental climate across Europe and into Asia are affected much more by temperature than rainfall (St. George 2014). Knowing these things has given rise to the discipline of *dendroclimatology*, where tree rings can be used to reconstruct the climate from times long before records were kept, going back hundreds and even thousands of years. This is what Fritz Schweingruber, a Swiss dendrochronologist, described as a 'tree's private diary'.

The pattern of wider and narrower rings across the years can also be used as a dating tool because *chronologies* can be built up. If a tree is living, then the year in which any ring was grown can be found by counting back from the bark on a cross-section of the tree or a core removed, as in Figure 175. This can be seen in tree 1 in Figure 179. Since trees of the same species growing in the same area tend to produce similar sequences of wide and narrow rings, it is possible to take a piece of wood from a dead tree (tree 2 in Fig. 179) and match the pattern of rings to that in tree 1. This can sometimes be easily done by eye, but since the pattern of rings is never quite the same between two trees, it is usually done using software that looks for the best match. In this way it is possible to add more and more pieces of wood to extend the chronology further back in time (trees 3 and 4 in the figure). The Northern Ireland oak chronology goes back over 7,000 years to 5452 BCE, and an oak chronology from the Rhine area of Germany goes back even further to 8480 BCE. Building these chronologies is possible because progressively older pieces of wood can be found in old trees, historic buildings and furniture and archaeological sites. They can also be found preserved in bogs and lake bottoms.

It can be difficult to join chronologies from different regions because differences in weather produce different ring patterns. Likewise, joining chronologies from different species can be equally difficult, since each species responds slightly differently to the same weather. But with greater computing power this is proving possible, since underlying patterns can be separated from

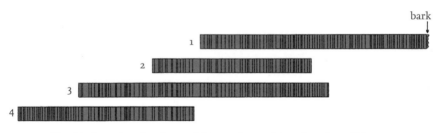

FIG 179. Dendrochronology involves matching patterns of tree rings from different trees. In a living tree, such as tree 1, the year a particular ring grew can be calculated by counting back from the bark. If the pattern of wide and narrow rings can be matched to pieces of dead wood (2–4), they can also be dated. In this way, long 'chronologies' can be built up and used to date pieces of wood of unknown history. The 'crossmatching' between cores is never quite as perfect as shown here; there is always some variation from tree to tree, usually requiring the use of computers to find the best match. From Thomas (2014).

local noise. The combined oak and pine chronology in Europe now goes back over 12,650 years to 10,644 BCE. Other long chronologies exist; the Bristlecone Pine chronology in western USA is almost 9,000 years long and it looks likely that it can be extended back to at least 10,359 years (Salzer *et al.* 2019). These long chronologies are extremely useful for reconstructing past climates and, of course, for dating pieces of wood of unknown age from old buildings and archaeological sites. They have also been used to check the accuracy of carbon dating.

Wood chemistry can also be used in dendrochronology. The amount of various chemicals incorporated into tree rings as they grow can be used to date events such as volcanic eruptions (which produce large quantities of sulphur into the atmosphere) but also other disturbances such as earthquakes, fires and human-made pollution. Increase in sulphur concentration found in a range of species over a wide geographical area helps to separate volcanic eruptions from any local variation produced by changes in the soil – for example, an increase in soil acidity can cause more sulphur to be taken up by a tree (Pearson *et al.* 2005). Eruptions can also lead to increases in rare earth metals in tree rings, as was found in Douglas Fir growing near the eruption of Mount St Helens in 1980 (Fig. 180). The value of this is that chemical signatures within a chronology can be used to fix the exact year of an eruption or other environmental event, particularly if they are prehistoric with no recorded date. This can also work the other way round; tropical trees that do not grow annual rings can be dated by looking for these distinctive chemical signatures in the wood.

CHANGES IN TREES WITH AGE: FROM JUVENILE TO ANCIENT

Growth of a tree slows with age – as seen by the increasingly narrower annual rings after a tree reaches maturity. Why does this happen? The answer lies in how a tree develops and changes with age. Trees can be categorised by their *chronological age*, which is how many years they have been alive. But another way of looking at a tree is to look at its *ontogenetical age*, or the different stages of its development – juvenile, mature, ancient and senescent (Dujesiefken *et al.* 2016). The two types of age are not necessarily the same thing, since a chronologically young tree growing in very difficult circumstances can be ontogenetically old – in the same way that a middle-aged person who has had a hard life can look much older.

In a *juvenile* tree, the upward and outward growth of the crown means it can hold more leaves each year, so the amount of sugary food it can produce by

FIG 180. The north side of Mount St Helens, which erupted in 1980, felling hundreds of square kilometres of forest. This photograph was taken in 2002 and the felled trees can still be seen on the hillside. The new trees regrowing and those surviving the eruption will be absorbing chemicals from the ash, giving their rings grown after the eruption a distinctive chemical composition. These chemical signatures can be used to date prehistoric eruptions. Coldwater Lake Visitor Center, Washington, USA.

photosynthesis goes up each year. At the same time, the woody skeleton is getting larger, and so each new shell of wood added under the bark (see Chapter 5) goes up in volume. This is like a set of Russian dolls, where each one added has to be larger than the last. The net result of an increasing need for wood, matched by increasing food production, is that the annual rings laid down tend to be roughly the same width, although there will be some year-to-year variation due to weather and other factors, as seen above.

Eventually the tree reaches *maturity*, when it is at its maximum height and the branches are as long as they can be without being at risk of breaking off. At this point it is holding the maximum number of leaves that it can, and so its income in terms of sugar from photosynthesis becomes fixed. Since the trunk and branches are still growing fatter, bigger shells of wood are needed each year to coat the tree under the bark. As the tree's income is now fixed, the result is that rings become progressively narrower each year as the same volume of wood is spread thinner and thinner. When a tree reaches maturity, it is also the time when it starts reproducing, so causing a further drain on the fixed amount of sugar it produces each year.

A tree may persist at maturity for many centuries. But as it gets older and the annual rings become ever narrower, something has to give eventually. What tends to happen at this point in *late maturity* is that the extra wood added each year is concentrated in the parts of the trunk linking vigorous branches and well-growing parts of the root system. This is when 'partial rings' are produced. As the parts of the trunk between these growth areas fall behind, the growing strips tend to become semi-autonomous units, able to survive more or less independently no matter what happens to the rest of the tree (Lonsdale 2013b).

The last but often longest *ancient* stage of the tree comes about when the crown of the tree declines or begins to die back as a result of pathogens or damage, and generally becomes more rounded and flatter, and often shorter, sometimes described as the tree 'growing downwards'. Due to this loss, the amount of sugar grown per year starts to similarly decline and the rings become even narrower. Growth can be increasingly concentrated in the living strips, so the stem becomes more fluted, creating a series of mini-trees within a whole tree that can eventually become separated into discrete stems as the dead wood between rots away, as was seen in the Fortingall and Payhembury Yews (Figs. 166 and 167) at the beginning of the chapter.

However, being modular, even at this stage trees can produce new shoots lower down the trunk and branches or from the base of the tree, using stored or newly created buds (see below). The growing of new juvenile shoots, or *rejuvenescence* (Del Tredici 2000), reverses the ontogenetic clock of a chronologically old tree. This leads to *retrenchment*, where older branches are lost but are replaced by new younger growth. A similar process is also undoubtedly happening below ground with the roots but, being hidden, it is much harder to be certain. As long as the retrenchment continues, and new branches hold a similar leaf area to the lost branches, the tree will persist.

At some point, however, the tree may enter a phase of terminal, *senescent* decline. Senescence is caused by genes associated with ageing being switched on. This leads to a physiological change in the tree, including the accumulation of detrimental mutations, a decline in reproduction, and a decline in the number of cells made. Growth also begins later in spring, so the growing season is shorter (Peñuelas 2005). There is also a declining ability to produce defensive chemicals against pests and pathogens, and compounds such as antioxidants and hormones that protect against environmental stresses. Interestingly, our oldest living trees do not appear to have a predetermined senescence built into their genes. These genes are not as readily switched on, which may be related to size rather than age. Slow-growing Maidenhair Trees up to 600 years old and Bristlecone Pines up to 4,713 years old have been found to show no symptoms of senility (Lanner & Connor 2001, Wang *et al.* 2020). Moreover, long-lived trees have more genes involved in repairing DNA than those that are short-lived (Blue *et al.* 2021). Our oldest trees really are potentially immortal.

ANCIENT AND VETERAN TREES

It is quite useful to give our chronologically and ontogenetically old trees some labels to help us appreciate what we have in our landscape. An *ancient tree* can be defined as a tree 'that has passed beyond maturity and is old, or aged, in comparison with other trees of the same species' (Anon 2008). So what in that case is a *veteran tree*? Simply put, a veteran tree is any tree that shows some of the characteristics of an ancient tree even if it is not very old: ontogenetically old even if not chronologically old. In effect, it looks old due to the way it has lived, caused by stress, physical damage to the canopy or roots, or disease.

The characteristics that are encompassed in being ancient include those in Table 13. In Yew trees, the pragmatic definitions are that veteran trees are from

TABLE 13. Characteristics that make up an ancient tree. The more of these a tree has, the more likely it is to be considered ancient.

Crown thinning or declining at the top or flattening in conifers

A large girth by comparison with other trees of the same species

Hollowing trunk which might have one or more openings to the outside

Stag-headed: dead, antler-like branches extending beyond the living crown (Fig. 183)

Fruit bodies of wood-rotting fungi (see Chapter 15)

Cavities such as where branches have broken away, sap runs or naturally forming water pools in branch hollows

Rougher or more creviced bark

An 'old' look which has high aesthetic appeal

Aerial roots growing down into the decaying trunk or branches

An irregular shape due to poor growth or old wounds that generally give a gnarled appearance

Rich in mosses, lichens or other epiphytes

Based on Anon (2008).

5–6.99 m in girth (approximately 1.6–2.2 m in diameter); ancient, 7–8.99 m in girth (2.2–2.9 m in diameter); and 'exceptional' trees more than 9 m in girth or 2.9 m in diameter (Hindson *et al.* 2019).

In the UK there are at least 123,000 ancient and veteran trees recorded in the Ancient Tree Inventory (Reid *et al.* 2021) and undoubtedly more to be added. Britain also has a number of *heritage trees*, defined as a tree that 'has contributed to or is connected to our history and culture' (Anon 2008). This includes such trees as the Major Oak in Sherwood Forest associated with Robin Hood, or those that play a special part in our landscape, such as the Meikleour Beech hedge in Perthshire, or those that have great rarity or botanical interest, such as being the first tree grown in Britain, either a new cultivar or an introduced species. We also have *notable trees* that stand out in their local environment usually by virtue of their size and provide the stock from which ancient trees will come.

Ancient and veteran trees should be conserved because they are rare, cannot be quickly replaced, often have historical or cultural links, and are home to a usually large diversity of other organisms, particularly epiphytes, invertebrates and fungi. Ancient trees, by definition, cannot be created other than by making sure older trees which will eventually become ancient are not lost from the landscape. Veteran trees, however, are defined by what they are, not their age, and can be produced by *veterinisation*. This can be done by encouraging some of the characteristics in Table 13 in various fun ways, as noted below.

THE VALUE OF DEAD WOOD

Ancient and veteran trees usually involve a lot of dead wood. Since dead wood sticking out from a tree or found on the ground is rather unappealing and unsightly, in the past it has been largely ignored or more often than not tidied away. However, it is now appreciated that dead wood has a high conservation value, and management of trees should be designed to keep and even encourage it. As an example of its value, European woodlands might contain around 40–50 species of trees and shrubs, yet there may be well over a thousand species of wood-inhabiting fungi.

Dead wood occurs in many sizes and places, including standing dead trees (*snags*), dead branches in the canopy, and trunks and branches lying on the ground. It also varies in the amount of decay, from sound wood to soft 'punky' wood into which a finger can be easily pushed (Fig. 181), all of which can be habitat for different species of animals, fungi and bacteria. There is also a big difference in moisture content and temperature between standing dead wood

FIG 181. (a) Coarse woody debris that is upright will be warmer and drier than that on the ground and offer habitats for different organisms. (b) The degree of decomposition also makes a difference, and even wood that is falling apart is valuable habitat. (a) Pedunculate Oaks *Quercus robur* at Moccas Park National Nature Reserve, Hereford, and (b) mixed conifer forest dominated by Jeffrey Pine *Pinus jeffreyi* near the Tioga Pass, Yosemite National Park, California.

and that lying on the ground. Bark kept on the outside may speed or slow decomposition of the wood, depending upon species and whether it is on the ground, and affect the fauna that can use the wood (Dossa *et al.* 2017). A useful term for this wide range of dead stuff is *coarse woody debris* (CWD), usually including all dead wood over a certain size, typically 2.5 cm in diameter. In a natural woodland, CWD normally accounts for one-third of the biomass, which in temperate woodlands typically ranges from 11–50 tonnes per hectare.

Coarse woody debris, both upright and on the ground, is an important habitat for a large number of invertebrates, including bark beetles, wood-boring beetles, termites and carpenter ants, many of which are specialist *saproxylic* insects (from the Greek *sapro*, meaning 'rotting', and *xylo*, meaning 'wood'; literally subsisting on and in dead wood). In Britain, more than 1,700 invertebrate species live in ancient trees, many using the dead wood (Butler *et al.* 2002). Since most of our woodlands do not have anywhere near a natural amount of CWD, and because trees once covered much more of our landscape than they now do, it is hardly surprising that many of the invertebrates using dead wood are threatened or endangered. Dead wood is also invaluable for epiphytes such as mosses and liverworts, since it makes a firm substrate, holds water and is often high enough above the ground for the epiphytes to escape the worst of the competition from woodland flora and to keep from being covered up by litter. Fallen, rotting trees can also act as nurse logs for tree seedlings for the same reasons.

There is thus a need to keep CWD in the canopies of trees where it is safe to do so (for people and the tree) and within aesthetic limits. We can also go one step further by encouraging dead wood in trees, even going as far as deliberate mutilation. Trees can be heavily and badly pruned, leaving long branch ends, or *stubs*, to encourage rot; the bark at the base of the trunk can be damaged with a sledgehammer or other forms of wounding; downward-pointing holes can be drilled into the trunk to encourage water pockets to build up and encourage rot; and branches can be ripped from the tree by winch. There are many ingenious ways of encouraging dead wood that quite rightly would normally be frowned upon but here are all in a good cause (Lonsdale 2013a). To quote George Peterken (1996, p. 420), dead wood 'can be achieved by sawing rot holes and ring barking, or by more exciting techniques such as fire, exploding the crown from the trunk or by holding a vandals' convention'!

Most of us would not want to inflict this sort of damage on our beloved trees, but we can make small changes to how we manage them. For example, when removing a branch, a more jagged wound (called *natural fracture pruning*) rather than a neat cut creates habitats for a whole range of dead-wood organisms. More dead wood can also perhaps be tolerated in a tree crown, resisting the temptation to tidy it up. The take-home message is that where dead wood can be maintained, it should be. And since CWD eventually rots away, there is the need to plan ahead to ensure that a steady succession of new CWD is maintained.

CWD is also proving valuable in freshwater environments. It slows the speed of water flow, creating pools of slower water and directly providing habitats. Some 147 species of invertebrates are strongly associated with submerged CWD (Godfrey 2003), including the Cranefly *Lipsothrix nigristigma*, which is subject to a UK Biodiversity Action Plan. This has led to management of waterways that would set the heart of any child aglow, such as winching trees over to fall into and across waterways; creating log jams from tree root plates and large branches wedged into the river bed; and ring-barking trees on river banks to kill the crown to create a future supply of CWD to drop into the waterway (Fig. 182). Of course, this also links to the benefits of introducing beavers into a river system.

WHAT KILLS A TREE?

If, as noted above, some trees are potentially immortal, what do they actually die of? First of all, there are many external agents that can kill a tree. Trees can be mechanically destroyed by extreme events such as lightning strikes, earthquakes or volcanic activity, like the devastation caused by the eruption of Mount St Helens in 1980 (Fig. 180). But trees are remarkably resistant unless forcibly uprooted; as noted in Chapter 2, trees such as the 'A-bombed trees' survived at the epicentre of the 1945 nuclear bomb dropped on Hiroshima. There can, however, be unseen extreme events. For example, radiation from the accident at the Chernobyl nuclear power station on 26 April 1986 killed all trees within 2.4 km (1.5 miles), creating a 'red forest' of dead pines and distorting the growth of trees outside this zone.

More gentle but persistent environmental problems can cause death either by a deficiency such as drought or an excess such as flooding. We humans can also make things worse by simple things such as disturbing roots by digging too close to the trunk, extensive use of herbicides and large-scale pollution. In urban areas we can add to this list the use of de-icing salt on roads and paths, soil compaction and old-fashioned vandalism. Newspaper headlines periodically declaim that 'winter road salt may kill off thousands of trees', and certainly in 1971, salt spray and salt in soils were estimated to be directly responsible for the deaths of over 700,000 trees each year in western Europe. However, better techniques of using and storing de-icing salt have drastically reduced this harm (Rose & Webber 2011).

FIG 182. (a) Before and (b) after management to introduce more coarse woody debris into the Churnet River, Dimmingsdale, Staffordshire. Photographs by Nick Mott, Staffordshire Wildlife Trust.

Pascal Pirone, a plant pathologist at the New York Botanical Garden, suggested in the late 1950s that large quantities of dog urine (high in sodium and potassium) would produce 'dog cankers' at the base of city trees, which could kill the trees (Pirone 1970). This is undoubtedly an overstatement but adds just another stress for the tree to endure. And that is the problem – all the individual stressors accumulate, weakening the tree until one or more strikes the final death knell. There may be a virulent pest or pathogen (Chapter 15) that looks to be the obvious cause of death, but often the poor tree has been weakened by a whole gamut of other things, all contributing to its death. The effect of these stressors may not show up immediately, taking several years to cumulatively weaken the tree. Subalpine Fir *Abies lasiocarpa* in the American Rocky Mountains can take up to 11 years to die as a result of a significant drought (Bigler *et al.* 2007). These are the outward causes of death, but what happens inside a dying tree really comes down to two main things – running out of food (sugary carbon), and hydraulic failure.

When it comes to natural death from 'old age', for trees it is primarily the size that is important rather than the age (Peñuelas 2005). The reason for this is down to the amount of sugar a tree can produce and the amount it needs to stay alive. Carbohydrates in the form of sugars and starch are often needed at different times to when they are produced. For example, a deciduous tree needs to grow a new set of leaves in spring when there is little photosynthesis going on. Some trees will also produce a bumper crop of seeds in a year, which need more sugar than the tree can produce in that year. So trees have mechanisms to accumulate reserves as *non-structural carbohydrates* that include sugars and starch (and also fats). These are primarily stored in the rays in the wood and are then used to fund the survival and growth of the tree when they are needed.

As carbohydrates are produced, they tend to be stored in the newest growth rings under the bark. Older growth rings will have carbohydrates stored from previous years but these do not get added to over the years, since they become increasingly difficult to access with age. The tree thus has a current account of *fast-cycling carbohydrates* in the newest rings that is used for the general running of the tree. But it also has a *slow-cycling* savings account of older carbohydrates that can be drawn upon to fund new growth or when the tree has a more exceptional need (Richardson *et al.* 2015). As an example, during a severe drought, the tree may be unable to photosynthesise, since it needs to keep the stomata tightly shut to prevent undue water loss. This lack of new sugar production is covered by its ability to call on the carbohydrates in its savings account of older wood. The outcome of this is that long-term stressed trees will use up progressively older and older carbohydrate reserves. In most situations, healthy trees are unlikely to run out of reserves. Hoch *et al.* (2003) estimated that temperate trees store carbohydrate reserves sufficient to be able to replace the entire crown four times over. These reserves are likely to be able to see a tree through a number of difficult years.

The problem for a bigger tree is when these reserves get progressively used up or become unavailable. As a tree gets bigger and enters the late mature and ancient stages of its life described above, the crown will tend to shrink as branches die back. This means that there will be fewer leaves and so less sugary income. Yet the new rings of wood have to be grown over the whole tree. These can become very narrow and may even be grown just partially around the trunk, but there is a point at which the outgoings will be more than the sugar grown each year, and the tree makes more demands on stored reserves.

Will the ancient tree eventually run out of stored reserves? There is evidence that trees will prioritise keeping a higher level of reserves even when growing in deep shade, but if something untoward happens, like repeatedly losing a set of leaves to defoliating insects, the reserves can be drained (Weber *et al.* 2019). Having said this, seedlings can certainly live on very low carbohydrate amounts for weeks (Weber *et al.* 2018). On the whole though, it is unusual that even a venerable old tree will run out of stored carbohydrates (Palacio *et al.* 2014). However, what is likely to happen is that these reserves may become increasingly unavailable and locked up in the older wood. So even though reserves are held in the wood, they are just not available to the tree, and all this leads to an inexorable spiral towards starvation.

At this point, something has to give, which usually means an increased loss of the topmost branches that are under most water stress. This leads to a *stag-headed tree*, where the top dead branches stick out of the crown like a set of deer's antlers (Fig. 183). Loss of sugary income will limit the reserves available for defences, making the tree more susceptible to pests and pathogens, leading to more dieback. Many trees can act to slow this process by *retrenching* – producing new branches from buds that have been stored unopened (called *epicormic buds*) for years on the trunk and lower branches, often in gnarled burrs on the trunk. Many trees can also produce completely new *adventitious buds* from the cambium under the bark. New epicormic branches hold extra leaves and go some way towards making up for those lost higher up. And since these new branches are thinner than the ones they replace, this reduces the amount of wood needed in new rings.

Epicormic branches tend to be fairly short-lived – 100 years in oaks, 60 years in Hornbeam and Beech and still less in birches and willows. Nevertheless, trees that have a plentiful supply of epicormic buds, such as oaks and Sweet Chestnut with big burrs, can keep up a steady supply of new epicormic shoots and slow the decline towards death for centuries (Fig. 184). This gives rise to the old saying that

FIG 183. (a) A Pedunculate Oak *Quercus robur* becoming stag-headed with age. (b) Even further dieback has left this squat Sweet Chestnut *Castanea sativa* with new epicormic shoots replacing the lost canopy. (a) Rural Staffordshire and (b) Greater Caucasus Mountains, near Balakan, Azerbaijan.

FIG 184. A Sweet Chestnut *Castanea sativa* with epicormic shoots of various ages arising from the base of the trunk and from burrs on the trunk.

'Oak takes 300 years to grow, 300 years to stay, 300 years to decline.' Some conifers, such as Giant Sequoia, Coastal Redwood, Sitka Spruce, Douglas Fir and Western Hemlock *Tsuga heterophylla* can produce epicormic shoots in the canopy and so prolong life for even longer. Trees with fewer epicormic buds, such as ashes and beeches and many conifers, have a shorter time of retrenchment and a shorter lifespan.

An important part of the retrenchment principal is that the lifespan of a tree depends more on the size of the tree than its age. The moral of the story is that if you keep a tree smaller by either growing it slowly or keeping it well pruned, even cutting it off at ground level, it will take longer to reach the ancient stage and will live much longer. Edmund Schulman (1954) noted that 'adversity begets longevity'. Indeed, there is evidence that in trees there is only so much living to do, and they either do it quickly and not for long, or more slower and for longer. This undoubtedly partly explains why the Bristlecone Pine lives for so long, growing on poor soil in a dry, cold environment with less than 30 cm of annual precipitation, most of which falls as snow, and with a short growing season measured in weeks. This tough environment also inhibits plant enemies such as fungi that could kill the tree (Piovesan & Biondi 2021). A corollary of this is that the lifespan of a tree can be extended by cutting it down to keep it small, such as by coppicing. As noted earlier, although Ash *Fraxinus excelsior* normally lives for 200–450 years, in Bradfield Woods, Suffolk, there is a coppiced Ash with a stump 5.6 m in diameter that is at least 1,000 years old. If only humans worked in the same way!

CHAPTER 13

Breeding and Genetic Engineering

PROPAGATING TREES

There are many ways in which new trees have traditionally been produced. One of the simplest is to take cuttings which can be rooted to create genetically identical, or cloned, trees. This method has been used in forestry to propagate a particularly good tree that might grow faster than its neighbours or have an exceptionally straight trunk. Cuttings taken from this can be used to create a *clonal plantation* of trees that are all genetically identical – but with the inevitable risk of pathogens spreading rapidly. Such clonal propagation is not new, and the huge number of Olive *Olea europaea* subspecies *europaea* trees planted around the Mediterranean are based on repeated selection of productive trees of the wild subspecies *sylvestris* and their clonal propagation to produce trees with narrower leaves, less thorny branches and bigger fruits. This has resulted in relatively few clones that are each genetically identical, with the downside that it makes them more prone to pathogens such as the bacterium *Xylella fastidiosa* (Chapter 15).

Propagation of clones can also be done by grafting where the top of one tree (the *scion*) is attached to the bottom of another (the *rootstock*). This has a number of uses:

- A vigorous scion can be kept small using a dwarfing rootstock.
- Or the other way round, a weak plant can benefit from a vigorous rootstock.
- A weeping variety can be made into a tall tree (as in Fig. 185).
- A susceptible plant can be protected from soil-borne pathogens.

The biggest historic development in apple production has undoubtedly been the use of dwarfing rootstocks (Fig. 186). Not only do they allow more trees to

FIG 185. Grafted form of Wych Elm *Ulmus glabra* 'Camperdownii'. This grows to be a large weeping tree so needs to get some height in the trunk to give the tree good shape. In this case the cultivar has been grafted fairly high up onto a standard Wych Elm stock, indicated by the horizontal line just above the black label. Botanic Garden of Smith College, Massachusetts.

be planted in an orchard but they also make it easier to pick the fruit from the short trees, and the physiological interaction between the scion and rootstock results in the scions flowering and producing fruit when younger. As shown, the rootstock can be chosen to produce a tree of almost any desired size. This is also seen in flowering cherries which were originally grafted onto Wild Cherry *Prunus avium* rootstocks but are now most commonly grafted onto a rootstock called Colt, described as 'semi-vigorous', which reduces trees to around 75 per cent of the height of those grown on Wild Cherry ('vigorous'), making them better suited for smaller spaces. Colt dwarfing rootstock was originally developed at East Malling Research Station in Kent to keep fruiting cherries down to 3 m tall, a more manageable size for picking. Colt is roughly comparable to the apple MM106 rootstock in Figure 186. As described in Chapter 7, a growing tree keeps

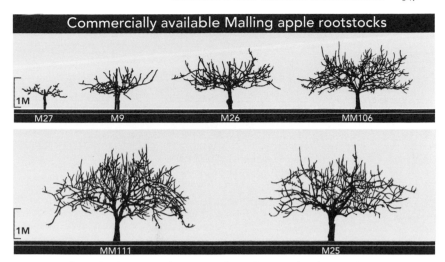

FIG 186. Effects of different dwarfing rootstocks on the size of apple trees above ground. Copyright NIAB EMR, used with permission.

the amount of roots and shoots in proportion to each other. A dwarfing rootstock works by the roots growing slowly and staying small, which forces the tree's shoots to stay small in proportion.

A particular problem with grafting is that the rootstock can produce vigorous shoots from its base or as suckers from its roots (as in Fig. 187) which, if not removed, will soon overcome a weak scion.

NEW CULTIVARS

Grafting can go a stage further to produce *different* trees, or new *cultivars*, a contraction of *cultivated varieties*. One method relies on a cutting 'remembering' how it was growing before (called *topophysis*: Robbins 1964). A cutting of a horizontally growing side branch will usually produce a weak, spreading tree, while a cutting from a top upright shoot will produce a more upright 'fastigiate' tree. In a similar way, cuttings from the top of a mature holly or False-acacia *Robinia pseudoacacia*, where there are few leaf prickles or thorns, can be used to create thornless varieties, as the scions remember that they were producing no prickles and the whole tree carries on with no or very few prickles.

We are also adept at using juvenile and adult growth stages in the same way. As explained in Chapter 10, many trees when young produce leaves that are a

FIG 187. An old Prunus 'Kanzan' that was grafted onto Wild Cherry *Prunus avium* rootstock. It is growing on a disused golf course, and once grass mowing was stopped, the many suckers of the rootstock flourished to make a circle around the parent tree. The Wild Cherry shoots will easily grow twice the height of 'Kanzan', so in time the ornamental pink cherry will be outcompeted and is unlikely to thrive unless the suckers are cut back.

different shape or size from those of an adult plant. A cutting taken from one of these stages has a memory of how it was growing before, and the whole new tree may stay growing that way. A vivid example is European Ivy *Hedera helix* which when young has horizontal, or *plagiotropic*, shoots that grow along the ground and up walls and tree trunks, rooting as they go and having five-lobed leaves. Then, often in response to higher light when they reach the top of their support, they change to adult plants which grow upright, or *orthotropic*, shoots with oval unlobed leaves and no aerial roots, and produce flowers and fruits. The adult stage of Ivy can be rooted (with difficulty!) or grafted to form a free-standing Ivy bush or small tree (*Hedera helix* variety *arborescens*) that flowers and produces fruit (Fig. 188). The adult plants are remarkably stable although the occasional juvenile branch can appear at the base, something which may also be induced by severe pruning or frost damage. Seeds from these bushes will, of course, initially produce the normal juvenile plant.

FIG 188. An Ivy bush *Hedera helix* variety *arborescens* created by rooting an adult flowering shoot. Biddulph Grange, Staffordshire (a National Trust property).

Some plants produced using this 'memory' may revert to normal growth after a few years, especially dwarf conifers – so beware when planting one that it may not stay as small as promised. Another problem of repeatedly taking cuttings is that there can be an increasing memory of age so that a physically young tree grown as a cutting or graft acts as if it is still part of an older tree. This can be quite useful, as scions taken from older trees will grow more slowly, tend to be less branched and will flower and fruit at a younger age, which is perfect if you are propagating apple trees for commercial production. But this can eventually be detrimental in clonal plants; it is known that after 500–20,000 years, pollen becomes progressively sterile in clonal trees, such as the American Aspen *Populus tremuloides*; perhaps too long for the horticulturalist to worry about! But the plants may also die at a younger age because they are physiologically older. The most notable case in the animal world was with Dolly the sheep, a clone of her mother, who was born in July 1996 but retained the 'age' of her parent and had lung disease and crippling arthritis by the time of her death in February 2003.

FIG 189. Copper Beech *Fagus sylvatica* 'Purpurea' can arise naturally from seed and has high
levels of red anthocyanin pigments in the leaves. The copper coloration needs high light levels
to develop, so the leaves inside the crown are much less red. Tyntesfield, Somerset (a National
Trust property).

New cultivars can also be produced by looking out for genetic mutations,
or 'sports'. Seeds will occasionally produce a variant seedling due to a mutation
in its genetic code. Copper Beech *Fagus sylvatica* 'Purpurea' (Fig. 189) arises this
way, although you might need to plant thousands of seeds from a normal green-
leaved beech in the hope of getting a Copper Beech, and even then the chance is
that it will be a less intense purple than the best available commercially, which
have been propagated from cuttings or grafts from the very best specimens. Most
seedling sports will only survive if protected from competition in the controlled
conditions of a garden or nursery, but a number of Copper Beeches are known to
be growing wild in continental Europe.

Another form of genetic mutation is a bud or branch sport that produces a
new, interesting branch (Fig. 190) from which a cutting or scion can be taken,
creating a whole new tree of the sport type (Fig. 191). This has been used for
centuries in trees such as flowering cherries, creating a plethora of subtly
different trees (Fig. 192).

FIG 190. *Prunus* 'Jacqueline' renowned for its early pink blossom. This specimen has produced two genetic sports at the top of the tree, which are more upright with smaller, fewer flowers. Cuttings of this could be grafted onto a rootstock to produce a new cultivar. Part of the Keele University National Collection of ornamental flowering cherries.

FIG 191. Genetic sports. (a) A cultivar of Sycamore *Acer pseudoplatanus* 'Brilliantissimum' that flushes orange in the spring before turning green. (b) An extreme sport of Large-leaved Lime *Tilia platyphyllos* 'Tiltstone Filigree' that arose at a Cheshire nursery. Most people would be hard-pressed from the leaf to recognise this as a lime!

OPPOSITE: **FIG 192.** The art of developing different cultivars. (a) Flowering cherries in Matsumae Park on the island of Hokkaido in northern Japan. With over 250 varieties of cherry, this park is renowned for its cherry festival each spring. (b) Many of these cherries have been bred by Mr Masatoshi Asari, shown here in 2016. Born in 1931, he spent his working life as a primary school teacher in Matsumae, but bred over one hundred new cultivars of cherry, some of which have the prefix Matsumae and many of which are still not known in the west. Mr Asari has been given the honorary title 'Sakura Mori', protector of the cherry trees.

Pink grapefruits, seedless grapes, navel oranges and 'Delicious' apples all come from such sports. A large tree can have anywhere between 10,000 and 100,000 buds, so even a relatively low mutation rate of around 1 in 10,000 per year will produce a large number of sports. Many of these will not survive the intense competition within a tree crown and will soon die, but some will be benign enough to survive and be worthy of propagation. Such mutations can produce dwarfness, particular shapes such as upright columns or plates of foliage or colour variation, such as the huge number of conifers that come in all colours from gold to deep blue.

Sports can also show juvenile leaf fixation where the juvenile foliage (Chapter 10) is retained in the mature tree. This is especially common in conifers and explains why Sawara Cypress *Chamaecyparis pisifera* is often seen in the form of 'Plumosa' (Fig. 193) or more commonly 'Squarrosa' – very common in English churchyards. Many of these sports tend to be unstable, and parts of the highly

FIG 193. Foliage of Sawara Cypress *Chamaecyparis pisifera*, native to Japan, with the juvenile fixed form of 'Plumosa' (left) and the normal adult foliage (right).

prized new tree may revert back to produce vigorous foliage of the original tree that should be cut out to prevent it taking over.

The domestication of the Apple *Malus domestica* shows these processes in action (Cornille *et al.* 2019). Our domesticated apple probably arose in East Asia some 4,000 years ago, originally based on *M. sieversii* from the Tian Shan mountains along the China–Kazakhstan border. It was spread along the Silk Road joining Asia to Europe, being bred with local apples on the way before our European Crab Apple *M. sylvestris* was included as a major part of the modern mix (Cornille *et al.* 2012). The Greek philosopher and botanist Theophrastus described six apple varieties some 2,300 years ago, the Romans probably introduced domesticated apples to Britain, and settlers took apples to North America, where further cultivars were developed. Development didn't stop there, and it is estimated that there are now between 6,000 and 10,000 apple cultivars worldwide that vary in a dazzling array of size, shape, colour of the skin and flesh, sweetness, and storage life. Modern apples are self-incompatible so cannot pollinate themselves, which means that the seeds from any one tree can be very variable depending upon which nearby varieties are providing the pollen. Thus, when a particularly good tree is found from a seed or from a genetic sport, it is 'fixed' by taking scions for grafting. And, of course, it can be kept and spread through further grafting.

CHIMAERAS

A chimaera is a Greek mythological fire-breathing monster with the head of a lion, body of a goat and tail of a dragon. Real-life chimaeras can be found in plants and animals and are defined as organisms composed of cells from more than one organism, or strictly speaking, cells of different genetic make-up.

In plants these happen most often as a *graft chimaera*, or *graft hybrid*. Usually this new growth has the core of one plant and a thin skin of another and both have an effect on what the new tree looks like (Frank & Chitwood 2016). A superb example of this is the Laburnum-Spanish Broom hybrid + *Laburnocytisus adamii* (the '+' before the name indicates a graft chimaera between two separate genera; if it was within the same genus, the '+' would be before the species name). All specimens we see today come from cuttings of the original tree that arose in France in 1825 in the garden of a nurseryman called Adam. He grafted a Spanish Broom (*Chamaecytisus purpureus*, which used to be called *Cytisus purpureus*) onto a Laburnum (*Laburnum anagyroides*). The Broom tends to be a low spreading bush that can be made more tree-like by grafting on top of a straight Laburnum stem.

In this case, a graft chimaera arose at the graft junction that had the Laburnum at its centre, which gives a more upright overall shape, but the outside layer, just one cell thick over the whole tree, is Spanish Broom, and so the leaves and twigs are Broom-like with just a hint of Laburnum in the shape. The flowers, however, can be Laburnum yellow or Broom purple (Fig. 194) or an interesting mixture, sometimes with a flower divided down the middle into half yellow and half purple. The flowers of graft chimaeras are either sterile or produce seeds of the tree at the core (in this case, Laburnum). While these chimaeras are stable, it is not unusual for the core species to break through the skin of its partner to produce branches of its own.

In some cases both a normal hybrid and a graft chimaera are possible; in the former the genes of both species are mixed in every cell, whereas the chimaera is a mix of cells of both species, each cell type with its original genes intact. As an example, hybrids between hawthorns (various *Crataegus* species) and Medlar *Mespilus germanica* are possible and are usually given the name × *Crataemespilus*

FIG 194. The Laburnum-Spanish Broom graft chimaera, or graft hybrid, + *Laburnocytisus adamii*, which in this specimen displays mostly yellow flowers of the core of Laburnum *Laburnum anagyroides* with some purple flowers of the Spanish Broom *Chamaecytisus purpureus* that forms a layer over the top of the Laburnum. National Botanic Gardens, Glasnevin, Ireland. Photograph by Chris Sanders.

FIG 195. A variegated form of Sweet Chestnut *Castanea sativa* 'Albomarginata', part of which has reverted to the non-variegated original. Being all-green, these can be more vigorous and take over the plant if not pruned out. Exmouth, Devon.

(the 'x' coming before the genus indicating a hybrid between different genera). But a chimaera named + *Crataegomespilus* also exists, resulting from grafting Medlar onto hawthorn rootstock (Thomas *et al.* 2021), which gives rise to branches from the graft union with a layer of Medlar outside a core of hawthorn, with white flowers tinted orange compared to the pure white flowers of the hybrid. These names may change in the future as there is some thought that the Medlar should be re-classified as a *Crataegus* species, and so the resulting hybrid and chimaera would also be *Crataegus* species.

The other form of chimaera is found in variegated plants, which starts naturally by a mutation in one bud. Buds are made up of 'initial' cells that each grow a discrete layer of tissue as the bud expands. If one of the initial cells mutates, then all cells in a layer will carry that mutation (and so be genetically different from the other layers – classifying the plant as a chimaera). If one of these changed layers happens to be colourless, then the sandwich of colourless and green layers can produce the yellow-edged or central variegated leaves commonly seen in pelargoniums and a number of trees, from maples to hollies and notably in *Prunus* 'Harlequin' (a sport of *P.* 'Okame'). Many of these variegated plants can 'revert' back to the original, as seen in the variegated form of Sweet Chestnut in Figure 195.

TREE BREEDING

Tree breeding in its most basic form has traditionally involved selecting the 'best' tree in a group – one having the most desirable characteristics, such as straightness of trunk, the brightest flowers, the best autumn colour or resistance to disease – and collecting seeds from that tree to grow the next generation. A step on from this is to breed together trees of the same species that have different useful characteristics in the hope that the resulting offspring will have them all in the same tree. As before, seeds are collected and grown, and the best, or *superior*, trees would be kept for their seeds. The main problem here is that seed collected from the superior tree is usually genetically mixed, as each seed has two parents and so not all seeds will produce equally good trees. One solution is to take pollen from a good tree and use it to fertilise the flowers of another good tree. This is fine, though time-consuming, for small trees but is less easy for large forest trees or on a large scale. Another solution is to produce a seed orchard where cuttings from a superior tree are grafted onto rootstocks and planted close together. This encourages pollen to be exchanged between the trees in the orchard, increasing the chance of a greater number of seeds being superior. But this is still expensive and time-consuming and so tends to be used only for high-quality fruit or nut trees and for aesthetically pleasing trees that can fetch a high price.

Once a superior tree has been developed, this can then be propagated by taking grafts from the original tree to produce clones, all genetically identical and looking the same. Alternatively, many more new plants can be produced by *plant tissue culture*, a method that was developed through the twentieth century (Thorpe 2007). Since every cell contains the DNA of the whole plant, sterile techniques have been developed to take cells from leaves or root tips and induce them to become embryonic cells by applying plant hormones and nutrients, technically called *somatic embryogenesis*. These embryonic cells can be separated and stored in liquid nitrogen or used straight away, with each cell growing into a whole plant as if it were a seed (Sánchez-Romero 2019).

New trees can, of course, be developed by hybridising related species. These hybrids can happen naturally, but we are not beyond deliberately producing them as they often show more vigorous growth (*hybrid vigour, or heterosis*) and they often bring together useful features of both parents into one plant. The hybrid Larch *Larix × marschlinsii* (formerly *L. × eurolepis*) is widely planted as a forestry tree in Europe because it has the good shape and fast growth of the European Larch *L. decidua* and the disease resistance of its other parent, the Japanese Larch *L. kaempferi*. Walking around urban Britain, you soon encounter what is probably the most notorious hybrid, Leyland Cypress *Cupressus × leylandii*, a hybrid between two species from the western seaboard of North America, the Nootka Cypress *Cupressus nootkatensis* and the Monterey Cypress *Cupressus macrocarpa* (Nootka Cypress was once in a different genus – *Chamaecyparis nootkatensis* – so the Leyland Cypress was originally called × *Cupressocyparis leylandii*). This original Leyland Cypress hybrid was found at Leighton Park in Wales in 1888, where both parents were growing near each other (in their natural habitats they don't come within 500 km of each other). Six plants were raised from seed collected from a Nootka Cypress by C.J. Leyland. In 1911 the reverse hybrid was discovered in two seedlings grown from a Monterey Cypress. The biggest Leyland Cypress is not surprisingly found in Britain and is currently 38.5 m tall and, since it shows hybrid vigour, is still growing well! It is possible that these might end up being amongst the tallest trees in Europe and possibly the world. Plant hybrids usually have fertile pollen and so can pollinate one or more of the parents. In tree families where hybridisation is common, such as the birch, pine, rose and willow families, this has led to *backcrosses* with the parents and with each other to produce an almost continuous range of intermediates, a *hybrid swarm*, making identification difficult to say the least.

Another way of improving trees is to use naturally occurring *polyploids* which have duplicated sets of chromosomes, often doubling or quadrupling the original number. Although rare amongst conifers, they have arisen naturally in a number of broadleaf trees. Polyploid forms of the North American Paper-bark Birch *Betula papyrifera* and the Japanese Alder *Alnus japonica* are known, which grow larger and faster than their normal relatives that have just one set of chromosomes.

The problem remains that the traditional tree breeding described above can take decades for each generation to be produced and for the best trees to be selected. This led to research in the early 2000s into whether it is possible to predict how well a tree will eventually perform from some feature that appears early and is easy to measure, such as the number or size of leaves. Indeed, in fast-growing hybrid poplars it was found, for example, that the size of the largest leaf on a branch is linked to the amount of wood grown each year, so the leaf size of young saplings could be measured in order to select the best trees to grow as a crop (Marron & Ceulemans 2006). A step on from this was to use genetic markers. For example, the Californian Valley Oak *Quercus lobata* is in decline due to the rapid rate of climate change in its native range (Fig. 196). A solution is to find trees growing well at low altitudes on the assumption that they will be best adapted to hotter, drier conditions, and these can be planted at higher altitudes ahead of climate change. Fortunately, *genetic markers* (a gene or short DNA sequence) associated with faster growth under warmer climates have been identified, so trees can be screened to determine their suitability before being planted (Browne *et al.* 2019). The problem, however, is that growth traits tend to be linked, and trees selected for ability to grow in high temperatures might also be the ones with poor disease resistance or some other problem. What was needed was a way of altering just the characteristic of interest. This had to wait until the development of genetic engineering.

FIG 196. Valley Oak *Quercus lobata* near the southern end of its natural distribution in the Coastal Ranges of California. With climate change, it is likely to be too hot and dry for it to grow here in the future, but individuals that are better at coping with hot conditions can be selected and planted to ensure these hills have some tree cover.

GENETIC ENGINEERING

The ability to alter selected desired characteristics (*traits*) of a tree became a reality when it proved possible to tinker directly with the genes inside a cell and produce new trees by genetic manipulation. The first problem is to find out where different genes are along the DNA strand. Each gene is responsible for producing a specific protein which does a particular job in the plant, helping to build it or make it work. For example, to make cellulose in a plant requires a chain of 10 genes, so to modify cellulose production, you need to know which genes these are and where they are on the DNA.

Every cell of a plant or animal contains a complete copy of its DNA which consists of genes strung along the chromosomes. We humans have 23 pairs of chromosomes while for comparison eucalypts have 10 pairs, poplars 19 pairs, Coffee *Coffea arabica* 22 pairs and the Black Mulberry *Morus nigra* an astonishing 154 pairs of chromosomes in each cell. This complete set of DNA in a cell (all the genes along all the chromosomes) is referred to as the *genome*. The first genome of a tree to be documented was the Black Cottonwood *Populus trichocarpa*, published in 2006 by Gerald Tuskan and his 108 colleagues from the USA, Canada and Europe. This has a comparatively small genome for a tree; eucalypts are twice the size, Pedunculate Oak four times, and spruces and pines have a genome 80 times the size of the Cottonwood (humans come in at six times the Cottonwood size). It is important to note that the size of the genome does not match how complex the organism is, since the simple single-celled freshwater amoeba has a genome more than 1,300 times larger than that of the Black Cottonwood. Part of the reason for the different genome sizes is that, over time, a genome can accumulate or lose parts of the *non-coding DNA*; that is, DNA outside of genes that does not hold the code for specific proteins. This has sometimes been called *junk DNA* but it is now known to occasionally prove useful to the plant in a variety of ways. The whole genomes of dozens of trees are now available, mostly temperate timber trees and commercially valuable fruit trees, as well as others such as Tea *Camellia sinensis*.

Once the genome is known, the position of genes can be worked out along the chromosomes by looking for *codons* – bits of DNA known to be at the start and end of genes. Computers can also be used to scan the DNA, looking for common genes found in most plants. Once the order of genes is known along the chromosomes, it is then possible to investigate what specific genes do and what happens if they are changed. One way to do this is to use a tree with a naturally occurring mutation, or sport, and compare how its genome differs from that of a normal individual – this is called *forward genetics*. Alternatively, the effect of a specific gene can be investigated by removing or silencing it using various methods and seeing what happens in a tree grown from the modified cells – *reverse genetics*.

Heather Coleman and colleagues published just such a reverse genetics study in 2008. They stopped a particular gene from working in a poplar hybrid and found that less lignin was produced in the wood and so proved that this gene is involved in the process of producing lignin. Moreover, in the same way that a blockage in a production line leads to a build-up of half-finished products, they looked for built-up compounds known to be involved in lignin production. From this they worked out which part in the production line of lignin this gene controlled. By repeating this process on different genes, they reached the point when they knew which genes were involved along the whole production line for lignin. As the position and role of different genes in a genome are worked out, we can begin to 'engineer' new trees.

Genetic engineering often involves *transgenesis*, the taking of genes from one organism and putting them into another organism (hence these genes are called *transgenes*). This creates a transgenic organism or a genetically modified organism (GMO or GM tree). Genes can be moved between anything that has DNA, unlike, for example, grafting, which can only usually be done between closely related trees. In the first transgenic trees this allowed genes that produce a naturally occurring insecticide to be moved from a soil-dwelling bacterium *Bacillus thuringiensis* into a poplar tree. This transgenic 'Bt poplar' was then able to produce the insecticide in the leaves, which gave it protection from insects.

Transgenesis raises a number of ethical concerns so there have been moves towards *cisgenesis* (*cis* meaning 'same'; *genesis* meaning 'origin'), using genetic material derived from the species itself or moved between plants that would be capable of hybridising naturally. In other words, this uses the gene pool available to traditional tree breeding. The benefit over traditional breeding is that a specific gene or genes of interest can be changed or moved without any other genes being altered. A good example would be the production of elm trees resistant to Dutch elm disease. A number of Asiatic elms have resistance to the disease, and it is possible that the gene or genes for resistance can be moved into European elms. Since many of these elms readily hybridise, such manipulation would come under the category of cisgenesis. This type of hybridisation has already been done by artificial cross-pollination, but can be more focused on changing disease resistance without inadvertently changing other variables such as tree shape or drought resistance. Chapter 15 gives details on progress so far.

Intragenesis takes a step in a different direction by allowing the introduction of new gene sequences created artificially outside the plant. Both cisgenic and intragenic plants are still classified as transgenic, but they can be a more socially and politically acceptable form of genetic engineering.

A variant is to use *transgrafting*. This uses a GM rootstock and non-GM scion. For example, a hybrid Walnut rootstock (*Juglans hindsii* × *J. regia*) is very good for growing walnuts on but it doesn't root very easily and so is difficult to produce commercially. So, a gene from a bacterium that induces rooting in plants was isolated and inserted into the Walnut rootstock, making it much easier to propagate (Vahdati *et al.* 2002). In other cases, the engineering will help the whole plant. Inside the cell, RNA is responsible for conveying the instructions for making a protein from the gene on the DNA to the protein-building apparatus. In transgrafted Wild Cherry *Prunus avium*, a gene giving resistance to the ringspot virus has been inserted into the rootstock (Song *et al.* 2015). The RNA carrying the gene's instructions moves across the graft junction into the scion, so the whole plant has viral resistance. The important thing about this is that in both these examples, the modified genes are only in the rootstock and not in the scion. Thus, in both the Walnut and Cherry, the plant benefits from genetic engineering but the fruits and nuts themselves are classified as non-transgenic.

Genetic engineering may also be able to do away with grafting altogether. As noted earlier, a dwarfing rootstock can be used to keep a fruit tree small so the fruit is easy to pick. The slow growth of the rootstock, which slows the growth of the scion grafted on top, is controlled genetically (Costes & García-Villanueva 2007). So there is the possibility that the gene or genes for dwarfing can be taken from the rootstock and put into the scion. For example, dwarfing in peach trees is known to result primarily from a mutation in a specific protein that recognises the plant hormone gibberellic acid. One of its roles in plants is to control growth, and if the hormone is not recognised, it has much less effect and the plant stays small. If this altered gene can be moved to other trees, we could have dwarf trees without the cost of grafting.

A big problem in genetic engineering has been how to physically move genes into the DNA of a new plant. It originally involved techniques such as using a *gene gun* that would fire minute inert metal particles coated in DNA at target cells (yes, really!). The particles would penetrate the cells and release their cargo of DNA into the cell, where it would hopefully become effective. This had the advantage that it could be done using cells still in the plant rather than having to isolate them in a Petri dish. Another technique is to insert new genes using a bacterium *Agrobacterium tumefaciens* which infects plants by inserting part of its own DNA into that of the host. This has been utilised in genetic manipulation by putting the gene or genes of interest into the bacterium DNA, adding the bacterium to the target host cell and allowing it to do its business. The main problem with these techniques is that the new material is inserted randomly into the host genome with often unpredictable effects, so it may take many goes to get it right.

However, we now have *gene editing* technologies (the most common one is called CRISPR, which stands for the equally opaque 'Clustered Regularly Interspaced Short Palindromic Repeats'). Gene editing allows a DNA chain to be precisely edited to alter, replace or silence a specific gene. This is where the concept of gene therapy in humans comes from, along with the desire for designer babies where we can remove genes associated with a disease and potentially even alter eye or hair colour. The ethical issue of doing this to trees is nowhere near as great as designing human babies but it still raises important questions that need to be dealt with.

This is perhaps exacerbated by the possibility of going back in time. Snippets of DNA have been recovered from subfossil pine wood around 13,000 years old from Switzerland, and the complete genome of Date Palms *Phoenix dactylifera* germinated from 2,000-year-old seeds has been worked out (Lendvay *et al.* 2018, Gros-Balthazard *et al.* 2021). It is thus possible to look at how genes have changed over time compared to modern trees, and we may even be able to reintroduce 'lost' genes into modern trees, or recreate past trees.

The next step in engineering trees comes under the realm of *proteomics*. Genes work by producing proteins that control what goes on inside a cell. While every cell in a tree has the same genome (set of genes), different types of cells have different roles in a tree (for example, cells in leaves are photosynthesising, and cells in the root are controlling water uptake). These different cells have different genes switched on and off and so produce different proteins they need for their role, which may be different from proteins produced by the same genes in other cells. Thus the *proteome* (the sum of all proteins in the tree) is infinitely more complex than the genome. To fully understand how a tree can be manipulated, the role and interaction of all these proteins needs to be understood. This gets us into the realm of *epigenetics*, discussed in Chapter 8 as part of the tree's immune system. Compounds produced by the plant can modify the expression or working of genes without changing the genetic code of the tree. These can modify what proteins the genes produce, making the proteome even larger. So stress on a plant, for example, will affect the proteome, changing which proteins are produced; these in turn can have epigenetic effects by altering which proteins *will* be produced in the future. While this is complex it does promise to allow some further fine-tuning of trees to suit our needs.

What is possible so far, and what about the future?

Genetic engineering is not cheap and so has been used to address significant problems faced in commercially valuable trees. The first transgenic trees were in production in the 1980s and were based on inserting insecticide genes giving resistance to insect attack (the Bt poplars mentioned above) or inserting genes into various poplar hybrids for herbicide resistance (so plantations could be sprayed to kill weeds without affecting the trees). Other genetic modifications have included Papaya *Carica papaya* made resistant to the ringspot virus in Hawaii and eucalypts in Brazil that grow faster (to produce biomass for power plants). Quicker growth has obvious economic benefits but can also help in carbon storage to combat climate change (see Chapter 2).

There have also been efforts to change the structure of a tree, and in particular to reduce the amount of lignin in the wood by suppressing one or more genes involved in lignin production. It has so far proved possible to produce poplars with less than half the normal lignin and with a third extra cellulose. This makes it easier and cheaper to make paper from the cellulose in the wood (lignin is a difficult molecule to remove from the cellulose) and to get greater yields of bioethanol, in some cases more than an 80 per cent increase compared to non-modified poplars. This is not without its downside, since reducing the amount of lignin tends to produce trees that are weaker and more easily damaged by weather, pests and pathogens. It also makes the wood more vulnerable to a breakdown of water transport in dry weather, resulting in reduced growth so the crop is smaller. Consequently, emphasis has changed to altering the chemical properties of lignin without reducing the overall amount. The aim here is to make it easier to separate the lignin from the cellulose when the wood is processed.

There have also been moves to modify trees in perhaps surprising ways. In trees where production of seed is the main aim, trees have been engineered to induce flowering earlier in their lives. Although some poplars and eucalypts will produce seeds in as little as three years, some commercially valuable timber trees such as spruces and pines can take 15 years or more in un-engineered trees. This works; in poplars and a number of other trees such as apples, citrus fruits and olives, advancement of flowering by several years has been managed.

Other trees have been modified to *reduce* the number of flowers or at least to produce sterile flowers. These are particularly useful in urban areas in reducing the amount of pollen, with the thanks of those who suffer hay fever, and reducing the amount of flower and fruit debris that can cause slips and falls or just be a nuisance when tracked into buildings. This debris can also attract insects, smell bad, block drains and be a general cause of justified grumbling. It is possible to plant just male trees of dioecious species, such as the male Maidenhair Tree (Fig. 197), but this still results in pollen and some flower debris. Of course, in some cases the flowers are wanted without the mess of fruit – such is the case with things like cherries, pears and crab apples. Flower and fruit production can be suppressed by spraying trees with hormones, but this is expensive for big trees, may have unintended consequences for other plants, and is understandably generally unacceptable to the public even if the spray is harmless to humans. Problems such as these are prime candidates for genetic engineering. The precise mode of action of many floral genes allows genetic modification to be very precise; for example, keeping the flowers but preventing the fruit from developing. Complete sterility is also possible for street trees. This can prove a useful addition to trees modified in other ways, since it minimises the chance of engineered genes escaping into wild populations of trees via pollen and seeds.

Trees have also been engineered to produce useful products. Genes for monellin, a sweet-tasting but low-calorie compound, have been taken from a tropical shrub and put into tomatoes and lettuces to make them sweeter with no added calories. In other cases, it is possible to engineer trees to produce more chemicals that we find useful. The most desirable of these are the terpenes which are used by trees in defence against invaders such as herbivores, insects and fungi (Chapter 8). Terpenes include things like eucalyptus oil, used in fragrances and pharmaceuticals, and are the main ingredient of conifer resins, hence turpentine based on terpenes. In conifers, the terpene content of leaves is around 2–4 per cent of leaf weight – enough to act as a good defence without being too taxing on the tree energy budget but very little if you're trying to harvest it.

FIG 197. Male Maidenhair Trees *Ginkgo biloba* used as street trees. Male trees are planted because although they produce some catkin debris, this is nothing compared to the odour of fallen fruits from female trees, which smell of rancid butter. San Sebastián, northern Spain.

So trees can be engineered to produce more terpenes either by increasing the number of glands that produce it, by increasing the amount stored or by causing the tree to put more energy into the biochemical pathways that produce terpenes (Peter 2018).

Although it appears perverse, these same terpenes, when released into the lower atmosphere (and then called volatile organic compounds – VOCs), react to form ozone. At street level this is a pollutant. There is thus a move to engineer urban trees to reduce VOC production. A reduction in global pollution can also be produced, since the trees most often used in intensive plantations such as poplars, willows, eucalypts and Oil Palm *Elaeis guineensis* emit high levels of VOCs. Engineering an increase or reduction in VOCs is not, however, straightforward. Although plants produce two main groups of VOCs – isoprene (44 per cent of global emissions) and monoterpenes (11 per cent) – temperate and tropical trees release hundreds of different VOCs in significant quantities, so engineering reduced levels is not necessarily easy.

A slightly different set of objectives in genetic engineering is to breed trees that can cope with a new or changing environment. For example, trees are being produced that can cope with drier, saltier, warmer or more toxic environments (Fig. 198). This gives us trees that can cope with the droughts and high soil salt content of areas where past removal of trees has resulted in open, dry growing conditions. Engineered sterility of trees might also help here, since it removes the burden of reproduction, which may allow better survival in a hostile environment. Given that trees are longer-lived than most other plants, it also gives us trees that can hopefully cope in areas of rapid climate change and so maintain a tree cover into the future. Other trees can be used to help clear up toxic environments. Engineered poplars have been produced that can break down volatile hydrocarbons such as trichloroethylene, vinyl chloride, carbon tetrachloride, benzene and chloroform, which are common pollutants. This is done by engineering the poplars to produce larger amounts of particular enzymes which help break down the pollutants.

Just how far can we go in engineering trees to suit our own needs? The largest constraint is understanding how much the character of a tree is determined by its genes and how much by the environment in which it is growing. Some aspects of tree growth are more tightly controlled by genetics, and other aspects are more influenced by the environment. For example, things such as phenology (the timing of yearly events; see Chapter 3) and wood density are highly controlled by genes. It is therefore realistic to expect to be able to engineer trees to flower early and have high-density wood. By contrast, things such as tree growth are not so tightly controlled by genes, which gives each tree more flexibility in how it grows and is clearly seen in trees being much shorter in windy areas, since they have reacted strongly to the environment. This *low heritability* of growth is why the breeding of poplars for fast growth has been so difficult in the past (Marron *et al.* 2007). In 2015, Brazil approved a trial planting of a GM eucalypt tree that grew 20 per cent faster than non-GM eucalypts in early trials (Ledford 2014), but in later trials in other parts of the country under different climatic conditions,

OPPOSITE: **FIG 198.** Trees are being bred to need less water. (a) In this arid lowland region of northern Libya, most crops need irrigation (as seen by the green squares). If crop trees can be bred to need less water, this will reduce pressure on valuable water supplies. (b) Even in higher-altitude areas, such as the Green Mountains near Benghazi, Libya, aridity is increasing due to climate change, leading to the death of the natural vegetation. In this case, breeding native plants to cope with less water would ensure the continuation of this already stressed habitat.

it failed to live up to these expectations, illustrating how complex the control of growth can be. The genetic control of branch angle is also being explored (Hill & Hollender 2019), since more horizontal branches produce more fruit, although this trait appears to be controlled by a complex number of factors, both genetic and environmental.

To make it more complicated still, some aspects of a tree's life are controlled by single genes (for example, resistance to white pine blister rust in a variety of pines), while others, including many aspects of growth, are controlled by a whole set of genes working together. This makes genetic engineering like trying to bribe corrupt officials: if there is only one to deal with, it may work, but if there is a whole committee of them, it becomes much more delicate and complicated to get the desired outcome. So it is with genes, since the more genes you tinker with, the more likely it is that there will be consequences other than the ones you want.

In case the use of all these forms of genetic engineering seems like playing God, it is worth pointing out that moving genes between organisms has been going on naturally between plants and their pests for millions of years, known as *horizontal gene transfer*. Both plants and insects have extensively borrowed genes from microbes. For example, the Coffee Berry Borer *Hypothenemus hampei*, an insect pest, has stolen microbial genes which it uses to extract more nutrition from hard-to-digest plant cell walls (Acuña *et al.* 2012). More recently it has been shown that some insects can also steal genes directly from plants (Xia *et al.* 2021). But this is not to underplay the ethical issues of releasing GM trees into the wild, which may have unintended consequences.

One of the main limits to genetic engineering is undoubtedly what can be tolerated by society. This can be a bigger problem in trees than in GM crops, since the engineered trees will live for decades and could migrate across international borders. GM trees have alternatively been called the saviours of future forests and 'Frankentrees', and their release into the environment involves complex arguments about the pros and cons. Nevertheless, in the USA, around 20,000 trees made up of around 100 different engineered tree varieties have been planted in field trials, the majority of them GM poplars, so in many ways the genie is already out of the bottle.

CHAPTER 14

Interaction with Helpful Organisms

GAINING MORE NUTRIENTS

There is a limit to how many roots a tree can produce and maintain, especially the fine roots which absorb most of the water and dissolved nutrients. To ensure that these roots are most productive, they tend to proliferate in areas where water and nutrients are most abundant (see Chapter 5). Fortunately, roots also have other ways of increasing their productivity.

Roots can gain extra resources from the soil by releasing chemicals into the thin layer of soil surrounding the fine roots, called the *rhizosphere*. The root exudates change the soil pH, which affects the solubility of nutrients and so can directly improve the take-up of nutrients. More importantly, the exudates contain a large variety of organic compounds including sugars, amino acids, organic acids and phenolics which have an effect on soil microbes (collectively called the *microbiome*). In a paper by Peter Bakker and colleagues in 2013, they write: 'This rhizosphere microbiome extends the functional repertoire of the plant beyond imagination.' The microbes help the roots extract extra nutrients from the soil and can help defend the roots against pathogens. The effect of the rhizosphere may only extend a few millimetres from the root but since there are so many fine roots, this can encompass around a quarter of the soil in which the tree is rooted. The exudates come at an energy cost to the tree, but at 1–5 per cent of the net sugar produced by photosynthesis, this is comparatively small compared to the benefits of this hugely important investment.

Gaining extra nitrogen

Another way for a tree to gain extra nutrition is for it to cooperate with another organism to their mutual benefit, the pair hence being called *mutualists*. Often this involves a close physical partnership, in which case the mutualism is referred to as *symbiosis*. One symbiotic relationship that trees use involves bacteria. Trees in the pea family (Fabaceae, or Leguminosae as it used to be called) share with their herbaceous relatives the ability to produce small swellings, or nodules, on their roots that contain bacteria of one of several *Rhizobium* species. These bacteria are capable of 'fixing' nitrogen, converting some of the 78 per cent of nitrogen in the atmosphere into a form that is useable by plants. The tree gains an extra supply of nitrogen, and in turn the bacteria gain a protected home inside the nodules and a supply of sugar from the tree.

A common example is False-acacia, native to North America (Fig. 199) and widely planted in Europe. A field experiment in Austria using five-year-old trees found that over two years, False-acacia contained 80 per cent more nitrogen than non-nitrogen-fixing species grown with them. In the False-acacia, 80 per cent of the nitrogen came from nitrogen-fixing bacteria and amounted to 110 kg of nitrogen per hectare each growing season (Danso *et al.* 1995). This helps to explain why False-acacia is so good at invading disturbed areas and why it can grow to 12 m in height in 10 years. Symbiotic relationships come at a cost to the tree, and trees are good stewards of their resources, so as levels of soil nitrogen increase, the number of nodules decreases. Symbiosis is not altruism; when the bacteria are not needed they are dispensed with.

Some other trees, particularly temperate broadleaf trees, use different partners. In these the nitrogen-fixing organisms are Actinomycetes, relatives of filamentous bacteria, found in the genus *Frankia*. Trees using *Frankia* bacteria produce a series of very short, fat, densely branched roots that hold the symbiont. A common European example is the alder, where the clusters of fat roots can reach the size of tennis balls and live for a decade. Other symbionts are also occasionally used. For example, the Australian Cycad *Macrozamia riedlei* has a photosynthetic blue-green alga (an *Anabaena* species) which invades the roots that grow near or above the soil surface, which can then photosynthesise and fix carbon and nitrogen.

Nitrogen fixation by trees is very beneficial in tropical agroforestry (Chapter 5) where the soils quickly lose their fertility and crop production drops. Planting trees such as *Acacia* and *Prosopis* species (which use *Rhizobium*) and the Australian She-oak *Casuarina equisetifolia* (*Frankia*) amongst the crops helps to maintain soil fertility over many years, while also reducing erosion and providing wood for fuel.

FIG 199. An old False-acacia *Robinia pseudoacacia*. The upper canopy shows the characteristic zigzagging of the branches. The central thinner branches are those of a Downy Birch *Betula pubescens* that has rooted in dead wood at the break of the two main stems.

Roots in unusual places

Trees are capable of producing roots on above-ground parts of the stem. These *aerial*, or *canopy*, *roots* can be used to help nutrition, particularly in tropical and temperate rainforests. Trees in these moist habitats often support rich growths of epiphytes such as orchids, bromeliads, ferns and mosses, which use the tree as a support nearer the sun but take nothing physically from it. Epiphytes rarely have roots reaching the soil and so have to gain all their nutrition from the atmosphere, from water running down the tree's trunk or from catching leaf litter as it falls. The litter creates little compost heaps which provide the epiphytes with nutrients as the compost decomposes. Some rainforest trees have developed the ability to produce canopy roots on the branches and trunk, which will infiltrate these humus holdings of the epiphytes, as well as any humus caught in hollows and branch forks, to help supplement their nutrition. The stilt-rooted mangroves (*Rhizophora* spp.) of the tropics use their epiphytes in a similar way but in a more symbiotic fashion (Fig. 200). The mangrove grows

FIG 200. The Mangrove plant *Rhizophora apiculata* uses stilt roots to provide stability in soft marine muds but also to gather nutrients from encrusting sponges when in deeper water. West coast of Penang, Malaysia.

fibrous roots into the marine sponges that grow epiphytically on their prop roots, from which they extract nitrogen, but in return the sponges extract sugar from the roots.

A tree of the Australian rainforest, *Ceratopetalum virchowii*, grows in areas with more than 6,000 mm of rain per year. This high rainfall is bad news for the tree, since it leaches nutrients from the branches and trunk as it runs over the surface. However, the tree also has canopy roots in clumps or enveloping the trunk like a fibrous coat but which do not root in the soil. These roots appear to act as a nutrient-recovery system, reabsorbing nutrients leached out higher up the canopy (Herwitz 1991).

A number of temperate trees in drier areas also produce canopy roots, including Beech *Fagus sylvatica*, but they tend not to interact with other organisms such as epiphytes to gain extra nutrition. Instead, the canopy roots explore pockets of humus that accumulate in the crown. However, despite the abundance of canopy roots in humus pockets, they are still greatly outnumbered by the roots below ground. It is likely that they do not make a large contribution to the tree's overall nutrition but may nevertheless be significant in long-term growth and survival (Hertel 2011).

A number of trees are also able to produce new roots inside their own rotting heartwood, which allows nutrients released from decaying wood by fungal rot to be recycled into new growth (Chapter 11). Wood is very poor in nutrients, particularly nitrogen, so this again may be a small contribution to the overall nutrient budget of the tree.

MYCORRHIZAS

By far the oldest plant symbiosis is the mycorrhizal association between roots and fungi, which works to the benefit of both. The word *mycorrhiza* is derived from the Greek *mycos* (fungus) and *rhiza* (root). This symbiosis can also be referred to as an *endophytic* relationship, since the fungi are partly inside the plant (*endo*, meaning 'within'; *phyto*, meaning 'plant'). Mycorrhizas are remarkably common, found in around 85 per cent of the world's vascular plants. In many trees, including beeches, oaks and pines, mycorrhizas are essential for the tree to prosper, but in others, including maples and birches, it is not essential and they can grow without them if they have to. In yet others, notably members of the Proteaceae (a common family of the southern hemisphere that includes proteas, banksias, grevilleas and the Macadamia Nut *Macadamia integrifolia*), mycorrhizas are rarely, if ever, found.

Two main types of mycorrhizas are found in woody plants. Of these, *ectomycorrhizas*, often referred to as ECMs or EMs, are rare overall in the plant kingdom, found in only 3 per cent of flowering plants, but are very common in conifers and 90 per cent of broadleaf trees of cool temperate and northern regions of the northern hemisphere, including birches (Table 14, Fig. 201). The other mycorrhizal type, *arbuscular mycorrhizas* (AMs; also called VAMs, AMFs or endomycorrhizas), evolved much earlier and are the commonest in flowering plants as a whole. They are also found in a selection of different trees, including some conifers and particularly trees in tropical forests (Table 14). The primary difference between these two mycorrhizal types is whether the fungus penetrates between the cells of the root (*ecto-*) or into the cell (*endo-*), as described below. Five other types of more specialised mycorrhizal associations have been identified, which in woody plants includes ericoid mycorrhiza, found in trees and shrubs in the heath family (Ericaceae).

In ECMs, the fungal mycelium, made up of small filaments called hyphae, forms a glove-like sheath around the outside of fine roots. Inside the sheath, the fungus penetrates *between* the root cells, forming a *Hartig net*, and branches to create a large surface area for the exchange of materials. Outside the sheath the fungus ramifies through the soil to form an extensive network of hyphae

FIG 201. Fruiting bodies of Fly Agaric *Amanita muscaria*. This fungus forms an ectomycorrhizal association with birches, seen at the back of the lawn.

TABLE 14. Types of mycorrhizas found in woody plants.

Ectomycorrhiza (ECM)

Found in:

– 90 per cent of temperate trees of the northern hemisphere, including most conifers

– Some southern hemisphere trees such as southern beeches (*Nothofagus*) and eucalypts (*Eucalyptus*)

– Less commonly in various families of the tropics, including the dipterocarps (Dipterocarpaceae) and in species-poor tropical forests

At least 6,000 species of fungi are involved, usually basidiomycetes, rarely ascomycetes.

Arbuscular mycorrhiza (AM; used to be called Endomycorrhiza)

Common in species-rich tropical forests and found in a wide variety of families:

– Sapindaceae (maples)

– Betulaceae (alders)

– Salicaceae (willows and poplars)

– Fabaceae (most acacias)

– Juglandaceae (walnuts & hickories)

– Ulmaceae (elms)

– Oleaceae (olives & ashes)

– Magnoliaceae (magnolias & tulip trees)

– Hamamelidaceae (sweet gums & witch hazels)

– Cupressaceae (cypresses & junipers plus redwoods & swamp cypress)

– Araucariaceae (monkey puzzles, kauris, Wollemi Pine & araucarias)

– Taxaceae (yews)

– Ginkgoaceae (Maidenhair Tree)

The fungi are from the small phylum Glomeromycota.

Modified from Thomas (2014).

which absorbs water and nutrients, particularly phosphorus but also nitrogen and other nutrients in short supply. These nutrients are given to the root in exchange for sugar and other compounds produced by the tree. Once a root is infected by the mycorrhizal fungus, it stops growing, loses its root hairs and in effect delegates exploration of the soil to the hyphae. The fungus seems to be less capable of colonising fast-growing or larger-diameter roots, which tend to stay non-mycorrhizal and have root hairs instead. This may partly explain the rarity of ECMs in the tropics where root growth is fast and continuous, but this is also undoubtedly contributed to by the comparative rarity of basidiomycete fungi in the tropics.

Roots with arbuscular mycorrhizas (AMs) can be hard to spot because they keep their root hairs, and the fungal hyphae are fewer and do not form an external sheath. Most of the hyphae are *inside* the cells of the roots (not between, as in ECMs) where the hyphae form highly branched structures called *arbuscles* for the exchange of materials.

The distribution of different mycorrhizal types can be tied to different soils. AMs are typically found on nitrogen-rich soils where phosphorus tends to be in short supply. The benefit of the fungus lies in the fact that the smaller hyphae can make very intimate connections to soil particles, able to target small mineral grains with the highest phosphorus content and extract the nutrients in a way that roots could not. The hyphae also spread out a great distance from the roots, making the effective size of the roots much larger, allowing greater volumes of soil to be exploited. This is very effective for the tree and comparatively cheap for the fungus to do. The cost to the tree is fairly large, typically 3–30 per cent of the sugar produced by the tree in a year. In return, however, AMs can increase plant growth by an average of 80 per cent compared to growth in unfertilised soils.

By contrast, ECMs are most frequent on colder soils where there is little nitrogen available. The main role of the fungus on these soils is to help in the breakdown of litter, absorbing phosphorus and nitrogen directly and passing it to the tree. The fungus still has to produce a large number of hyphae. In pines, for example, this can reach 1,000–8,000 cm of hyphae per centimetre of root (Lipson & Näsholm 2001), and there can be as much as 100 m of hyphae within one cubic centimetre of soil, accessing soil pores that plant roots are unable to penetrate. Because of this, the production of ECMs is often a more costly relationship than using AMs and typically takes around 20 per cent of sugar production from the tree. On these impoverished soils it is still a worthwhile investment to gain scarce nutrients. This can be supplemented by a bit of carnivory. The ECM fungus *Laccaria bicolor* is known to trap springtails, which it digests and then passes their nitrogen to the tree. Eastern White Pine *Pinus strobus* has been found to gain up to 25 per cent of its nitrogen from this source (Klironomos & Hart 2001). This does not quite make the tree a carnivore, but the pine seeds may be one of the few foods available to vegetarians that are produced by consuming animals!

It is worth pointing out that within both ECM and AM trees, individuals may be infected with several species of fungi at the same time or show a succession of different species as the tree ages. Some further muddy the waters

FIG 203. Stinkhorn *Phallus impudicus* renowned for its smell of rotting meat

FIG 202. Tulip trees (in this case a variegated variety of *Liriodendron tulipifera* called 'Purgatory') are capable of forming both arbuscular mycorrhizas (AMs) and ectomycorrhizas (ECMs) at the same time. Bluebell Nurseries, Leicestershire.

which attracts flies to the grey-green spore mass. This species can sometimes be a mycorrhizal fungus but even when not it is still beneficial to trees by helping decompose wood, releasing nutrients. Beech forest *Fagus sylvatica*, Transylvania, Romania.

by forming ECM and AM relationships, sometimes both at once or sequentially. This includes many in the Salicaceae family, such as willows and poplars, but also junipers, eucalypts and tulip trees (Fig. 202). These multiple connections across trees make sense for the fungi. Many mycorrhizal fungi can only live in a symbiotic relationship so it is not surprising that they reduce the problem caused by a host tree dying by spreading their connections to many host trees.

Several nutritional aids can be used together; for example, Red Alder *Alnus rubra* native to North America has root nodules with *Frankia* actinomycetes and is also commonly ectomycorrhizal. This brings its nitrogen fixation rates up to 300 kg per hectare per year, a level as high as found in any of the pea family. Plus, some fungi such as the Stinkhorn *Phallus impudicus* (Fig. 203) can sometimes be mycorrhizal but are also involved in decomposing dead wood and thus releasing nutrients that are then available to tree roots and their mycorrhizal partners.

Mycorrhizal fungi do more than supplement a tree's nutrition. They can also protect a plant against soil-borne pathogens such as *Phytophthora* (Chapter 15). Moreover, since the mycorrhizal fungi are related to pathogenic versions or can be pathogenic themselves, the mycorrhiza is a mild form of pathogen that can immunise the tree against this disease in the process of *systemic acquired resistance*, described in more detail in Chapter 8. The trees may also benefit from the mycorrhizal fungi providing some protection from toxic levels of salt or heavy metals in the soil.

Mycorrhizal fungi also have an interaction with the rhizosphere. Amongst the microbes found around the outside of the root are *mycorrhiza helper bacteria* that attract the mycorrhizal fungi to the root. The mycorrhizal fungi in turn modify the rhizosphere by fostering bacteria that enhance plant growth and also help suppress pests and pathogens (Cameron *et al.* 2013). This creates a *mycorrhiza-induced resistance* that further reinforces the tree's immune system.

Commercial mycorrhizal inoculants

It is commonly suggested that we should use commercially available mycorrhizal inoculants when we are planting new trees (and other plants) on the grounds that the mycorrhizas are needed for the plant to be healthy. The mantra is that by adding the powder to the planting hole, you are ensuring the tree will quickly build mycorrhizal links and will flourish. It will help the tree, or at least it will not do any harm, so is worth using to be on the safe side.

However, the fungal species contained in the inoculant and where they are sourced are usually not given and they are very unlikely to be local or even from the same country. It used to be argued that fungal spores are so small that they are blown around the world and all fungi are the same, so it didn't matter. But this is not the case, and there is increasing concern that using commercial inoculants is leading to the introduction of invasive fungal species that can be every bit as devastating as invasive plants and animals. At least 200 species of ECM fungus are known to have been introduced, primarily into the southern hemisphere on planted pines and eucalypts (Vellinga *et al.* 2009). To be fair, this is not just down to the inoculants, since mycorrhizal fungi are also being moved in soil or inside roots of imported plants.

What does this mean for tree planting? Most studies show that even in the most barren urban planting hole, there are likely to be sufficient mycorrhizal fungi of the right kind. If you are worried that they might be lacking, it is far better to take a small amount of soil from beneath a nearby successful tree to act as an inoculant. This is likely to have the right fungi and creates less risk of introducing new potentially invasive fungi to the system, and is, of course, free.

WOOD WIDE WEB

It is well established that the stumps of conifers, which generally do not have the ability to produce new shoots, can be kept alive via their roots being joined with those of surrounding living trees. All of the stump's needs are passed to it through the root-grafts between root systems. These junctions between root systems of neighbouring trees are also widely used by living trees, since it has been shown using dye injections and radioactive tracers that sugars and other compounds are readily transferred between trees. But fungal connections can also be important in this exchange. In the chaparral shrublands of California (Fig. 204), seedlings of Douglas Fir *Pseudotsuga menziesii* have been found to only successfully establish in clumps of various manzanita *Arctostaphylos* species with which they share a number of ECM fungi. But the seedlings do not establish in clumps of another bush called Chamise *Adenostoma fasciculatum* which is

FIG 204. Chaparral on the coast of California. These are dominated by shrubs capable of coping with the Mediterranean climate of mild, wet winters (as seen here) and hot, dry summers. At higher altitudes away from the coast, trees such as Douglas Fir *Pseudotsuga menziesii* are able to invade, assisted by their mycorrhizal partners. Pacific Coast Highway, California, USA.

AM (Horton *et al.* 1999), suggesting that sugars and nutrients are being shared through the fungal network.

Further evidence for this was found in uneven-aged groups of Douglas Fir in central British Columbia. Big old trees, referred to as 'hub' trees, had more ECM connections than younger trees. Bingham and Simard (2012) found that one particular hub tree was directly linked to 47 other trees within a 30 m × 30 m area of forest, and was likely linked to at least another 250 trees in the wider forest. They also found that the survival of Douglas Fir seedlings was four times higher when they had access to the mycorrhizal network of older Douglas Fir trees compared to when they were planted in nylon mesh bags that did not allow the fungi through. This suggests that sharing mycorrhizal connections is indeed very important, certainly to young plants trying to get started in intensively competitive habitats. The extensive fungal connections between trees happens in our own woodlands as well. For example, in a mixed central European forest, 75 ECM fungal species were found on Beech roots, and 29 per cent of these ECM species were shared with one other tree species and 10 per cent with two other tree species.

With so many fungal links it is not surprising that nutrients and sugars are being moved around through this underground network which has aptly been named the *wood wide web*. Some of the movement appears to be controlled by the trees, as there is evidence that a tree can recognise which neighbouring trees it is genetically related to (probably through root exudates or mycorrhizas), and the flow of materials can be controlled to preferentially aid offspring and relatives.

There is also evidence that the wood wide web is involved in moving not just nutrients and water between relatives of the same species but also moving sugars around between different species, allowing trees to 'help' those that are young and shaded. Research led by Suzanne Simard of the University of British Columbia, Canada, looked at Douglas Fir seedlings growing in the shade of Paper-bark Birch and joined together by the same ECM fungi, forming an interlinked 'guild'. The researchers found that carbon (the main ingredient of sugars) could flow both ways between the two trees. Douglas Fir seedlings that were shaded gained carbon from the Birch, and in winter, when the Birch was leafless, the flow of carbon was from the Douglas Fir to the Birch. Seedlings of Western Red-cedar *Thuja plicata*, which are AM rather than ECM and so not part of the same connection, gained only very small amounts of carbon, suggesting that the carbon transfer was indeed through the fungal connections (Simard *et al.* 1997).

This suggests that the sugars are simply flowing through the fungal connections from areas of high concentration to areas of low concentration along a concentration gradient; that is, from rich *sources* to needy *sinks*. But it's not that simple. Tamir Klein and colleagues (Klein *et al.* 2016) found that when carbon dioxide with a mix of isotopes was given to a century-old Norway Spruce *Picea abies*, this carbon turned up in the roots of nearby Beech, Scots Pine and European Larch, flowing through the ECM fungal network. The astonishing thing was just how much carbon was being shared – up to 40 per cent of the carbon in any given root came from a neighbouring tree. This amounts to trees exchanging around 280 kg of carbon within a hectare of forest each year, equivalent to 4 per cent of the forest's total carbon uptake. Interestingly, the movement of carbon did not follow obvious source-sink gradients, flowing from trees with more carbon to those with less, but instead mixed between trees in both directions. This might appear as if the trees are 'helping' each other in complex ways, but it is much more likely that the mycorrhizal fungi are modifying the distribution and flow of carbon for their own needs.

To make the story even richer, trees are transferring not just nutrients and carbon but also sharing other compounds, including *infochemicals* that can serve as an early warning of attack for their neighbours. As described in Chapter 8, some of this is done above ground through the air by trees releasing volatile chemicals, particularly jasmonates, to repel further attack by insects or microbes, but this is picked up by surrounding plants of the same or even different species. They in turn increase their defences against the insects or microbes. It is increasingly understood that this exchange of infochemicals is also happening below ground. The exchange happens not just between tree roots but through the mycorrhizal 'fungal super highways' (Barto *et al.* 2012). Signals that include plant hormones, jasmonates, and even genetic material are moving between trees over a time period of hours or a few days.

This is sometimes explained as trees being intelligent, 'talking' to each other and even acting in an altruistic manner. If intelligence is defined as being able to respond to stimuli and make choices in reaction then, yes, trees are intelligent, but it is a specific plant kind of intelligence (Trewavas 2014). For example, a tree will grow upright to avoid gravitational forces that could snap it, but if light is only coming from the side, it will grow more towards that despite this putting pressure on the woody frame; the tree is weighing up opposing stimuli and reacting. However, trees do not have brains and have no consciousness, will or intent and cannot be altruistic; they are adapted through natural selection to do things which help their survival. Infochemicals such as jasmonates are a way of getting the message quickly to other parts of the same tree, which just happens to benefit other nearby plants, so it is perhaps better to think of the trees as eavesdropping on each other – perceiving chemicals which they themselves produce and react to – rather than actively 'talking to' and 'helping' each other. Neither are the mycorrhizal fungi being altruistic in helping trees. Infochemicals are indeed being transferred by the fungi but it is likely that the fungi that pass the signalling chemicals between trees will survive better by protecting their sources of food, and the genes in the fungi that allow this to happen are thus perpetuated. So, the fungi help trees to communicate but are doing so for their own purposes, and it is possible that they may preferentially send these signals to their best sugar producers, whether the receiving trees are related to the tree sending the message or not.

The view that trees and fungi are cooperating in a positive, harmonious unit is rather attractive, and there are many publications that take this stance. However, I think the reality is that the symbiotic relationship between trees and mycorrhizal fungi is an uneasy truce at best, which has evolved because it is normally in the best interests of each of the partners. For the reasoning behind this we can look at what happens when environmental conditions change.

Mycorrhizas are usually only found on trees that need them. Individuals with good sources of nutrients tend not to have mycorrhizas or the mycorrhizas are quite sparse, saving some sugar that can be used for tree growth. This makes good economic sense for the tree – only spend out on sugar when it brings a benefit. But getting rid of the fungal partner is not easy, and if the plant reduces sugar release to the fungus, the relationship can become exploitative or even parasitic, with the fungus attacking the tree. The same thing can happen if the tree is weakened by, for example, being shaded or being defoliated by insects and sugar production is reduced. This shifts the balance of power such that, rather than both partners gaining, the fungus can get the upper hand and be a net drain on the tree, progressively changing from a helpful symbiont to a potentially harmful parasite or even a deadly pathogen. This becomes more complicated when multiple fungus species are involved, and working out whether a relationship is mutualistic or parasitic becomes very difficult.

The effectiveness of mycorrhizas may also be changing due to large-scale nitrogen pollution. This has been seen to reduce the abundance of the AM fungi by 24 per cent, presumably because the tree can obtain more nitrogen for itself from the soil and so shifts sugar to its own growth rather than the AM. However, the smaller AM network will also mean less phosphorus for the plant, and the tree may selectively invest in AM fungi that are most effective in delivering phosphorus under these high-nitrogen conditions (Treseder *et al.* 2018). Trees enter into mycorrhizal relationships because it is to their benefit most of the time, rather than because they are being altruistic in playing their part in the wood wide web.

Many trees acting as one

It is worth noting at this point that trees of the same species growing in close proximity take on the shape of a single tree (Fig. 205). Being so close, they share many root grafts and thus infochemicals below ground, as well as above ground, including the main plant hormone signals that lead them to inadvertently act more as one organism. Each tree may benefit from being in the group by increased stability against wind or sharing a larger collective root system. The benefits are the outcome of growing in close proximity and the increased chance of root grafts. This is likely to be an evolved strategy to aid the survival of each individual tree, but this is not the trees altruistically 'choosing' to help one another within the group.

FIG 205. Eleven Common Lime trees *Tilia* × *europaea* growing sufficiently close together that the crowns of each merge together and appear as one.

ALLELOPATHY

Trees have the ability to interact with other plants below ground by the release of chemical compounds from the roots. This is referred to as *allelopathy* and is common in a wide range of trees, particularly those in the walnut family, Juglandaceae, which produce the compound juglone from their roots. It has also been suggested that caffeine in various coffee *Coffea* species acts in the same way (Silva *et al.* 2013). These compounds released into the soil depress the germination of seeds and the growth of susceptible plants. Mycorrhizas appear to be actively involved in moving these compounds around in the soil and helping to protect them from being broken down as soon as they are released into the soil from the root. For example, tomatoes are very sensitive to the juglone produced by the walnut. Achatz *et al.* (2014) planted tomato plants in pots beneath a Black Walnut. The pots had 30 μm wide holes – wide enough for fungal hyphae to grow through but not roots. Some of the pots were periodically rotated to break the fungal connections, and others were left with the hyphae intact. Pots with fungal connections had four times as much juglone in the soil as those without connections, showing that it was transported by the fungus. The tomato plants exposed to juglone were two-thirds the weight of those in the rotated pots. This provides the possibility of plants targeting a rival species with toxic chemicals by favouring the growth of fungi to which they can both connect.

PARASITES ON TREES

There are a number of woodland plants that hitch a ride on the system of tree roots and mycorrhizal fungi. Whether you view them as a helpful interaction depends upon whose point of view you take! But on the whole, many plants that are parasitic on trees are fairly innocuous and to be treasured, and so they are included in this chapter. Parasitic plants have no chlorophyll and so cannot photosynthesise. They gain all their nutrition and other needs from other organisms while giving nothing in return. This includes the European Toothwort *Lathraea squamaria* (Fig. 206c), which is parasitic on the roots of Hazel *Corylus avellana*, Alder *Alnus glutinosa* and occasionally Beech.

Others, such as the Bird's-nest Orchid *Neottia nidus-avis* and yellow bird's-nest *Monotropa* species (Fig. 206a and b), are classified as *myco-heterotrophs* (from the Greek *myco*, meaning 'fungus', and *heterotroph*, meaning 'feeding on others'). These are, as the name suggests, effectively parasitic on fungi in the soil. This is classed as parasitism because the fungi appear to get nothing in return. Others tend

FIG 206. (a) Bird's-nest Orchid *Neottia nidus-avis* and (b) North American Indian Pipe *Monotropa uniflora* (the British species is Yellow Bird's-nest *M. hypopitys*) are myco-heterotrophs, living off soil fungi. (c) By contrast, Toothwort *Lathraea squamaria* is directly parasitic on the roots of various trees. (a) Sissach, Switzerland, (b) Harvard Forest, Massachusetts, and (c) Needwood Forest, Staffordshire.

to blur the margins between parasitising fungi or tree roots. For example, the Coralroot Orchid *Corallorhiza trifida* (classified as a myco-heterotroph) can gain food from woody plants such as Silver Birch *Betula pendula* and Creeping Willow *Salix repens* transmitted via the mycorrhizal network (McKendrick *et al.* 2000) and would therefore seem to be at least partially parasitic on the trees.

Seeds of root parasites can be triggered to germinate by detecting compounds released into the rhizosphere. In particular, *strigolactones* are plant hormones that stimulate root branching but also encourage the development of AM fungi in the soil. They will also trigger seed germination in parasitic plants such as broomrapes *Orobanche* species, signalling to them that roots are nearby. AM fungi once developed can change the nature of root exudates, particularly reducing the level of strigolactones. So the mycorrhizal association can help defend the tree against these parasites. Again, this is not the fungus 'helping' the tree but an adaptation to protect the sugar it is getting from its host.

Trees can also be host to above-ground parasites, notably mistletoes. These are often thought to be fairly innocuous but can be quite damaging. As such they are included in the next chapter.

TREES AS PARASITES

There are a few instances where trees themselves benefit from being parasites. The Western Australian Christmas Tree *Nuytsia floribunda* (Fig. 207) is a beautiful flowering tree with vivid orange flowers, which varies from a low shrub up to a tree 12 m tall. This is a *hemiparasite* (really a woody mistletoe) taking just water and nutrients from the host and growing its own sugar by being green and photosynthetic. Since it is green, it looks like a normal woody plant above ground. Below ground, however, it taps into the roots of a diverse range of other plants, which saves it from having to grow its own extensive roots. As such it has been described as the world's largest parasite (Hopper 2010).

Root parasites like *Nuytsia* attach themselves to the roots of the host using *haustoria*, specialised roots that can penetrate its host. In the case of *Nuytsia*, the haustorium is unique, as it has a pair of sharp blades like secateurs (Calladine & Pate 2000). These are used to cut off the host root, and the haustorium plugs itself into the plumbing (xylem) of the host, stealing water, dissolved nutrients, amino acids and other compounds. It is not fussy about which host it attacks and has even been found to cut through plastic-covered wires such as telephone cables and try to attach itself! The tree tends to have a number of deep tap roots which provide water in the dry season, while the haustoria attached to shallow

FIG 207. A hemiparasitic tree, the Western Australian Christmas Tree *Nuytsia floribunda*. Photograph by Gnangarra from Wikimedia.org. Reproduced under the Creative Commons Attribution 2.5 Australia.

roots are quite short-lived, and so these connections are most likely best at providing abundant water in the wet season (Hopper 2010). The haustoria die in the dry season, which stops water being drawn from *Nuytsia* back into the hosts. Longer-lived haustoria persist, however, on deep roots of hosts to help fund dry-season survival.

The only tree that is thought to be a true parasite is also unusual in that it is not a flowering plant (angiosperm) – *Parasitaxus usta* is a rare conifer in the family Podocarpaceae, native to the Pacific island of New Caledonia (Fig. 208). It grows up to 1.8 m tall with red shoots and scale leaves, resembling a staghorn coral, hence the local name 'coral wood'. Although the leaves contain chloroplasts, they do no photosynthetic work (Qu *et al.* 2019), and all the tree's needs appear to come directly from the roots of another podocarp, *Falcatifolium taxoides*. Unlike other root parasites, it does not produce haustoria but instead has fungal hyphae at the parasite–host junction, so carbon transport from the host to *Parasitaxus* is probably done using the fungus as an intermediary. This

FIG 208. A parasitic conifer, *Parasitaxus usta*, native to New Caledonia. From iNaturalist (https://www.inaturalist.org/photos/55058052). Reproduced under the Creative Commons Attribution 4.0 International license.

puts it somewhere between being a true root parasite and a myco-heterotrophic plant, parasitic on the mycorrhizal fungi rather than directly on its host plant (Field & Brodribb 2005).

CARNIVOROUS TREES?

Some trees are meat-eaters. The South African shrubs *Roridula dentata* and *R. gorgonias* are described as being *protocarnivorous*. They have leaves covered in sticky glandular hairs which trap insects but they lack the digestive enzymes or specialised absorptive glands that carnivorous plants use to take up nutrients from the prey. Instead they use *indirect digestion*. Two species of Hemipteran bug (*Pameridea roridulae* and *P. marthothii*) live on the leaves and eat the caught insects;

FIG 209. A Lesser Noddy *Anous tenuirostris* caught in the sticky seeds of *Pisonia grandis* and unable to fly. They are often caught in the seeds because they favour making their nests in the *Pisonia* out of the leaves stuck together with droppings. Fairy terns often lay their eggs directly onto bare branches and so suffer less from the sticky seeds. The Seychelles. Photograph by Matthew Tosdevin.

they then defecate on the leaves and the nitrogen is absorbed directly through the leaf cuticle (Anderson & Midgley 2002). This indirect carnivory is similar to the mycorrhizal fungus *Laccaria bicolor* mentioned above, which catches springtails and passes the nitrogen to its host tree, Eastern White Pine.

In other cases, what appears to be carnivory probably is not. The aptly named Bird-catcher Tree *Pisonia grandis* is common in seabird colonies on small tropical islands. Its fruits are surrounded by a persistent calyx that exudes a very sticky resin which sticks strongly to feathers. This is an obvious method for long-distance transport of the seeds between remote islands, carried by birds. But some birds become so laden with sticky seeds that they cannot fly, and so they die (Fig. 209). The dead bodies drop to the ground with the attached seeds, but this could still be beneficial, since the seedlings would gain nutrients from the rotting carcass – another example of indirect carnivory. But it seems that seedlings developing from these carcasses actually do less well than others dropped further away, primarily due to disturbance caused by crabs and other scavengers feeding on the dead bird bodies. A study in the Seychelles found that 64 per cent of birds found were unable to fly, and it was presumed they would die (Burger 2005).

The death of such large numbers of birds appears to be collateral damage for a very good mechanism of seed dispersal. This explains why some groups such as Nature Seychelles have begun replacing *P. grandis* with other native trees.

BENEFICIAL ANIMALS

As seen in Chapter 4, pollination – moving pollen from one flower to another – is often carried out by animals such as insects and even birds and mammals. This, of course, is another mutualism where both partners win. There are costs for the tree in terms of lost pollen and the production of nectar, but also huge gains in having pollen targeted at other flowers and often specifically flowers of the same species. The pollinating animal in turn gains a reliable food supply for itself and its young. There are also risks within this mutualism. If a tree species and a pollinator rely solely on each other, then they are vulnerable if one of the partners declines or disappears. This risk is reduced by mutualisms usually being more diverse. Most trees that use animal pollination can be pollinated by a range of different species, and these in turn visit a range of different plants.

Problems can still arise with the wholesale loss of groups of insects. Southern Argentina has seen a large decline in native bumblebees. A study there found that apple orchards where native bumblebees were re-introduced, produced twice the number of apples even when honeybees were hundreds of times more abundant (Pérez-Méndez *et al.* 2020). Similarly, in parts of China, overuse of insecticides has resulted in a serious decline and even complete loss of pollinating bees. Many studies have shown that artificial pollination with a paintbrush or feather can produce just as many apples as insect pollination. But in these cases, apple production is only maintained by a huge investment of labour in hand-pollinating flowers, which in China involves every family member on a farm (Partap *et al.* 2001). Other parts of the world may not yet be at this stage, but in Britain, 23 species of bees and flower-visiting wasps became extinct between 1850 and 1998 (Ollerton *et al.* 2014), and similar stories of decline are seen in other western countries. We owe a huge debt to pollinating insects, and their long-term survival is in our best interests.

Defence
A wide range of trees around the world, particularly tropical trees, are *myrmecophytes*, using ants as their main form of defence against other organisms. The ants patrol around the tree, repelling birds and other animals by stinging,

and they also prune away epiphytes and lianas using their sharp jaws. In some cases they also cut back anything growing within a 10 m radius of the tree trunk. In this mutualism, the ants in turn are given board and lodgings. This includes trees in the genus *Macaranga* (family Euphorbiaceae) in Southeast Asia where the ants live inside hollow stems, and acacias (Fabaceae), primarily in Africa, which have swollen thorns used for nesting (called *domatia*, from the Latin *domus*, meaning 'home'). Food is given as sugar-rich nectar from *extrafloral nectaries* (nectaries outside of flowers) found at the base of leaf stalks, and as protein, fats and carbohydrates supplied in small knobs (called *Beltian bodies* – named after the nineteenth-century naturalist Thomas Belt) found at the ends of the leaves in some acacias.

These ants can even be a defence against the biggest animals. The Whistling-thorn Acacia *Vachellia drepanolobium* (Fig. 210) is a common tree in the savannah of East Africa, often forming monocultures in the uplands. Various large herbivores such as giraffes and elephants roam these areas but seldom feed on or cause

FIG 210. The Whistling-thorn Acacia *Vachellia drepanolobium* in East Africa. In amongst ordinary sharp thorns are those with bulbous swellings 2–3 cm across and into which *Crematogaster nigriceps* ants burrow and make their nests. Photograph by Pharaoh Han, reproduced under Creative Commons Attribution-Share Alike 3.0 Generic, courtesy of Wikimedia Commons.

damage to the Acacia. This is because of the ant bodyguards, which aggressively bite the sensitive interior of elephant trunks or any part of other animals that try to graze on the foliage.

But like any mutualism, the balance between the partners is delicate, and there is distinct competition between ant species. For example, in Kenya there are four species of ants competing for exclusive use of individual Whistling-thorn Acacias. Of these, *Crematogaster mimosae* is found in around half of the trees. Another ant, *C. nigriceps*, will cut away tree buds on side branches, apparently to reduce sideways growth and the chance of touching other trees, which would allow the more aggressive *C. mimosae* to invade. Yet another ant, *Tetraponera penzigi*, does not use the nectar produced by the trees but instead destroys the nectar glands to make a tree less appealing to other ant species.

A study in the same part of Kenya showed that there is a subtle relationship between the trees and the ants. By excluding large herbivores from some trees, it was found that over the next 10 years the Acacias reduced the amount of nectar and number of swollen spines produced – the Acacia no longer needed such strong protection and reduced the resources for ants, to the detriment of the resident ants. In response, the usually dominant *C. mimosae* became less common, found on only 30 per cent of trees, and their behaviour changed. To supplement their diet they increasingly farmed sap-sucking scale insects for their honeydew, to the detriment of the tree. In contrast, the number of trees holding *C. sjostedti* increased, as these ants' main food is other invertebrates rather than the resources provided by the tree. Moreover, they nest in cavities produced by beetles rather than hollow thorns, so on their trees the occurrence of boring beetles was much higher. As a result, the mutualistic relationship between Whistling-thorn Acacias and their ants breaks down in the absence of large herbivores, and trees become less healthy as a result (Palmer *et al.* 2008). This shows that such relationships often oscillate between mutually beneficial partnerships and opportunistic parasitism.

Some of these features are found in other trees even in temperate areas, such as poplars, cherries, hawthorns, elders, viburnums, roses and even oaks. The Black Cherry *Prunus serotina* of North America is an excellent example. It produces extrafloral nectaries at the base of its leaves during the first three weeks after the leaves emerge. These attract large numbers of the ant *Formica obscuripes* from nearby colonies that feed on the nectar. They also eat invertebrates and so offer protection to the newly produced, vulnerable leaves, including protection from the Eastern Tent Caterpillar *Malacosoma americanum*, a major defoliator of Black Cherry. Once the leaves are mature and less vulnerable, the nectar dries up as the tree no longer needs the services of the ants (Tilman 1978).

Other temperate plants provide their defenders with a more complete service. Many trees have small tufts of hairs or pockets of tissue on the leaves at the base of major veins. These domatia offer homes to small invertebrates which help protect the leaves. The European evergreen shrub Laurustinus *Viburnum tinus* has been found to house up to 10 species of mites that are predators or microbivores (eating microbes), offering protection for the leaves in return for lodgings.

DEAD BEES UNDER LIME TREES

Bees are renowned as good pollinators of lime trees, which in turn supply the bees with copious nectar (Fig. 211). However, dead bees, particularly bumblebees *Bombus* species, are often found under lime trees, particularly Silver Lime *Tilia tomentosa* and Caucasian Lime *T.* × *euchlora*. In fact, these limes are often considered poisonous to bees. Why would a tree harm its pollinators in this way?

FIG 211. (a) Flowers of the Large-leaved Lime *Tilia platyphyllos* which are very attractive to bees. (b) Add in the huge numbers of flowers on a tree, and they become very important sources of food for bees. Poznań, Poland.

A study in Germany by Illies and Mühlen (2007) found that Silver Lime produces abundant nectar (850 µl from 540 flowers over 19 days) that is fairly rich in sugar (30 per cent sugar, compared to 20–50 per cent in most temperate flowers) and produces nectar for a long time (around 20 days compared to half that in the Large-leaved Lime *T. platyphyllos* – Fig. 211). So the tree is beloved by bees but is seemingly fatal. In the same study, Silver Limes were seen to be visited by equal numbers of bumblebees and Honeybees *Apis mellifera*, but over a thousand dead bumblebees were collected under a single Silver Lime, six times more than the number of dead Honeybees, and indeed often it was only bumblebees that were found to be dead. In the 1970s it was suggested that this was because of high amounts of the sugar mannose in the nectar, a sugar that bees are poor at digesting. However, more recent studies have not found any mannose in the nectar. It is possible that other components of the nectar, pollen or scent of Silver Limes are toxic, although there is little evidence. It is also possible that the occurrence of caffeine in the nectar could deceive the bees into continuing to forage even when there is little nectar (Koch & Stevenson 2017). Further study shows that the number of dead bees increases towards the end of the flowering period and is also highest in dry years. The cause of death thus appears to be that bumblebees become conditioned to visiting abundant nectar sources and their visitation rate is not correlated to the amount of nectar on offer as it is with Honeybees – less nectar, fewer visits; the bumblebees just keep turning up. The bumblebees continue to visit at the end of the season, or earlier in dry years, when little nectar is left, and they effectively die from starvation rather than poisoning (Kirk & Howes 2012).

EPIPHYTES

A number of organisms that live on trees are in reality fairly benign. For example, mosses, lichens and green algae (including the ubiquitous green *Pleurococcus* and orange *Trentepohlia* species) commonly hitch a ride on the outside of the bark of trees, particularly in unpolluted areas with high humidity (Fig. 212).

OPPOSITE: **FIG 212.** Epiphytes on trunks. (a) A multitude of lichens on a Beech *Fagus sylvatica* in Kent, (b) mosses and ferns on a Pedunculate Oak *Quercus robur* in Wistman's Wood on Dartmoor, and (c) higher plants in the form of bromeliads together with lichens in the cloud forest of Honduras. (d) A cautionary tale: I pontificated about what the unusual brown epiphyte was on this Beech as I marched towards it with a group of students, until they pointed out that the nearby telegraph poles and fence had the same epiphytes – it was the result of haphazard muck-spreading by the farmer!

Although they can be quite luxuriant, they are doing little damage, but see Figure 217 in Chapter 15. It may look as if they are blocking up the lenticels – the corky breathing holes that allow oxygen and other gases to pass in and out of the tree – but this is very rarely the case. At worst they may be considered aesthetically unpleasing on trees with ornamental bark, and some gardeners will scrub their prize specimens with soapy water or even pressure-wash the bark. This is much harder with larger trees but can still be done on the main trunk and branches which are within the gaze of most garden visitors.

The epiphyte that often causes concern in Europe is the common woodland Ivy *Hedera helix* (Fig. 213). It is rooted in the ground, grows its own sugar by photosynthesis and so does little except use the tree for support. In effect it saves the cost of growing its own self-supporting trunk by leaning on a tree. Since

FIG 213. Luxuriant growth of Ivy *Hedera helix* on a Pedunculate Oak *Quercus robur*. A problem for the tree, of no consequence, or beneficial wildlife habitat that should be encouraged?

it is not parasitic, it does little damage to a healthy tree. In many ways, Ivy is beneficial, since its flowers provide nectar late in the year, its fruits are important food for birds in late winter/early spring, and its dense growth provides shelter for many birds and insects. As such it should be left on healthy trees unless it is causing an aesthetic problem.

However, Ivy can become a problem on old, weak trees where it adds significant weight to the crown and where the sail area created by the evergreen ivy can make the tree more prone to windthrow in the winter when winds are strongest and surrounding deciduous trees are otherwise leafless (Fig. 213). Vigorous flowering shoots of Ivy that grow upwards on free-standing shoots can also smother the sparse leaves on slow-growing old trees, outcompeting them for light, particularly on shoots close to the main trunk. In these cases, the removal of Ivy is justifiable. Dense Ivy can also be a problem by making it harder to see problems such as cavities or other structural defects in the trunk and branches.

Other epiphytes can be more problematic. Lianas, or woody vines, occur throughout the world but are most abundant in tropical forests (Fig. 214) where

FIG 214. Lianas joining together the various layers of rainforest trees and shrubs into a solid unit. Royal Belum National Park, Malaysia.

they can have a serious detrimental effect due to competition for light. The lianas hold their leaves above those of their hosts, reducing growth and seed production. Moreover, lianas can quickly invade tree-fall gaps and prevent the regeneration of trees (Putz 1984). Like European Ivy, they avoid growing their own trunk by draping themselves over trees, such that in tropical forests lianas may represent less than 5 per cent of above-ground biomass but may have more than 40 per cent of the leaf area (Kainer *et al.* 2014). The competition they present to the trees is such that cutting lianas in tropical forests used for timber production leads to greater growth of timber. In Belize it has been found that nine years after cutting lianas, the 'liberated' timber trees (in this case *Swietenia macrophylla*) grew 38–63 per cent faster than infested trees (Mills *et al.* 2019). Removing lianas can also increase non-timber products. For example, a 10-year experiment cutting 454 lianas from 78 Brazil Nut trees *Bertholletia excelsa* in Amazonia led to treated trees producing 77 per cent more nuts than control trees 10 years after liana removal (Kainer *et al.* 2014). Removing lianas also reduces 'canopy connectivity', where tree crowns are joined by thick woody lianas. This reduction lowers incidental tree damage when a tree is felled or falls down; otherwise the connected trees come down like dominoes.

CHAPTER 15

Pests and Pathogens

AERIAL PARASITIC PLANTS ON TREES

Woodland plants that are parasitic on tree roots and which tend to be fairly benign are covered in the previous chapter. But trees can also be host to above-ground parasites, notably mistletoes. The European Mistletoe *Viscum album* and the broadleaf mistletoes of the Americas, *Phoradendron* species, are green and so produce their own food by photosynthesis, ostensibly taking just water and nutrients from their hosts. The seeds of these mistletoes germinate under the bark and develop root-like structures, *haustoria*, that grow into the sapwood of the host, effectively plumbing themselves into the water-conducting tissue (xylem). Since they grow their own sugar, these can be classified as partial parasites (*hemiparasites*) as they just take water.

But European Mistletoe is not as benign as once thought. It is known that heavy infestations of Mistletoe on apple trees and poplars leads to a shortened lifespan, but it is now clear that it will also reduce the growth and seed production of the host. This is exemplified by an increasing problem in Scots Pine in mainland Europe due to global warming. A study in the mountains of northeast Germany by Chris Kollas and colleagues (2017) showed that Mistletoe infection on Scots Pine increased from 1–11 per cent between 2009 and 2015, and the Mistletoe had moved onto new trees 200 m uphill. This is creating a number of problems. It is now clear that Mistletoe steals not just water but can also steal about 30 per cent of its sugar from the host, and so is really acting as a true parasite. At the same time, Mistletoe is a profligate user of water, showing hardly any regulation of water use. Under severe drought stress when Scots Pine has tightly closed its stomata, Mistletoe is still losing water at near its maximum rate.

The loss of sugar to Mistletoe and the extra water stress inevitably affects the host tree, reducing needle size, reducing seed production, slowing growth, and even hastening death.

Other mistletoe species are genuinely complete parasites, taking not just water and nutrients but everything else they need, including sugar. These include the small, often yellow, dwarf mistletoes, such as *Arceuthobium* species, in North and Central America, Asia and Africa. In western North America and Asia, these are particularly damaging to conifers where they cause distorted and twisted growth of branches, and heavy infestations can lead to reduced growth and even death.

In Europe, similar congested growths referred to as witches' brooms are created of densely branched small twigs clustered in the crown, which resemble a besom. These can be induced by a range of different organisms: sometimes by fungi, such as *Taphrina betulina* in birches, or by mites. Unless these are very numerous, they are unlikely to be causing the host tree any great loss in vitality and are harmless.

Stranglers

Strangler figs are usually epiphytic to start with, as the seeds germinate high in the crown of a host tree after being dropped there by birds. The young seedling has the advantage that it is high in its host canopy with plenty of light and can live on scavenged nutrients and water just like any other epiphyte. As it grows, it sends roots down to the ground, which give a more certain supply of water and nutrients. Using this, the strangler's shoots are able to grow out above the crown of the host, outcompeting it for light. The roots descending on the outside of the host's trunk will branch and fuse together into a three-dimensional cage that constricts the growth of the trunk, literally strangling the host, leading to its death. But stranglers are not always figs – see Figure 215. It is also important to note that species of figs that are normally strangling can grow as self-supporting independent trees if the seeds germinate on the ground. In this case they grow their own trunk and use their many aerial roots to bolster their stability and explore further for water and nutrients (Fig. 216).

MICROBES: SAPROPHYTES, PARASITES AND PATHOGENS

Like animals, plants carry a large load of bacteria, yeasts and fungi which form communities on the surface and inside the plant, both between and inside cells. Some are decidedly beneficial, such as the mycorrhizas described in the previous

FIG 215. Stranglers. (a) A strangler, Northern Rātā *Metrosideros robusta*, in New Zealand. The seeds germinate in the high light of the forest canopy and are initially epiphytic but soon send roots down to the ground. The roots branch and fuse to encircle the tree, as is beginning to happen in the photograph. These will eventually kill the host either by stopping outward growth of the trunk or out-competing it for light. (b) Stranglers (in this case a fig *Ficus* species) can also continue to produce new roots from the crown to create a 3D network, helping in support and in the competition for water and nutrients. (a) Near Greymouth, South Island, New Zealand, and (b) Singapore Botanic Gardens.

chapter. Some are epiphytes, hitching a ride on the outside of the plant and gaining their living by what they can scavenge off the plant (Fig. 217). Others are *saprophytes*, eating dead plant material. This includes the microbes on the outside of leaves (the *phylloplane community*) that feed on exudates from inside the leaves or on animal products such as faeces left on the leaf surface. They may also be feeding on areas of dead leaf tissue, but since these are dead already, they cause little new damage.

This saprophytic lifestyle is also common on a larger scale in the decomposition of dead wood. Fungi in wood decompose cellulose, hemicellulose and lignin in various proportions, depending on the fungus type. Many fungi that we see on a walk through the woods are not pathogenic but are involved in the important job

FIG 216. Strangler figs can germinate on the ground and grow into independent trees without a host to hold them up. (a) Preah Khan and (b) Ta Prohm temples in the Angkor area of Cambodia. The doorway in (b) was immortalised in the 2001 film *Lara Croft: Tomb Raider*.

of breaking down dead wood and releasing nutrients. Moreover, they can be useful to us humans in many different ways. For example, the Beefsteak Fungus *Fistulina hepatica* (Fig. 218a) produces brown rot in oak, which is highly prized for its colour in fine woodworking. It is also one of the few brown rot fungi that causes *spalting* in broadleafed trees (Fig. 218b). In spalted wood, different fungi occupy a volume of wood and limit competition with their neighbours by producing a *zone line*, a black line (actually two – one by each fungus) made up of hard, dark mycelium. The different colour of each zone can be due to pigments produced by the fungi and the degree of rot. Spalted wood is keenly sought by woodturners.

FIG 217. A leaf still alive and attached to the tree with epiphytic growth of algae and lichens over the surface. This creates a rich phylloplane community of bacteria, algae and other microbes on the surface that may do no damage or may eat away at dead spots in the leaf. This community can also include fungi that begin decomposition of the leaf while still on the tree. Tropical cloud forest of Honduras.

FIG 218. (a) Beefsteak Fungus *Fistulina hepatica* on a Pedunculate Oak *Quercus robur*. (b) Spalted beech wood where each zone is colonised by a different species of fungus, separated by black zone lines. (a) Wistman's Wood, Dartmoor, and (b) Transylvania, Romania.

Saprophytic fungal rots become most harmful when they threaten the structural strength of a tree's woody skeleton, as described in Chapter 11. Although the wood they are attacking is wholly dead, if enough is rotted away, the tree will not be able to withstand wind and other stresses and will collapse. To reduce the amount of internal rot, there have been attempts to keep wounds 'dry' by filling cavities with a variety of materials, including cement (Fig. 219). However, the filling is stiff and since it does not move with the tree is liable to crack and fall out, and may even cause more damage to the tree by resisting its natural bending in the wind. A similar method is to fit a thin metal sheet or mesh into a wound and cover this with plaster. This is more flexible but will still eventually crack and fall out. Moreover, with any of these fills, once they crack, water seeps in and they are more likely to trap water inside and keep the wood

FIG 219. (a) An impressive attempt to fill a large cavity in a Japanese Elm *Ulmus japonica* using concrete painted on the outside to look less obtrusive. Sadly, as the tree has flexed in the wind, the concrete has cracked and part has fallen out. (b) An open wound on a Maidenhair Tree *Ginkgo biloba* left by some interesting chainsaw work has been covered with arboricultural paint in an attempt to stop rot developing. In both cases it is now recognised that leaving the wounds open and allowing them to dry is the best policy. Hokkaido, Japan.

wetter. Current advice is to leave wounds open to allow them to dry naturally. The same is true of the use of arboricultural paint on sound wood left by tree surgery (Fig. 219). The paints will do more harm than good by trapping water and fungal spores underneath, which can hasten rot rather than prevent it.

Other microbes invade the living tissue of the tree and so act as *parasites*, which is readily seen in many trees where the leaves become mottled with dead areas as the growing season progresses. Some of these are more of an aesthetic issue rather than a danger to the tree. For example, the tar spot fungus *Rhytisma acerinum*, which creates black spots on leaves of Sycamore *Acer pseudoplatanus*, does little to harm the long-term growth of the tree, despite leaves being heavily infested in years with humid weather. A similar effect is caused on the various horse-chestnuts *Aesculus* species in Europe, North America and Southeast Asia by *Guignardia* leaf blotch, caused by the fungus *Guignardia aesculi*. This produces reddish necrotic areas with bright yellow borders usually on the tips or edges of the leaves, but seems to cause little significant damage. The *Guignardia* leaf blotch and a number of different fungi are not necessarily lethal to a tree but may do enough damage that they make the tree unsightly, which can lead to it being removed for aesthetic reasons (Thomas *et al.* 2019). The fungus *Sirococcus tsugae* introduced into the UK in 2013 is causing just this problem on Atlas Cedar *Cedrus atlantica* (Fig. 220).

BACTERIA AS PATHOGENS

Some bacteria are *pathogenic* and the cause of serious diseases leading to death. This includes chestnut bleeding canker (Fig. 221) caused by the bacterium *Pseudomonas syringae* pv. *aesculi* (Pae) on horse-chestnuts ('pv' is short for *pathovar* or *pathogenic variety*). The pathogen probably originated from the Himalayas where it infects leaves of the Indian Horse-chestnut *Aesculus indica*. Pae in Britain is almost genetically identical to that in mainland Europe and so it is all likely to have come from a single, recent introduction into western Europe, possibly from India. The bacterium spreads mainly in windblown rain. By 2007, over 70 per cent of horse-chestnut trees surveyed in England, 42 per cent in Scotland and 36 per cent in Wales had symptoms of bleeding canker (Thomas *et al.* 2019).

The disease is recognisable by a yellowing of the leaves, leading to crown dieback (but this can be hidden by the twin blights of *Guignardia* leaf blotch and the leaf mines of the horse-chestnut leaf miner – see below). The most distinctive feature of Pae is patches of bleeding cankers (hence the name) on the bark, where necrotic lesions exude a rust-coloured or black sticky substance, not unlike those

FIG 220. Atlas Cedar *Cedrus atlantica* infected with the fungus *Sirococcus tsugae* which causes the death of leaves and shoots. (a) The cedar in the foreground is losing many of its needles but is likely to survive. In the background is a Deodar C. *deodar*, currently unaffected in Britain although they are susceptible in North America. (b) The affected needles turn a characteristic pink before going brown and producing the fungal fruiting bodies. The fungus also attacks hemlocks *Tsuga* species, hence the species name of the fungus, and infected hemlocks have been found in southwest Britain.

FIG 221. Chestnut bleeding canker on Horse-chestnut *Aesculus hippocastanum* caused by the bacterium *Pseudomonas syringae* pv. *aesculi* (Pae).

caused by *Phytophthora* species, and hence why the dieback was originally thought to be yet another problem caused by *Phytophthora*. These lesions penetrate the cambium and the phloem, and so the biggest threat to the tree is when the cankers become so numerous that they merge around the trunk, thus girdling the tree and leading to death. The size of Pae cankers can be larger in trees with leaf miner infestations, probably due to the suppression of tree defences as a result of leaf miner defoliation (Thomas *et al.* 2019).

Fortunately, a high proportion of horse-chestnuts show some resistance to the canker, and trees are known to have survived for a decade or more with half of their bark area affected. A study in southern England found that Pae was responsible for the death of 11 per cent of horse-chestnuts and 27 per cent of Red Horse-chestnuts *Aesculus carnea*, and surviving trees showed a decrease in growth of 22 per cent between 2003 and 2012 and a decline in crown density of 4–5 per cent (Straw & Williams 2013). Death of major branches can be a problem from a safety point of view even if the tree survives. But infected trees are weakened, and secondary agents such as *Phytophthora* may gain access to a diseased tree and be the ultimate cause of death. All is not lost: Pae infections have been successfully treated with potassium and silicon phosphites that lessen the impact or even prevent infection.

Potentially far more damaging is the bacterium *Xylella fastidiosa*. There are four reported subspecies and a number of varieties, known mostly from the Americas. In Europe, subspecies *pauca* was first confirmed in southern Italy in 2013 in olive trees, leading to the destruction of tens of thousands of trees in an attempt to stop its spread. By 2017 *Xylella* was found in Spain on cherry and almond trees and by 2019 in Portugal on lavender *Lavandula*. This was subspecies *multiplex*, which is able to survive in a cooler climate and the one most likely to thrive in Britain. More than 360 host species across 75 different plant families have been identified so far, including grapevines, olives, citrus fruits, coffee, *Prunus* species including cherries and peaches, and other broadleaf trees such as oaks, elms and planes. The potential effect on our diet, wildlife and aesthetics of urban areas is truly staggering, and much energy is being spent trying to stop its establishment and spread. This includes restriction on imports of affected plants and increased vigilance to find and eradicate any infections found. This makes it even more imperative not to bring plants or bits of plants home from holidays.

Two biological aspects will act to hamper the efforts of keeping *Xylella* from establishing in Britain. Firstly, the bacterium lives unusually in the xylem of its hosts and is therefore readily spread by insects that feed on the xylem sap. This includes 13 species of xylem-feeding bugs in Britain that are known to feed on hosts susceptible to the pathogen, including the very common Meadow Spittlebug or Froghopper *Philaenus spumarius* that produces the familiar cuckoo spit. Although these Froghoppers rarely fly more than 100 m, they can be blown much further by the wind, carrying the bacterium inside their bodies, and so spread is easy and rapid. Secondly, there may be no initial symptoms in infected plants, and when they do appear they are quite unspecific, with the main symptom being a scorching of the leaves from their edges, which is very similar to that of a number of other pathogens and also similar to scorching produced by the sun, drought, salt spray, other air pollutants and nutritional deficiencies. This may be followed by wilting of the leaves and leaf loss, crown dieback and death. So detecting *Xylella* is not easy and may be as difficult to stop in Britain, should it arrive, as it currently is in mainland Europe.

FUNGI

Most pathogens are, however, fungi that are able to cause a diverse range of diseases in trees. Sometimes, though, it is difficult to be sure whether a fungus is parasitic on living tissue and unlikely to kill the tree, or a more virulent pathogen that will lead to death. Whether it is a parasite or pathogen will often depend on a balance between the fungus and the tree, which can shift depending on conditions. For example, a fungus can parasitise a tree without causing great harm, but if the tree is weakened or under stress by factors such as flooding or drought, the fungus can overwhelm the defences of the tree and become a pathogen that may kill it. The fungus may then live saprophytically off the dead wood while looking for another host tree to infect.

A classic example is the common honey fungus disease caused by 40 or so species of *Armillaria*. Many species of honey fungus are just saprophytic, but some, including *Armillaria mellea*, are also harmful pathogens. It has been suggested that no tree in the world grows to its full potential due to the ubiquitous effects of honey fungus, although this is undoubtedly somewhat overstated. In fact *A. mellea* has been found to be less virulent in woodlands where other competing fungi reduce its effectiveness as a pathogen (Fig. 222). Moreover, the exact relationship between honey fungus and host can vary; if the host is vigorous and growing well, then honey fungus can change from a lethal pathogen to a less-damaging parasite. So the presence of honey fungus in a garden is not necessarily a sign of impending doom.

FIG 222. Honey Fungus *Armillaria mellea* growing on the base of a Pedunculate Oak *Quercus robur*.

Dutch elm disease

Pathogenic fungi can be truly formidable, exemplified by Dutch elm disease (DED), which has killed millions of elms across Europe, North America and western Asia and is still active, continuing to alter landscapes across the world.

The disease is caused by a fungus which was first identified in 1921 by Dutch plant pathologists, hence the name. In fact, it probably originated in the Himalayas and was brought to Britain in 1927 on imported diseased logs. The first European epidemic from the 1920s to the 1940s was caused by the fungus *Ophiostoma ulmi* which killed around a third of elms. But sometime after this, a more aggressive strain, *O. novo-ulmi*, arose, which was imported to Europe on Canadian elm logs. This new strain ravaged elms across Europe in a second epidemic in the 1970s, killing some 28 million elms in Britain alone and largely removing the elm as a mature tree in the landscape. Death was equally devastating elsewhere: in 1970 the 30,000 elms in Paris, half of them in the Bois de Vincennes, were reduced to less than 1,000 surviving trees, while another 40 million trees were killed in North America.

I can remember the heart-rending loss of large numbers of stately hedgerow elms in the 1970s in southeast England. Most affected were the 'suckering' English Elm *Ulmus procera* and Field Elm *U. minor*; more than 90 per cent of mature Field Elms were killed. Wych Elm *U. glabra* was (and is) less affected and is now the most numerous elm in central and eastern Europe. Part of the resistance of Wych Elm comes from the genetic diversity in populations, since they are mostly of seed origin, whereas the suckering elms are genetically similar clones. Once the disease has cracked the defences of one suckering elm, it can spread rapidly through the whole population, including through root grafts that develop between adjacent elms.

The *Ophiostoma* fungus is moved between groups of trees by elm bark beetles, particularly the Large Elm Bark Beetle *Scolytus scolytus* and to a lesser extent by the Small Elm Bark Beetle *S. multistriatus* (and the Native Elm Bark Beetle *Hylurgopinus rufipes* in North America). As the female beetles burrow into the bark, they eat the sugary inner bark (phloem) and score the sapwood (Fig. 223), allowing the fungal spores carried by the beetles to infect the wood. The fungal hyphae block the water-conducting tubes of the outer ring of wood, and since elm is ring-porous (Chapter 6), conducting all of its water through this ring, this results in the death of the crown from lack of water.

Wych Elm and the European White Elm *U. laevis* are less attractive to these beetles than the Field Elm and Siberian Elm *U. pumila*, and so although they are more susceptible to the fungus, this is another reason why they have better survival. This is also helped by Wych Elm being more of a woodland tree, often

FIG 223. Wych Elm *Ulmus glabra* wood scored by the galleries of the Large Elm Bark Beetle *Scolytus scolytus*. The female beetle eats through the nutritious phloem and outer sapwood, laying eggs as she goes, creating the central channel. The newly hatched larvae move out sideways, eating as they go, creating this characteristic fan. At the end of their enlarging channels the larvae pupate, and the new adults burrow out through the bark.

hidden amongst other trees rather than growing exposed in hedgerows filled with elms, and so better escaping the attention of the beetles. Moreover, Wych Elms are favoured by another fungus, *Diaporthe eres* (formerly called *Phomopsis oblonga*), which may compete with the beetle for habitat within the wood (Thomas *et al.* 2018).

Dutch elm disease is still prevalent in Europe, causing successive waves of death at 15–25 year intervals. If the tree is not killed outright by DED, new stems regrow from the base of Wych Elm and as root suckers from other elms until they reach 5–9 m tall, with stems around 10 cm in diameter with bark 5–8 mm thick, at which point they become susceptible to the Elm Bark Beetles, starting another cycle of infection. Wych Elm is the only British elm that reproduces by seed and fortunately by the time the trees are big enough to be re-infected, they have already been producing viable seed and so can contribute to the long-term survival of Wych Elm populations by spreading to new areas. Genetic variation is also maintained, which may increase the resistance of populations to the disease.

Our other common elms have always reproduced by suckers, which can gradually spread along hedgerows and open areas.

In Scotland and northern Europe, Wych Elm has been growing beyond the reach of DED, primarily because it is too cold for the Elm Bark Beetles to be effective. In 2011, the northern limit of the beetles was 63° north, so around a third of Wych Elms along the Norwegian coast were out of reach of DED (Solheim *et al.* 2011). However, global warming is causing a race to the north between the elms and the beetles. It is reasonable to expect the range of Wych Elm to expand northwards as the climate warms. That same warming will, however, also allow the Elm Bark Beetles to pursue the elms up through the latitudes, and thus the disease may well keep pace with the range expansion of the elms or, more likely, outpace it. Dutch elm disease was first found in Scotland in 1938, in the Borders area, and in the west part of the Central Lowlands in 1947. It is now widespread around Loch Ness and moving further north and west each year. When the elms and the beetles eventually get to the northern shores of Scotland, the elms will have no place to move further. At that point, abundant large elms will cease to be commonplace in the British landscape.

However, we do still have some magnificent large elms. Isolated mature elms have survived in the landscape, presumably by some inbuilt resistance to the beetles or the fungus. Other mature elms (such as those in and around Brighton) have been kept alive by management, including the use of sanitation zones in which all infected trees are felled, and by the use of insecticides and injected fungicides. Vaccines have been developed that help the trees to produce their own antifungal compounds, but these need to be injected every year and so are only likely to be used in the protection of high-value trees.

A long-term approach commonly used in combating tree diseases is to wait until the susceptible individuals have been killed and to use the remaining trees, which are assumed to be resistant to the pathogen, in breeding programmes. This is happening with elms, producing resistant strains of Field and Wych Elms. An example of this is the 'Great British Elm Experiment' run through the Conservation Foundation (2017), which produces new trees from apparently resistant elms and distributes them to schools, community groups, local authorities and private landowners for planting. The main problem with this approach is that when 'resistant' trees are inoculated with the disease, it is seen that very few trees are fully DED resistant (Coleman, 2009), and so the long-term survival of these trees is by no means certain.

Another approach is to hybridise European elms with disease-resistant Asiatic elms in the hope of getting elms that look like our elms but with the disease resistance of their Asiatic cousins. This was done with gusto after the first epidemic in the 1940s, producing resistant hybrid trees such as *Ulmus* 'Commelin', 'Groeneveld', 'Plantyn' and 'Doedens', which were planted in their millions. Sadly, they mostly had Wych Elm as one of their parents (since it produces seed and is thus hybridisable), and they were partially or completely susceptible to the new *O. novo-ulmi*. Although they can still be found, they only really prosper where they have some protection from the disease, such as isolation from other elms.

More recent hybrids are showing promise. For example, *Ulmus* 'New Horizon' is a hybrid between an American cultivar of the Japanese Elm (*Ulmus davidiana* var. *japonica* 'Reseda') and a clone of the Siberian Elm (*U. pumila*) that was patented in the USA in 1994. This has proved very resistant ever since to DED and a number of other diseases and pests. In a group planted at Keele University in Staffordshire in 2012, one tree died in 2020 of what may possibly have been DED. But even if true, mortality is still much lower than in native elms on the campus going through a new wave of DED deaths. Although it is not quite the same

FIG 224. A hybrid Elm, *Ulmus* 'New Horizon', that is resistant to Dutch elm disease. The beauty of this is that it looks almost like a native British elm with the distinctive 'herring bone' pattern of small twigs and young flower clusters just emerging.

shape as a native European Elm (Fig. 224), it is a terrific addition to the arsenal of trees that can be relatively safely planted. Moreover, the White-letter Hairstreak butterfly *Satyrium w-album*, a specialist of elms, has been found breeding on 'New Horizon' and the closely related 'Sapporo Autumn Gold'.

It is not fully understood how elms naturally resist the DED fungus so it is difficult to isolate the genes responsible and move them between elms. But transgenic elms have been created using anti-fungal genes from other sources, inserting the genes using *Agrobacterium* (Chapter 13). In this way resistant varieties of American and English Elms (*U. americana* and *U. procera*) have been produced (Gartland *et al.* 2000, Newhouse *et al.* 2007) that appear suitable for native elm insects. There are, however, social and political problems of releasing these into our landscapes, as discussed in Chapter 13.

Chestnut blight

Dutch elm disease is, sadly, not alone in causing large-scale tree death. Chestnut blight caused by the fungus *Cryphonectria parasitica* has killed 99.9 per cent of American Chestnuts *Castanea dentata* (about 3.5 *billion* trees) in eastern North America along the Appalachian Mountains. This happened in less than a century after the fungus was introduced from Asia between 1882 and 1904. Chestnuts used to be a dominant tree, making up 40–50 per cent of the canopy over an area of 8 million square kilometres, and so its loss is truly staggering, leading to a wholesale change in forest composition and the extinction of at least seven species of moths that fed exclusively on the Chestnut. The trees can still be found as small saplings (Fig. 225) but these do not survive long enough to flower and fruit.

FIG 225. A small sapling of American Chestnut *Castanea dentata*. Since the introduction of chestnut blight into North America, this once abundant magnificent tree is rarely seen in the wild other than as small shoots. Harvard Forest, Massachusetts.

The chestnut blight pathogen has been spread throughout the world, mostly by the movement of infected timber. It is now found from Australia to Europe and was first discovered in Britain in 2011 in Warwickshire. Fortunately, while the European Chestnut *C. sativa*, and a handful of other trees, including our two oak species, are susceptible, they are seemingly less affected than the American Chestnut, and there are cases of trees recovering. The recovered trees appear to have been infected with a strain of the virus made less virulent (*hypovirulent*) by it being infected by something resembling a fungal virus. This agent has been released into the chestnut blight populations of North America but unfortunately it offers a smaller degree of protection to the American Chestnut than to its European cousin.

The most promising recovery option for the American Chestnut is the ongoing breeding programme, taking trees that show some resistance to the pathogen and hybridising them with highly resistant Asiatic species of chestnuts, including the Japanese or Korean Chestnut *C. crenata* but especially the Chinese Chestnut *C. mollissima*. The most resistant offspring have been backcrossed with the American Chestnut with the aim of producing trees that look like the American original but with the resistance of the Asian trees.

A potentially quicker and better solution is to see if genetic engineering can implant the resistance into the American Chestnut. Success will hinge on whether disease resistance is controlled by a few genes or by many hundreds or even thousands of them. A significant breakthrough has, however, been made by William Powell of the State University of New York and his collaborators. It is known that the chestnut blight fungus kills chestnut cells by secreting oxalic acid. Powell's team have transferred a gene from wheat that produces an enzyme which breaks down oxalic acid (an oxalate oxidase). In the chestnut, this neutralises the toxicity of the fungus, allowing tree and fungus to coexist. The question remains whether this modified tree will remain resistant and whether it will be accepted for release into the wild (Powell *et al.* 2019).

Ash dieback

A disease that is of concern across Europe is ash dieback (Fig. 226), caused by the fungus *Hymenoscyphus fraxineus*, originally identified as the asexual stage, *Chalara fraxinea*. It was first discovered in northeast Poland and Lithuania in 1992, from where it spread rapidly across much of Europe, being first officially identified in Britain in Buckinghamshire in February 2012 but was so widespread that it was certainly present before then. The fungus infects through the leaves and in some cases may get no further before the leaves are shed in the autumn. But in many cases the fungus invades the vascular tissue of the twig, and once in, readily spreads through the tree, killing leaves and shoots progressively back into the crown, hence the name 'dieback'.

Seedlings die within a few years of infection, but older trees may take a decade or so to succumb. Trees over 20 years old are progressively weakened by ash dieback, so they may actually die from secondary effects such as honey fungus or environmental stresses (Thomas 2016). This disease is particularly tragic for the British countryside, since ash is the second most abundant tree in small woodlands after oak, and it dominates woodlands on rich soils. Jon Stokes of the Tree Council has estimated that we have over 1.7 billion ash trees with a trunk more than 4 cm in diameter and something like 2 billion seedlings smaller than 4 cm, both inside and outside of woodland. Sadly, the Narrow-leaved Ash *Fraxinus angustifolia* 'Raywood', a cultivar renowned for its bright red autumn foliage, is also susceptible to ash dieback. Other members of the olive family (Oleaceae) are also known to be susceptible, including *Phillyrea latifolia*, *P. angustifolia* and *Chionanthus virginicus* and potentially species of *Olea*, *Syringa*, *Forsythia*, *Jasmina*, *Osmanthus* and *Ligustrum*.

Fortunately, around 5–7 per cent of our ash trees are resistant to the pathogen and even if infected do not die. This gives us a solid resource from which to breed and spread ash back into the countryside. To make this easier, the genes responsible for resistance have been identified (Harper *et al.* 2016), so trees can be screened for resistance far more rapidly than waiting to see whether a tree dies or not. As with elms and the American Chestnut, it may also be possible to breed our ash with resistant trees from Asia (where the pathogen is native), such as the Chinese ashes *Fraxinus chinensis* and *F. bungeana* or those from North America, including the Oregon Ash *F. latifolia* and Arizona Ash *F. velutina*. The spread of the pathogen can also be reduced locally by collecting and burning leaf litter in the autumn, since the fungus overwinters on the old leaf stalk (*rachis*) and is a source of invading spores in the spring. This, of course, is most feasible in urban areas where the leaves can be easily swept up.

The main fly (or beetle) in the ointment for our ash trees is a small iridescent green beetle around a centimetre long – the appropriately named Emerald Ash Borer *Agrilus planipennis*, native to Asia. It has killed tens of millions of ash trees in North America since 2002 and is considered by some to be the most destructive forest pest ever seen in North America (Thomas 2016). The adults feed on ash leaves, doing comparatively little damage, but the larvae bore into the trunk, breaking the phloem link between roots and crown, ring-barking the tree so effectively that it causes 99 per cent mortality in ash populations.

The beetle was recorded in Moscow in 2003, where it has killed more than a million trees, and is now moving west into Europe at the rate of 30–40 km per year, seemingly hitching rides on vehicles along main roads. If it arrives in Britain, it is certain that the small percentage of ash trees surviving ash dieback will succumb to the beetle, in which case the future of ash as a mature tree in

FIG 226. Ash dieback caused by the fungus *Hymenoscyphus fraxineus*. (a) A street tree showing the 'dieback' from the outside of the crown inwards. The tree was dead two years after the photograph was taken. (b) In this woodland edge in Kent, a number of ash trees are dead and others have a very thin canopy and have undoubtedly died since the picture was taken in 2016. (a) Newcastle-under-Lyme, Staffordshire, and (b) Kent Downs Area of Outstanding Natural Beauty.

Britain and most of Europe will be over, and it will occur only as small saplings and bushes, much like the American Chestnut. This will obviously have an effect on the animals and plants that live in or on ash. It has been recorded that 1,058 other species are associated with ash (Chapter 2). Of these, 44 are found only on ash, including 4 lichens, 11 fungi and 19 invertebrates (Mitchell *et al.* 2014b).

PHYTOPHTHORA

Another worldwide set of pathogens, the most virulent of which also come from Asia, are the *Phytophthora* species (from the Greek *phyto*, meaning 'plant', and *phthora*, meaning 'destruction'; hence 'plant-destroyer'). These are fungus-like organisms called water moulds, or Oomycetes. The most virulent is *P. ramorum*, found in over 100 different plant species. It is the cause of *Sudden Oak Death* in western USA, first identified in the mid-1990s, which is still killing numerous Tanoaks *Notholithocarpus densiflorus* and various oak species, including California Black Oak *Quercus kelloggii* and Coast Live Oak *Q. agrifolia*. A strain of *P. ramorum* is now found in Europe and is showing the same propensity to move between host species. It was first found in Britain in 2003 in the introduced Rhododendron *Rhododendron ponticum* and has now spread to camellias, magnolias, species of *Pieris*, viburnums and larches (with implications for forestry), and can potentially spread to oaks, maples, sweet chestnuts, beeches and a number of shrubs.

Oaks can be divided globally into eight genetic groups. Fortunately, our two British oaks, Pedunculate Oak *Q. robur* and Sessile Oak *Q. petraea*, are members of the white oak group which shows some resistance to *Phytophthora*. So the good news is that our oaks should be less vulnerable to the pathogen. Other *Phytophthora* species are causing their own problems in other parts of the world. For example, the large Kauris *Agathis australis* of New Zealand are being threatened by *P. agathidicida* (Fig. 227).

Disease caused by *Phytophthora* is difficult to treat, since the spores remain viable in the soil for many years, allowing for reinfection of their hosts. Despite this, phosphites injected into the stem and sprayed on the bark give some control of *P. ramorum* (Garbelotto *et al.* 2007). Other studies have shown that biochar and mulches also show promise as treatments to reduce the severity of disease caused by various *Phytophthora* species (Zwart & Kim 2012, Percival 2013).

Deciding whether *Phytophthora* is affecting European oaks is made more complex because some species of the pathogen are fairly innocuous, and although

FIG 227. Dead and dying Kauris *Agathis australis*, the result of *Phytophthora agathidicida*. Waipoua Forest, New Zealand.

trees show typical symptoms of small bleeding cankers that ooze a dark sticky liquid, they are not being seriously affected by the disease. Moreover, bleeding cankers can be caused by bacterial infections, producing similar weeping patches to *Phytophthora*. These bacteria are at least partially responsible for *Acute Oak Decline* (AOD) which has been affecting European trees for decades and is particularly noticeable in the warm southeast of Britain. The causes of AOD appear to be mainly bacterial infections of the phloem linked with the galleries of the Two-spotted Oak Buprestid Beetle *Agrilus biguttatus*. In combination, these can seriously weaken and kill an oak in less than six years after infection. The affected trees might finally die from a range of causes, but death is ultimately caused by the weakening effects of AOD.

A similar syndrome is *Chronic Oak Dieback/Decline* which has been causing a slow, long-term decline in European oaks for many decades. It is seen as a progressive dieback of the canopy, from the twigs back to the larger branches and eventually the trunk. Some trees recover, surviving as stag-headed individuals, but many eventually die. The causes include the Two-spotted Oak Buprestid Beetle of AOD but interacting with many other possible factors. These include infection of the leaves by Oak Mildew fungus *Erysiphe alphitoides*, aided by defoliation by moth caterpillars, particularly the Green Oak Tortrix *Tortrix viridana* and Winter Moth *Operophtera brumata* following a warm spring. Honey fungus is also involved but this may just be the final straw that kills weakened trees rather than a major cause of the decline. There are rumours that *Phytophthora* is involved in weakening the oaks. Environmental problems can also contribute, particularly exceptionally cold winters in mainland Europe, spells of drought and high levels of nitrogen pollution that put trees under extra stress, and soil compaction affecting water supply. The extra nitrogen in leaves makes them more attractive to microbes and insects. Moreover, as described in Chapter 7, the increased nitrogen in the soil can create nutrient imbalances in the tree, leading to a reduction in the levels of phosphorus and other nutrients like calcium, magnesium and potassium. Excess nitrogen also leads to soil acidification, putting the tree under further stress (Brown *et al.* 2018). All of this can make it very hard to determine exactly which agent is the main problem causing decline in a tree.

BIOSECURITY

A common theme running through the above accounts of formidable pathogens is that many of them originate in Asia, including parts of China. This area of the world has long been a hotbed of evolutionary battle of tree defences against pests and pathogens, such that a number of very virulent pests and pathogens have evolved there. With increased international trade, it is inevitable that more of these pests and pathogens are shipped around the world. And it need not be just in living plants. Beetle larvae have been transported inside the wood used to make pallets, packing crates and even furniture, and some pathogens (like Dutch elm disease) have been transported on cut logs. In the case of ash dieback, it could also be argued that the porous nature of national boundaries within the European Union (the UK was still a member at that point), and the perceived lack of political will to prioritise this problem or prevent the international movement of ash seedlings, led to biosecurity measures being ineffective. Biosecurity thus becomes a political rather than an ecological problem, requiring the investment of sufficient money to allow adequate vigilance and surveillance to reduce the risk of importing new pests and pathogens, and enough resources to allow containment and eradication if something does slip into the country.

INSECTS

Trees have so many leaves and form such a large food supply that inevitably a wide range of insects has evolved to munch their way through this cornucopia. When you look at a woodland, the effect of these insects on the trees is usually quite small, although there are always exceptions – Figure 228. Removal of up to 30 per cent of the leaves by defoliating insects in any one year will generally affect growth by very little, and even losing 50 per cent of leaves will usually have little impact, as long as there is no other damage to the tree. However, repeat defoliation year after year is more of a problem, since the tree will need to draw on stored food reserves, which gives it fewer reserves for emergencies in the future. An experiment that involved spraying areas of forest in northern Russia with insecticide over four growing seasons found that the background insect herbivory was reduced from 4 per cent to 1 per cent loss of leaf area in a sprayed area (Shestakov *et al.* 2020). This might seem insignificant, but at the end of the four years, the sprayed sites showed nearly double the increase in biomass compared to unsprayed plots, so the small, constant loss of leaves to insects is important.

FIG 228. Extensive insect damage to leaves in lowland Honduras.

Moths

Prominent amongst leaf-eating insects are the caterpillars of moths, but despite their reputation, only a few species are capable of defoliating a whole tree. Those that can defoliate an oak include the Winter Moth and the Green Oak Tortrix that contribute to Chronic Oak Dieback/Decline. I remember walking through oak woodland in Devon as an undergraduate, and the loudest noise was the falling frass from the many caterpillars of Winter Moth high in the canopy, and walking through a forest of silken threads each with a green caterpillar at the end, heading for the soil to pupate.

This infamous pair has been joined by a new defoliator – the Oak Processionary Moth *Thaumetopoea processionea*, native to central and southern Europe. This has reached infestation levels in many European countries and was first discovered breeding in Britain in 2005 in west London. It has since spread throughout London and surrounding counties and has also been found in isolated incidences throughout Great Britain, arriving on imports from the Netherlands and Germany. Like the Winter Moth, the caterpillars can spread through the air by 'ballooning' – blown away on silken threads by the wind.

Oak Processionary Moth caterpillars are capable of stripping the leaves from a variety of trees, including our two native oak species, as well as Turkey Oak *Quercus cerris* and hornbeams, sweet chestnuts, beeches, hazel and birches, although they cannot complete their development on non-oaks. Their name comes from the habit of the caterpillars 'processing' to fresh food supplies in head-to-tail chains, often in the shape of an arrowhead behind a leader. Oaks are well used to coping with defoliation by growing a new set of leaves in August (lammas growth – see Chapter 5) but this is another stress on trees that may exacerbate *Chronic Oak Dieback/Decline*. Moreover, the caterpillars hit our consciousness, since they possess urticating hairs that can cause an itchy rash, eye irritation, breathing difficulties and, occasionally, an allergic reaction in us and our pets. The hairs are released when the caterpillars are threatened or disturbed, and then blow in the wind and stick to trunks and grass where they can be disturbed and picked up by people.

Other moths cause aesthetic problems rather than endangering the life of a tree. An example is the Horse-chestnut Leaf Miner *Cameraria ohridella* (Fig. 229). The caterpillar of this moth is a leaf miner, eating the internal tissue within the leaf blade and leaving brown lines of dead epidermis on both sides of the leaf (Thomas *et al.* 2019). It is native on the last remnants of natural Horse-chestnut *Aesculus hippocastanum* populations in the Balkans and Greece, but has been spreading through Europe on the many planted horse-chestnuts. It was first seen in Britain in Wimbledon, London, in 2002, but has now spread to most parts of

FIG 229. Horse-chestnut Leaf Miner *Cameraria ohridella*. (a) Tunnels in leaves and (b) the small adult moth, around half a centimetre long. Photographs by David Emley BEM.

England and Wales and is still on the move north. It can also occasionally mine Sycamore and Norway Maple *Acer platanoides* when horse-chestnut is unavailable or already heavily infested. Up to five generations of the moth per year affect the leaves, with the final generation overwintering as pupae in the leaf litter. Leaf damage can rise dramatically in the first three years of infestation, reaching 200 mines per leaf and removing up to 75 per cent of leaf area so that the whole crown turns brown or can even be completely defoliated. This damage shortens the lifespan of the leaf by a third, so the leaves begin falling from the beginning of July.

The cumulative loss of green leaf area over the summer could be quite devastating, but fortunately, when the loss of green tissue reaches its highest, the photosynthetic efficiency of the leaves has decreased anyway and so the loss in photosynthesis is restricted to something around 65 per cent that of uninfected leaves. This inevitably reduces the vigour and growth of the trees, but there is no compelling evidence that damage by the Leaf Miner leads to long-term health problems or tree death. However, it does affect tree reproduction because the Leaf Miner has been found to reduce seed weight by 40–50 per cent, and in some cases germination is up to 32 per cent lower (Takos *et al.* 2008, Percival *et al.* 2011). Fortunately, there is no evidence that seedling survival in the first 2–3 years is affected.

Moths can also be damaging to fruits. Apples are prone to damage by the Codling Moth *Cydia pomonella* (Fig. 230), but the moth can also be found to a lesser extent in fruits of pear, walnut, apricot, peaches, plums, cherries and

FIG 230. Damage to the seeds and core of an apple caused by larvae of the Codling Moth *Cydia pomonella*. The flesh is also damaged as larvae exit the fruit. Without the seed, the apples are shed earlier than those with intact seed.

chestnuts. The caterpillars burrow into the core of the ripening fruit and eat the developing seeds, usually riddling the apple flesh with frass-filled channels – a right mess. By killing the seeds, the fruit is shed early, so the first apples to fall are those with the moth, which is a way for the tree to save energy by not wasting it on fruits that have no seed. It just means that the first windfalls, eagerly awaited, are always a disappointment.

Historically, Codling Moth was treated with chemicals such as lead arsenate and DDT with all the resultant problems we now know about. Organophosphate sprays can be used to coincide with the hatching caterpillars searching for a fruit to burrow into. While some of the larvae pupate in the soil, many will find cracks or loose flakes of bark in the crown to hide beneath, so traditional sticky bands put around the tree to catch crawling females (as used for the Winter Moth and other pests) will not stop the Codling Moth. If the problem is severe or annoying enough, there are now pheromone traps to catch the males, and pathogenic nematodes can be sprayed on the tree in the autumn to infect the caterpillars as they leave the apples. In our garden we just put up with the disappointment of a few lost apples. It is worth noting that another *Cydia* species, in this case the Large Beech Piercer *Cydia fagiglandana*, is reducing beech seed production, described in Chapter 4.

Other insects

A large number of true bugs (Hemiptera) feed on the largesse of trees. This includes numerous aphids and scale insects that feed on the sugary phloem sap of vulnerable trees, including limes and some maples such as Sycamore. Since these insects mainly want the small amount of nitrogen in the sap, they excrete large quantities of sugar-rich sap which lands on the windscreens and paintwork of cars and on benches, making a sticky and recalcitrant mess. The sticky liquid is rapidly invaded by sooty moulds, creating an unsightly black mess on the trunk and lower leaves that is more an aesthetic problem than harmful to the tree.

Other aphids can be much more damaging, particularly woolly aphids which are covered in a waxy wool that gives them protection from predators. The commonest in Britain is the Woolly Apple Aphid *Eriosoma lanigerum*, found on apples and pyracanthas. While unsightly, it does little damage by itself but the wounds it produces can be infected by the fungus *Neonectria ditissima* (also called *Neonectria galligena* or *Nectria ditissima*), causing apple canker that can kill whole branches. Other woolly aphids include the Woolly Beech Aphid *Phyllaphis fagi* found on various beech species around the world.

Closely related are the various adelgids that are causing extensive damage when introduced to hosts outside of their native range. This includes the Balsam Woolly Adelgid *Adelges piceae*, native to central Europe, which is causing great damage to fir species in North America (particularly Balsam Fir *Abies balsamea*, hence the common name, but also Fraser Fir *A. fraseri*) where it is a particular problem for Christmas tree growers. It causes curling of leaves, shoot swellings, dieback and death within 3–4 years. Equally damaging is the Hemlock Woolly Adelgid *A. tsugae* (Fig. 231), native to Asia. It was introduced onto the eastern

FIG 231. (a) Hemlock Woolly Adelgid *Adelges tsugae* on the underside of the twigs of Eastern Hemlock *Tsuga canadensis* at Harvard Forest, Massachusetts. (b) The effect of the adelgid is causing high mortality of trees across the New England landscape. Photographs by (a) David Orwig and (b) Will Blozan.

coast of North America in the early 1950s from Japan, where it is causing widespread death of Eastern Hemlock *Tsuga canadensis* and the Carolina Hemlock *T. caroliniana* which are proving more susceptible to the pest than western or Asian hemlocks. The small insects, less than 1 mm long, congregate on young branches and inject a toxic saliva while taking up the sap. Within a few months, the needles and associated buds die, leading to the progressive loss of branches. The weakened tree is then susceptible to other pathogens and pests, eventually leading to death.

The Eastern Hemlock covers some 10,000 km^2 from southern Canada to northern Georgia. Since Eastern Hemlock, like most conifers, cannot sprout from old wood, it will not recover from severe loss of its needles. Once an area is infected with the Woolly Adelgid, Hemlocks die over 1–2 decades with up to a 99 per cent mortality rate. This is proving to be as devastating as the fungal diseases Dutch elm disease and chestnut blight.

Control is difficult, since the adelgid is spread by the wind and easily carried by birds and other animals. It can also be difficult to see the adelgid in a tree until it is well established, and the most obvious method of control using insecticides is problematic over whole trees and landscapes. The best option for management of the pest looks to be biological control offered by a range of insects (Motley *et al.* 2017). Field trials are showing promise, but modelling by Aaron Ellison and colleagues from Harvard Forest (2018) suggests that Eastern Hemlock is still likely to be completely lost from all of its range by 2050 except for the extreme north of the USA and southern Canada.

Ants can also play their part in defoliation, particularly leaf-cutter ants (Fig. 232). They cut sections of leaf which they carry back to their nest and add to a fungal culture they tend. This mutualism sees free food delivered to the fungus which is nicely protected underground, and in return the ants can eat the leaves once the fungus has decomposed the cellulose. The ants can tell when a particular leaf type is not decomposing very well or when chemicals in the leaf are affecting the fungus and will adjust what they bring back.

Wood and bark borers

Many species of insects burrow into wood, making it their home and source of food. In warmer parts of the world, more than 2,000 species of termites (Isoptera) are renowned as wood borers, with around 100 species capable of causing significant damage to buildings. In temperate areas, most wood borers are beetles (Coleoptera) and moths (Lepidoptera), with a few sawflies and wood wasps (Hymenoptera). Most of these live in dead wood that is too dry for fungi (Fig. 233). For example, the Common Furniture Beetle *Anobium punctatum* can survive in

FIG 232. Leaf-cutter ants (probably an *Atta* species) cutting away leaf sections which they add to a fungal culture in their nest. Significant quantities of leaves can be cut up and transported, including, as shown, a pile of leaves collected as part of a research project, which largely disappeared over a day! Cusuco National Park, Honduras.

wood down to 12 per cent moisture, which is why furniture in damp houses may be riddled with woodworm but is safe from fungal rot. Due to dead wood being removed from woodlands, many of these beetles are rare and in need of conservation (see Chapter 12 for the value of dead wood). As an example, the Two-spotted Oak Buprestid *Agrilus biguttatus* is a harmful pest in mainland Europe,

FIG 233. (a) Beetle exit holes in an old oak fencepost. Beetles can survive in wood at lower moisture contents than most fungi. (b) If conditions are just right, the wood can be reduced to no more than powder without any fungal rot, as is the case of this Eudoia log (*Tetradium daniellii* var. *hupehensis*, formerly *Euodia hupehensis*) that I was storing in a damp – but not too damp – garage.

linked to Acute Oak Decline, but is a rare British Red Data Book species mostly associated with mature oaks more than 30 cm in diameter.

Of the insects that cause damage to living trees, the majority live in the wood as larvae and the damage is caused by their tunnelling. Most notorious of the wood-boring beetles are the large and colourful longhorn beetles native to Asia, including the Asian Longhorn Beetle *Anoplophora glabripennis* and Citrus Longhorn Beetle *A. chinensis*. The former can be up to 4 cm long with antennae the same length again. They have both been introduced to North America and Europe, and have been found in the UK sporadically since 1994, in each case being eradicated before they could spread. The larvae are also large, up to 5 cm long, and can remain in the tree for four or more years, so in this case it is the physical damage that is the problem, caused by the many tunnels through the wood, which can be up to 1 cm wide as the larvae grow in size. This also, of course, reduces the quality of timber, but the larvae can be inadvertently spread when carried inside furniture! The problem is made worse by the beetles being catholic in taste, readily invading birches, elms, horse-chestnuts, maples, willows and poplars, and to a lesser extent alder, ash, beech, plane, and fruit trees such as apples, pears and *Prunus* species. It has been calculated that, depending upon the species composition of the trees, between 15 and 98 per cent of urban trees will be lost in major Nordic cities due to these two beetles – a huge and expensive impact (Sjöman & Östberg 2019).

A number of wood-boring insects cause significant problems for the host trees because they have ring-porous wood (see Chapter 6) which is very dependent upon the outer ring of wood for conducting water, and this is the ring most frequently damaged as the larvae burrow through the nutrient-rich phloem and score the sapwood. In this category come the Emerald Ash Borer and elm bark beetles described above, the latter made worse by also introducing the Dutch elm disease fungus.

Some of the beetles bring with them fungi which help them to overcome the defences of the tree. The European Spruce Bark Beetle *Ips typographus* is a serious pest of spruce and other conifers, causing large-scale death of trees (Fig. 234). As the adult beetle burrows into the tree, it brings with it the blue-stain fungus *Ceratocystis polonica* (sometimes called *Ophiostoma polonicum*). This fungus can kill the tree and also leaves blue streaks in the wood, which reduces its commercial value. The beetle tends to target weakened or diseased trees where resin production is likely to be less, but the fungus also reduces resin production within the tree, making it easier for subsequent beetles to burrow into the wood without being physically washed away or killed by the resin flow, and so increasing the beetle load of a tree and the likelihood of death.

FIG 234. Tree death caused by the European Spruce Bark Beetle *Ips typographus* in the Tatra mountains on the Polish–Slovakian border. Although the beetle specialises in Norway Spruce *Picea abies*, it will infect a number of other conifers.

The Mountain Pine Beetle *Dendroctonus ponderosae* of North America does the same, bringing with it the blue-stain fungus *Grosmannia clavigera*, which may also slow down resin production. Additionally, the fungus may help the beetle larvae to feed by breaking down the wood, making it more digestible. The Mountain Pine Beetle is native to western North America and is an important natural component of the forest ecosystem, helping to kill old trees and making openings for seedlings to grow. However, a series of very mild winters and hot, dry summers since the late 1990s has led to the beetle reaching epidemic numbers and killing millions of pines over a large area of western North America, affecting more than 36 million hectares of forest and killing 70–90 per cent of the pines. It affects a number of different pines but particularly the northern Lodgepole Pine *Pinus contorta* and the more southerly Ponderosa Pine *P. ponderosa*.

LARGER ANIMALS

Birds can be surprisingly damaging to trees, especially when they are looking for nesting material. I have stood under a lime tree watching pigeons repeatedly landing in the tree and tugging at small branches until they came away. Many other birds build nests of woody material and will raid trees, from magpies through to eagles. In a woodland this is unlikely to cause any great problems for the trees but it becomes more serious in isolated trees that bear the attention of many local birds. We have a Gutta Percha *Eucommia ulmoides* tree in our garden, and last year I wondered why the upper leaves were being left as a skeleton of main veins until I saw a group of pigeons sitting in the canopy eating the leaves (Fig. 235); they also broke a good number of smaller branches in their attempt to reach the leaves. I have also seen it happen with ash.

Bullfinches can be a major pest in fruit orchards, attacking the developing buds in the spring. Pears and plums are particularly susceptible, but apples, gooseberries and currants can also be affected. Culling licences were regularly issued in the past, but since bullfinch populations are in decline, a number of other deterrents, such as noise makers and removing scrub close to the orchard, have been used. Bullfinches are frequent visitors to our garden, and we rarely have fruit on our plum tree.

Small mammals such as mice, voles and rabbits can cause considerable damage to bark, especially on young trees. The Red Squirrel *Sciurus vulgaris*, native to Eurasia, is renowned for stripping the bark of broadleaf trees between April and July. However, this is minor compared to the damage caused by the Grey Squirrel *S. carolinensis*, introduced into Britain from North America in the 1880s. This stripping (Fig. 236) is partly to get access to the nutritious phloem but is

FIG 235. Pigeon damage to a Gutta Percha *Eucommia ulmoides*. Upright shoots have had their leaves stripped, and branches have been broken by the scrambling of the birds.

FIG 236. Squirrel damage (a) to a Sycamore *Acer pseudoplatanus*. Bark has been removed from the right-hand stem. Both trunks are showing distorted growth due to damage in previous years. (b) Trees will often snap where the bark has been removed and no wood can grow, so weakening the trunk, as in this Beech *Fagus sylvatica*.

also aggressive posturing to rival squirrels. Well-tended, vigorous trees are most favoured over self-sown trees in a dense woodland; these vigorous trees have thinner bark and a higher sap content, and so it is often specimen trees that are targeted. So common is this bark stripping that ashes, limes and Wild Cherry may eventually become more common in Britain, other things being equal, because of their relatively low palatability to squirrels compared to the more palatable Beech *Fagus sylvatica* and maples, including Sycamore.

Inevitably, deer with their longer reach can cause extensive damage to trees by browsing lower branches and repeatedly devouring young saplings such that they die. Further damage can be caused by the deer debarking trees when rubbing the velvet off their young antlers, although the same trees tend to be used repeatedly so this damage is usually quite localised. But bad luck for the chosen tree! The same problem has been found in parks and gardens, caused by dogs chewing or scratching the bark from young trees. The solution to the damage caused by large animals is either to exclude them from an area by fencing, or to give each tree its individual planting tube (Fig. 237) or some other means of protection, such as chestnut paling fences or, for dogs, hessian impregnated with chilli powder.

FIG 237. Protection from grazing animals. Tuley tubes also act as miniature greenhouses, speeding the growth of young trees. They are not without potential problems caused by too high temperatures inside and rapid weed growth up the tube, which swamps the tree.

Even larger animals can cause problems for trees. In North America, the Black Bear *Ursus americanus* uses its claws to grip when climbing trees, causing distinctive patterns of scar tissue of parallel lines on the trunks of poplars, one line for each claw. This creates a thrill when finding them rather than being a serious problem for the tree. Bears can, however, damage vigorously growing conifers, particularly those 15–40 years old, as they peel away the bark to eat the sugar-rich phloem in spring when other food is scarce. This, of course, makes the trees vulnerable to girdling and windthrow and lays them open to pathogens and pests. In the Pacific Northwest of North America this damage has been estimated to result in a loss of timber production of $56 per hectare – small but significant over large areas (Taylor *et al.* 2009). In North America, the main control in the past was culling the bears but this has ceased in more enlightened times and the loss is just accepted.

In Japan, where the Asian Black Bear *Ursus thibetanus* causes similar damage, particularly to Japanese Red-cedar *Cryptomeria japonica* and Japanese Cypress *Chamaecyparis obtusa*, culling almost led to extinction of the bear. Now, fortunately, other forms of control are used, including protecting the trees using biodegradable plastic netting (Fig. 238) which gives protection for more

FIG 238. (a) Bark stripping of a tree by Asian Black Bears *Ursus thibetanus* in central Honshu island, Japan, done using teeth and claws to get at the nutritious inner bark. (b) To prevent this, valuable plantation trees can be wrapped in biodegradable plastic netting. Photographs by Shinsuke Koike.

than 10 years and reduces damage from around 28 per cent in unprotected trees to less than 6 per cent in those with netting (Kobashikawa *et al.* 2019). But this is obviously only practical in small areas, and aesthetically acceptable only in plantations!

Thirteen different primate species around the world have also been found to strip bark from trees, presumably again to get access to sugar-rich phloem or, in the case of the White Colobus Monkey *Colobus guereza*, to get access to the sodium in the phloem of eucalypts (Di Bitetti 2019). The urban equivalent is the overzealous use of strimmers to cut vegetation close to the trunk of young trees, thus neatly removing the bark and killing the tree (Fig. 239).

FIG 239. A young cherry cultivar that has been repeatedly hit by a strimmer line, effectively removing the thin bark round the whole trunk below the hand, and so killing it. Older trees with thicker bark are less readily damaged.

CHAPTER 16

What Is the Future of Our Trees?

THE PRESSURE OF PEOPLE

It is widely agreed that we are in a new geological epoch, the Anthropocene, marked by human activity becoming the dominant influence on climate and the environment. This influence is creating the world's sixth major extinction of species since life began (Ceballos *et al.* 2020), with extinction rates of trees running at three orders of magnitude higher than might be expected in the absence of humans. The human population of the planet has increased from 3 billion in 1960 to 7.8 billion in 2021 and these extra people are inevitably putting pressure on our trees and forests. The extra people are concentrated in the world's cities – over half of the world's population lives in cities (Fig. 240), and it has been estimated that by 2050, this will increase to 68 per cent as another 2.5 billion people are added to the world's urban population. Europeans are already city dwellers, with 80 per cent of us living in cities, but Asia and Africa will catch up, since 90 per cent of the increase in cities by 2050 will be in these continents (United Nations 2019).

Calculations have shown that for the first time, in 2020, the weight of human made things, or *anthropogenic mass*, overtook the weight of living biomass on Earth that currently stands at 1,100 gigatons. And each week the new anthropogenic mass produced equals the weight of all humans on the planet (Elhacham *et al.* 2020). Although we are more concentrated in cities than ever before, higher populations mean that rural areas need to supply the resources and can still suffer from our influence. For example, in Great Britain (England, Scotland and Wales), 25 per cent of land is within 79 m of a road and only around 12 per cent, mainly in the uplands, is more than a kilometre from a road (Phillips

FIG 240. (a) New York City from the Empire State Building. In 2018, with almost 19 million people in the city and suburbs, New York ranked as one of the world's 39 megacities (cities of more than 10 million people). (b) However, many smaller cities, such as the beautiful Sighişoara in central Romania with a population of under 30,000 and reputed to be the birthplace of Vlad the Impaler on whom Dracula is based, are still growing due to people moving in from surrounding rural areas, attracted by better-paid jobs. The result is that cities of all sizes are growing.

et al. 2021). Whilst pollution will obviously be highest along the busiest roads, low levels of pollution – particularly nitrogen – will be pervasive and have a detrimental effect on trees, not to mention on human health.

Inevitably, the resources needed for this growth in people and our stuff is putting pressure on the world's trees and forests. It has been estimated that since the origins of agriculture around 12,000 years ago, 46 per cent of the world's estimated original 5.8 trillion trees have gone. And since the beginning of the industrial era, we have lost 32 per cent of the world's forest area (Fig. 241). A quarter of this loss of global forest is due to converting forest to produce commodities such as beef, soy, palm oil and wood (Curtis *et al.* 2018). Moreover, the quality of what we have left has also suffered. Grantham *et al.* (2020) calculated that only 40 per cent of forest area (17.4 million square kilometres, or 6.7 million square miles) is high in quality and without human modification. This is mostly found in Canada, Russia, the Amazon, Central Africa and New Guinea, and since only 27 per cent is within protected areas, and many of these countries are using their forests as a resource, it is likely that the quality of what we have left will continue to decline. High-quality natural forests contain higher levels of biodiversity and store more carbon than plantations, primarily because natural forests are

FIG 241. Deforestation of rainforest for subsistence farming in Honduras.

denser with a more complex structure and higher carbon accumulation in the undisturbed soil. These features take centuries or millennia to develop, so natural forests should be conserved where possible (Waring *et al.* 2020).

Individual tree species are also under threat. Over 15,500 (30 per cent) of the world's estimated 60,065 tree species are threatened with extinction due to the expansion of agriculture and livestock farming as well as forest degradation. Around 440 of these species (0.75 per cent) are critically endangered, with fewer than 50 individuals remaining in the wild, and at least 142 species have been recorded as extinct in the past five years (BGCI 2021). Within Europe, the *European Red List of Trees* (Rivers *et al.* 2019) concluded that 42 per cent of the 454 native tree species are regionally threatened with extinction. Among Europe's endemic trees – those that are not native anywhere else on Earth – 58 per cent are classified as threatened, and another 15 per cent (66 species) as critically endangered and thus just one step away from becoming extinct.

We in Britain are not immune, and there are currently 15 endangered tree species listed as 'priority species' in the UK biodiversity action plan. Ten of these are endemic microspecies of whitebeam *Sorbus*, some of which have fewer than 20 individuals. These are mostly on rocky outcrops or screes, habitats which are becoming rare themselves in Britain, so their future should not be taken for granted. One might argue that since these microspecies are largely apomictic clones of individual trees (Chapter 1), the loss of one or more is not that significant. But every microspecies lost is still a loss of diversity and genetic resources that is part of the drain on our countryside's richness. Other endangered species in Britain include the Woolly Willow *Salix lanata*, restricted to rock ledges in the central Scottish Highlands where it can escape the pressure of grazing animals. The Juniper *Juniperus communis* (one of Britain's three native conifers along with Yew *Taxus baccata* and Scots Pine *Pinus sylvestris*) is also classified as vulnerable to extinction. Juniper has an unusual distribution, since it is found in the uplands of Wales and Scotland and also on the chalk downs of southern England (Thomas *et al.* 2007). However, in all three areas it is declining. For example, Ward and King (2006) found that 56 per cent of Juniper populations in Sussex in southern England became extinct between 1970 and 2003, and bush numbers fell from 4,767 to 1,465, a 69 per cent loss.

THE VALUE OF TREES

We are integrally linked with trees and they are very important to us physically and mentally (look back at Chapter 2 for the details). Millions of people are reliant upon food, fuel and medicine derived from trees, and trees are important in conserving biodiversity. The high species diversity is not just important in itself – it also adds to the resilience of ecosystems. It has been found, for example, that deforestation leads to a greater incidence of human disease by, for example, favouring the spread of malarial mosquitoes (Morand & Lajaunie 2021). The largest 1 per cent of trees in diameter in a forest contain 50 per cent of the above-ground live biomass and so are very important to the functioning of the forest in terms of physical structure, carbon storage and providing habitat and food for many other plants and other organisms (Fig. 242). But these largest trees tend to be amongst the oldest trees, and once lost are impossible to replace in any meaningful timescale – and they are disappearing rapidly in many parts of the world (Lindenmayer *et al.* 2012).

FIG 242. A large, old Pedunculate Oak *Quercus robur* of great value not only for carbon storage but also for biodiversity and its aesthetic quality, which is an important part of human wellbeing. Moccas National Nature Reserve, Herefordshire.

REASONS BEHIND TREE DECLINE

The causes behind the decline of trees are many and varied. Things we value in the west such as coffee, cocoa, cheap soybeans and beef are fuelling deforestation. It has been calculated that consumption patterns of the G7 countries drive an average loss of 3.9 trees per person per year (Hoang & Kanemoto 2021). While we gain tree cover by extensive planting in the north, our needs are leading to a net loss of natural forest around the world, especially of older, richer forests in the tropics.

Then we add in the many biotic stressors we put on trees, especially from the more rapid and widespread movement of pests and pathogens around the planet as a result of an increase in globalisation and difficulties in enforcing biosecurity checks in the growing international plant trade (Chapter 15). Global warming is making things worse by allowing introduced and native pests and pathogens to persist and spread where they were previously controlled by colder winters. The *European Red List of Trees* (Rivers *et al.* 2019) states that invasive species, both those we already face and those waiting in the wings, are the largest threat to European trees. On top of all this, we can also add in abiotic stressors which can be locally important, from dog urine to road salt, or more regional, such as large-scale pollution, particularly nitrogen, and global, such as climate change.

Climate change (Chapter 2) is a large-scale, long-term threat to trees and biodiversity in general, not just from warmer temperatures but also from other changes in climate, such as more frequent and extensive floods, droughts and heat waves (Menezes-Silva *et al.* 2018, Breshears *et al.* 2021). It has been estimated that up to a third of all species are vulnerable to climate change–related extinction, and we could be facing widespread forest dieback. As the climate warms, species can adapt, or move higher in elevation, or move northwards to regions more suited to their climatic needs. This could mean an expansion of forests in the north with conifers replacing up to half of what is now frozen tundra. However, it is not clear just how many tree species will be able to disperse quickly enough to keep up with predicted changes in climate. Migrating forests will not just need an agreeable temperature, they will also need suitable soil, moisture and topography, and the unsuitability of any of these to seed movement and tree establishment will act to slow down the rate at which forests migrate.

It was calculated in 2009 that the global average speed at which temperatures are moving towards the poles was 0.42 km per year (km/yr). At the same time, the speed at which different forest types can move was calculated (Loarie *et al.* 2009). The potential speed of movement of boreal forests (0.43 km/yr) and dry tropical forest (0.42 km/yr) matches that of climate change, and so, everything else permitting, these forests should be able to move northwards and cope with the rate of change. However, other forests will move more slowly; tropical rainforests (0.33 km/yr) and temperate broadleaf forests (0.35 km/yr) will lag, but temperate coniferous forests (0.11 km/yr) will be even further left behind. This may lead to the loss of some tree species, and even the death of entire forests. Nothing much will change at first as the climate continues to warm, but the most susceptible species will go into decline and show *extinction debt*, where they are doomed to die out but haven't yet physically died. By the time the trees die, it is too late to put things right. Species that are more adaptable are likely to survive, and there is evidence from North America that after the last ice age, trees with smaller seeds and those that could form ectomycorrhizal associations with more fungal species (Chapter 14) were able to move more rapidly (Pither *et al.* 2018), and this is likely to be the case in the future. So it is probable that our future woodlands and forests will not appear as they do now, since some species will stay and survive, new species will move in and other species will be lost, helped on their way by an unsuitable climate and by introduced pests and pathogens.

THE GOOD NEWS!

Fortunately, all is not yet lost. It is imperative that we slow deforestation (golden rule No. 1, Fig. 245 below), especially in tropical regions which are storehouses of biodiversity and carbon. We also need to improve the quality of the forests we have left. An important initiative along these lines is the global REDD programme (Reducing Emissions from Deforestation and Forest Degradation) which creates a financial value for the carbon stored in forests, paid for by the richer and more carbon-polluting nations. There was some criticism that this programme resulted in a blinkered view of forests such that carbon was the only important issue. This was taken on board in REDD+ which recognises the multiuse value of forests and includes payment for biodiversity conservation and sustainable management, including the needs of indigenous peoples. Inevitably there have been mixed results so far (Visseren-Hamakers *et al.* 2012, Duchelle *et al.* 2018) due to the problems of integrating management with the needs of people on the ground and with national, regional and local government agendas, set against a backdrop of fluctuations in the carbon credit price and the obstruction of renegade leaders. But the evidence 'so far paints a moderately encouraging picture' (Duchelle *et al.* 2018), and I am optimistic that we can slow deforestation while protecting the needs of countries and indigenous peoples.

This can be integrated with improved management of existing forests to maintain or improve their carbon storage and biodiversity. An additional step has been termed *proforestation*, helping existing forests to reach their full potential as 'intact ecosystems' so that they hold as much carbon as possible and provide abundant ecosystem services (Moomaw *et al.* 2019). The aim is to achieve this by reducing the cutting of wood or stopping it altogether. The loss of revenue from wood is more than compensated for by the increased value of other ecosystem services (Chapter 2). But the wood we use has to come from somewhere.

Most positive of all, it is clear that people are passionate about the trees around them, and there is a groundswell of interest in preserving significant or just lovely trees. This can be for aesthetic value or age (Fig. 243) or because of the tree's perceived value to society (Fig. 244). This growing interest is reflected in groups such as The Arborealists, an artistic movement that portrays trees and treed landscapes (Peterken *et al.* 2020). Moreover, more people are wanting trees in their streets and small urban *microforests* within walking distance. There is a growing trend to plant tiny urban woodlands using the *Miyawaki method*, named after its inventor, the Japanese botanist Akira Miyawaki. This requires a piece of land that can be just 3 × 4 m and so is ideal for a community project. The soil is

FIG 243. Preserving significant trees in our landscape. (a) A Dawn Redwood *Metasequoia glyptostroboides* in autumn colour photographed in 2015 at Smith College Botanic Garden, Massachusetts. (b) Next to the tree is a plaque that celebrates the tree and its preservation.

first improved with a covering of organic matter using whatever is available. As many tree and shrub species as is practical are then planted very densely, typically 2–7 trees per square metre (Manuel 2020). The ensuing competition ensures that the strongest trees survive and grow tall very quickly such that within a decade or two, there is a dense woodland with a surprisingly high biodiversity.

In addition to protecting what we already have, we also need to add new woodlands (*reforestation* on old forest sites and *afforestation* on new sites such as old agricultural land) using the ten golden rules in Figure 245. It is likely that we will need to manage our woodlands differently to the way we do now as weather extremes become more common (Field *et al.* 2020). We also need to add more trees on a local scale. This might be through planting or allowing natural regeneration, as discussed in golden rule Nos. 5 and 6. But as noted in Chapter 2, planting trees is not a magic bullet for solving climate change. It helps but

FIG 244. A 'superior' forestry tree, Scots Pine *Pinus sylvestris*, marked by the plaque being pointed at. This designation is to ensure that it is protected for the future. Białowieża Forest, Poland.

OPPOSITE: **FIG 245**. Ten golden rules for a successful reforestation project. The order of the rules matches the order in which tasks should be considered during the planning and running of a project, although some are interdependent and should be considered in parallel. From Di Sacco *et al.* (2021).

does not relieve us of the burden of taking action to reduce emissions of carbon dioxide and other greenhouse gases. Of course, there are also many other benefits of planting trees for wildlife and humans in urban areas, also discussed in Chapter 2. This is something that we can all contribute to, whether it is helping plant new large-scale woodlands through a Wildlife Trust or the Woodland Trust or planting a tree in a garden. Every little helps. Even scattered trees can make a difference; Prevedello *et al.* (2018) looked at 62 published studies and found that the local abundance of arthropods, vertebrates and woody plants was 60–430 per cent greater in areas with scattered trees compared with open treeless areas, and overall species richness was 50–100 per cent higher.

In the light of climate change and the lag in natural movement of forests, we can help future-proof our wooded landscape by moving trees north; for example, planting trees from warmer, drier parts of Europe – most likely France and Italy – into Britain so that they are preadapted to the climate as it changes over the next decades (golden rule No. 7). This can be as *assisted gene flow* if moving seed within the natural range of a species, such as moving Pedunculate Oak from central France to Britain, or as *assisted migration* if moving a species beyond its current natural range, such as moving Holm Oaks *Quercus ilex* from Spain to Britain (Aitken & Bemmels 2015). Movement of trees to new areas also overcomes problems of habitat fragmentation where gaps between woodlands can be too large for trees to bridge on their own, leading over the long term to small populations at risk of local extinction and increased inbreeding (Breed *et al.* 2018).

The tricky part is ensuring that *provenance selection* matches the predicted future climate. This is where *big data* proves useful, using huge datasets to help address these issues, often under the banner of 'the right tree for the right place for the right reason'.[6] For example, by looking at the range of conditions in which a tree species currently grows, either within just its natural range or the full range over which it has successfully been introduced, we can work out whether that species will be suitable for a particular place, and more importantly, whether it will still be suitable in the future. Figure 246 shows that our native Silver Birch *Betula pendula* is clearly not a suitable tree to grow in Jakarta or Cairo, but moreover will face conditions far outside its comfort zone in London in 2050 and is unlikely to survive. This sort of analysis allows us to work out which trees will

6 In forestry this is sometimes cynically adapted to a narrower definition of 'a tree that pays is a tree that stays'; i.e. a tree that can be sold for profit is the one that is grown.

Ten golden rules for reforestation

1 Protect existing forest first
- Reforestation doesn't easily compensate for the losses of deforestation
- Old- and second-growth, degraded and restored forests are all valuable

2 Work together
- Involve local communities with interactive participation in every project phase

3 Aim to maximise biodiversity recovery to meet multiple goals
- Restoring biodiversity will maximise carbon sequestration and help deliver socio-economic benefits

4 Select appropriate areas for reforestation
- Only target previously forested lands
- Connect or expand existing forest
- Do not displace activities that will cause deforestation elsewhere

5 Use natural regeneration wherever possible
- It can be cheaper and more efficient than tree planting, if conditions are suitable
- Works best on lightly degraded sites or those close to existing forest

OR

6 Plant species to maximise biodiversity
- Always plant a mix of species
- Use as many natives as possible
- Include rare, endemic and endangered species
- Promote mutualistic interactions
- Avoid invasive species

AND

7 Use resilient plant material
- Incorporate appropriate genetic variability
- Pay attention to provenance

8 Plan ahead for infrastructure
- Use the locally available infrastructure, capacity and supply chain or build it into the project
- Refer to seed standards to ensure maximum seed quality and process efficiency
- Provide training and use local knowledge

9 Learn by doing
- Research existing data and perform trials
- Adapt management accordingly
- Monitor the results beyond project life
- Use appropriate indicators according to project goals

10 Make it pay
- Ensure the project's economic sustainability
- Income can come from carbon credits, non-timber products, watershed and cultural services
- Make sure the economic benefits reach rural and poor local communities

New forest established

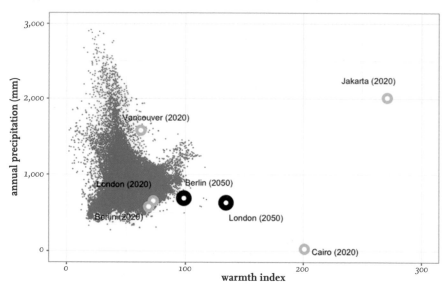

FIG 246. Distribution of Silver Birch *Betula pendula* against a warmth index (based on temperature during the growing season) and annual precipitation for places where it is a native tree. Also marked are five cities around the world plus, for Berlin and London, the predicted climate in 2050. From Sjöman & Watkins (2020).

grow in the future, both native and those introduced from elsewhere. A similar dataset has been compiled for magnolias around the world (Watkins *et al.* 2020), and it is clear that different species have significantly different requirements of moisture and warmth. If this information is fed to nurseries and those buying trees, it would allow the right tree to be chosen for a specific site.

Moving trees is not without its problems, since there are arguments for only planting with a local seed source (*local genetic provenancing* – Chapter 1) to maintain the distinctive local genetic type which is expected to be best adapted to the local conditions. However, against this, trees have been moved around Europe for centuries already, readily mixing genetic types. Inevitably, with the problem of migration rates outlined above, assisted gene flow and assisted migration are likely to be the best ways of maintaining our forests, and assisted migration is certainly proving necessary for forestry plantations. Undoubtedly, the best of both worlds is to leave some woodlands to regenerate or to plant them with local stock, and to plant more southern provenances in others (Gömöry *et al.* 2020).

In urban areas there is less of an ethical problem in planting non-local stock or indeed non-native species, since these are often human-made communities

with no pretension of being 'natural' woodlands. In these places, golden rule
No. 6 can be stretched to include non-natives. However, despite the lack of
restrictions, we tend to be quite conservative in what we plant; just eight tree
genera dominate planted urban trees in many parts of the world (Fig. 247).
This leaves urban areas at risk from climate change losses and from pests and
pathogens sweeping through genetically uniform areas. The solution to such
problems is to diversify. We should be planting a much wider range of trees in
urban areas and a greater genetic diversity within those species, obviously making
sure that the growing conditions are suitable and they are likely to survive.
Thankfully, there is increasingly good information available on choosing the
right trees for the right place, such as *Tree Species Selection for Green Infrastructure*
(Hirons & Sjöman 2019), readily available on the web.

The biggest problem faced by urban trees is arguably water stress, so
we can look particularly at how trees growing in local botanic gardens and
arboreta are reacting to water stress. This helps us to pick suitable tree species
and even genotypes for urban planting that are resilient to the challenges of
climate change (Hirons *et al.* 2020). We can also potentially breed trees with
paler leaves that contain less chlorophyll, which will reflect more light rather
than absorbing it, keeping our urban areas cooler (Genesio *et al.* 2021). When
introducing new species into urban areas or forestry, care must obviously be
taken that they are not going to be invasive outside of their initial planting
sites or encourage the establishment or spread of pests and pathogens (Ennos
et al. 2019). Diversifying urban trees will help safeguard against wholesale loss
of trees and will do so much to brighten up and diversify our streets. The trees

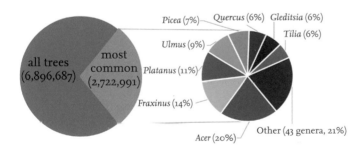

FIG 247. The OpenTrees dataset (opentrees.org) was used to find which genera of trees were
most common in cities around the world. Of these, the 10 most frequent genera per location
('most common') accounted for over 2.7 million trees, of which eight genera make up almost 80
per cent, including *Acer* (maple), *Fraxinus* (ash), *Platanus* (plane), *Ulmus* (elm), *Picea* (spruce), *Quercus*
(oak), *Gleditsia* (honey locust) and *Tilia* (lime, basswood or linden). From Stevenson *et al.* (2020).

may require more watering and other maintenance, but surely they are worth the expense and effort.

A growing trend is for urban areas to become *smart cities* where data is collected by sensors and citizens and then used to manage the city efficiently (Nitoslawski *et al.* 2019). In terms of green infrastructure, this could include water-stress sensors in trees linked to automatic irrigation systems. It could also allow data on tree performance in relation to local environmental conditions to be collected so that future tree planting could be optimised. This *Internet of Nature* could be expanded to incorporate all aspects of the city to improve the quality of life of people as well as the greenery. It could, for example, include suggested management options for an area of trees to be shared with citizens via virtual reality, allowing a collective choice to be made (Galle *et al.* 2019). This all fits within golden rule No. 2 (involving local people) and rule Nos. 8–10 (Fig. 245).

A positive development is the growing international movement to value trees. This is exemplified by the 17 *Sustainable Development Goals* of the United Nations, which includes the first two worthy goals of no world poverty and zero hunger but also includes promoting good health and wellbeing (Goal 3) and protecting, restoring and promoting life on land (Goal 15). Moreover, in 2011, the Bonn Challenge (www.bonnchallenge.org) was set up with the aim of restoring 350 million hectares of forest globally by 2030. As I write, there have been more than 70 pledges from more than 60 countries committed to restoring 210 million hectares of degraded and deforested lands. There have also been many tree-planting programmes proposed around the world, discussed in Chapter 2. On top of these international initiatives, many national programmes are being developed. For example, the UK has the vision of its 25 *Year Environment Plan,* which includes the goal of increasing woodland in England from 10 per cent of land area in 2019 to 12 per cent by 2060, which involves planting 180,000 hectares by the end of 2042 (the whole of the UK had 13 per cent of land area under woodland in 2019). This should not only put trees in the ground but also has positive political knock-on effects. For example, as I write, the *Tree-lined Streets Bill* is passing through the House of Commons which, if enacted, will require building developers to ensure that the streets of major new developments are lined with trees. Not all initiatives will work, but there is a growing consensus that trees and green spaces are valuable and need to be protected and enhanced.

In conclusion, although there are many pressures on trees, there is also plenty of optimism that the problems can be overcome in a variety of local to global ways that will ensure that trees will always grace our landscapes, both in the wild and in our towns and cities. This will require work – from governments and from individuals – but the reward of enjoying our trees will make this fully worthwhile.

References

Achatz, M., Morris, E.K., Müller F., Hilker, M. & Rillig, M.C. (2014). Soil hypha-mediated movement of allelochemicals: arbuscular mycorrhizae extend the bioactive zone of juglone. *Functional Ecology* 28, 1020–1029.

Acuña, R., Padilla, B.E., Flórez-Ramos, C.P. *et al* (2012). Adaptive horizontal transfer of a bacterial gene to an invasive insect pest of coffee. *PNAS* 109, 4197–4202.

Adams, J. (1905). The occurrence of yew in a peat bog in Queen's County. *Irish Naturalist* 14, 34 (plus facing plate).

Aerts, R. (1996). Nutrient resorption from senescing leaves of perennials: Are there general patterns? *Journal of Ecology* 84, 597–608.

Ågren, G.I., Axelsson, B., Flower-Ellis, J.G.K. *et al*. (1980). Annual carbon budget for a young scots pine. *Ecological Bulletins* 32, 307–313.

Ågren, G.I. & Weih, M. (2020). Multi-dimensional plant element stoichiometry – looking beyond carbon, nitrogen, and phosphorus. *Frontiers in Plant Science* 11, article 23.

Aitken, S.N. & Bemmels, J.B. (2015). Time to get moving: assisted gene flow of forest trees. *Evolutionary Applications* 9, 271–290.

Alexander, K., Butler, J. & Green, T. (2006). The value of different tree and shrub species to wildlife. *British Wildlife* 18(1), 18–28.

Aloni, R. (2010). The induction of vascular tissue by auxin. In P.J. Davies (ed.), *Plant Hormones: Biosynthesis, Signal Transduction, Action!*, 3rd edition. Springer, Berlin, pp. 485–518.

Alonso, C., Ramos-Cruz, D. & Becker, C. (2019). The role of plant epigenetics in biotic interactions. *New Phytologist* 221, 731–737.

Améglio, T., Ewers, F.W., Cochard, H. *et al*. (2001). Winter stem xylem pressure in walnut trees: effects of carbohydrates, cooling and freezing. *Tree Physiology* 21, 387–394.

Anderegg, W.R.L., Trugman, A.T., Badgley, G. *et al*. (2020). Climate-driven risks to the climate mitigation potential of forests. *Science* 368, eaaz7005.

Anderson, B. & Midgley, J.J. (2002). It takes two to tango but three is a tangle: mutualists and cheaters on the carnivorous plant *Roridula. Oecologia* 132, 369–373.

Angst, Š., Cajthaml, T., Angst, G. *et al*. (2017). Retention of dead standing plant biomass (marcescence) increases subsequent litter decomposition in the soil organic layer. *Plant Soil* 418, 571–579.

Anon (1948). *Woody-Plant Seed Manual*. Forest Service, US Department of Agriculture, Miscellaneous Publication No. 654.

Anon (2008). *Ancient Tree Guide No.4. What are Ancient, Veteran and other Trees of Special Interest?* Ancient Tree Forum, Woodland Trust, Grantham.

Archetti, M. (2009a). Phylogenetic analysis reveals a scattered distribution of autumn colours. *Annals of Botany* 103, 703–713.

Archetti, M. (2009b). Classification of hypotheses on the evolution of autumn colours. *Oikos* 118, 328–333.

Archetti, M., Richardson, A.D., O'Keefe, J. & Delpierre, N. (2013). Predicting climate change impacts on the amount and duration of autumn colors in a New England forest. *PLoS ONE* 8, e57373.

Arnon, D.I. & Stout, P.R. (1939). The essentiality of certain elements in minute quantity for plants with special reference to copper. *Plant Physiology* 14, 371–375.

Aschan, G. & Pfanz, H. (2003). Non-foliar photosynthesis – a strategy of additional carbon acquisition. *Flora* 198, 81–97.

Bakker, P.A.H.M., Berendsen, R.L., Doornbos, R.F., Wintermans, P.C.A. & Pieterse, C.M.J. (2013). The rhizosphere revisited: root microbiomics. *Frontiers in Plant Science* 4, article 165.

Ball, E., Hann, J., Kluge, M. *et al.* (1991). Ecophysiological comportment of the tropical CAM-tree *Clusia* in the field II. Modes of photosynthesis in trees and seedlings. *New Phytologist* 117, 483–491.

Barnes, B.V. (1966). The clonal growth habit of American aspens. *Ecology* 47, 439–447.

Bar-On, Y.M., Phillips, R. & Milo, R. (2018). The biomass distribution on Earth. *PNAS* 115, 6506–6511.

Barto, E.K., Weidenhamer, J.D., Cipollini, D. & Rillig, M.C. (2012). Fungal superhighways: do common mycorrhizal networks enhance below ground communication? *Trends in Plant Science* 17, 633–637.

Basler, D. & Körner, C. (2012). Photoperiod sensitivity of bud burst in 14 temperate forest tree species. *Agricultural and Forest Meteorology* 165, 73–81.

Bastin, J.-F., Finegold, Y., Garcia, C. *et al.* (2019). The global tree restoration potential. *Science* 365, 76–79.

Bayala, J. & Prieto, I. (2020). Water acquisition, sharing and redistribution by roots: applications to agroforestry systems. *Plant Soil* 453, 17–28.

Beech, E., Rivers, M., Oldfield, S. & Smith, P.P. (2017) GlobalTreeSearch: The first complete global database of tree species and country distributions, *Journal of Sustainable Forestry* 36, 454–489.

Beerling, D. (2007). *The Emerald Planet*. Oxford University Press, Oxford.

Belcher, C.M., Mills, B.J.W., Vitali, R. *et al.* (2021) The rise of angiosperms strengthened fire feedbacks and improved the regulation of atmospheric oxygen. *Nature Communications* 12, e503.

Bennie, J., Davies, T.W., Cruse, D. & Gaston, K.J. (2016). Ecological effects of artificial light at night on wild plants. *Journal of Ecology* 104, 611–620.

Berdanier, A.B., Miniat, C.F. & Clark, J.S. (2016). Predictive models for radial sap flux variation in coniferous, diffuse-porous and ring-porous temperate trees. *Tree Physiology* 36, 932–941.

Bevan-Jones, R. (2004). *The Ancient Yew: A History of Taxus baccata*. Windgather Press, Bollington.

BGCI (2021). State of the World's Trees. Botanic Gardens Conservation International, Richmond. Retrieved from https://www.bgci.org/wp/wp-content/uploads/2021/08/FINAL-GTAReportMedRes-1.pdf

Bigler, C., Gavin, D.G., Gunning, C. & Veblen, T.T. (2007). Drought induces lagged tree mortality in a subalpine forest in the Rocky Mountains. *Oikos* 116, 1983–1994.

Bingham, M.A. & Simard, S. (2012). Ectomycorrhizal networks of *Pseudotsuga menziesii* var. *glauca* trees facilitate establishment of conspecific seedlings under drought. *Ecosystems* 15, 188–199.

Bloemen J., Vergeynst L.L., Overlaet-Michiels L. & Steppe K. (2016). How important is woody tissue photosynthesis in poplar during drought stress? *Trees* 30, 63–72.

Blue, Y.A., Kusumi, J. & Satake, A. (2021). Copy number analyses of DNA repair genes reveal the role of poly(ADP-ribose) polymerase (PARP) in tree longevity. *iScience* 24, article 102779.

Bobbink, R., Hicks, K., Galloway, J. *et al.* (2010). Global assessment of nitrogen deposition effects on terrestrial plant diversity: a synthesis. *Ecological Applications* 20, 30–59.

Bogdziewicz, M., Kelly, D., Tanentzap, A.J. *et al.* (2020a). Climate change strengthens selection for mast seeding in European beech. *Current Biology* 30, 1–7.

Bogdziewicz, M., Kelly, D., Thomas, P.A., Lageard, J.G.A. & Hacket-Pain, A. (2020b). Climate warming disrupts mast seeding and its fitness benefits in European beech. *Nature Plants* 6, 88–94.

Bolmgren, K. & Cowan, P. (2008). Time–size tradeoffs: a phylogenetic comparative study of flowering time, plant height and seed mass in a north-temperate flora. *Oikos* 117, 424–429.

Borchert, R., Renner, S.S., Calle, Z. *et al.* (2005). Photoperiodic induction of synchronous flowering near the Equator. *Nature* 433, 627–629.

Bowker, G.E. & Crenshaw, H.C. (2007). Electrostatic forces in wind-pollination – Part 2: Simulations of pollen capture. *Atmospheric Environment* 41, 1596–1603.

Brandt, M., Tucker, C.J., Kariryaa, A. *et al.* (2020). An unexpectedly large count of trees in the West African Sahara and Sahel. *Nature* 587, 78–82.

Bräuning, A., De Ridder, M., Zafirov, N. *et al.* (2016). Tree-ring features: indicators of extreme event impacts. *IAWA Journal* 37, 206–231.

Breed, M.F., Harrison, P.A., Bischoff, A. *et al.* (2018). Priority actions to improve provenance decision-making. *BioScience* 68, 510–516.

Brelsford, C.C., Nybakken, L., Kotilainen, T.K. & Robson, T.M. (2019). The influence of spectral composition on spring and autumn phenology in trees. *Tree Physiology* 39, 925–950.

Breshears, D.D., Fontaine, J.B., Ruthrof, K.X. *et al.* (2021). Underappreciated plant vulnerabilities to heat waves. *New Phytologist* 231, 32–39.

Brown, A.A., Arnold, R.K., Fons, W.L., Sauer, F.M. & Reifsnyder, W.E. (1953). *Report on Project 3.3. Blast Damage to Trees – Isolated Conifers.*

Forest Service, US Department of Agriculture, Washington, DC.

Brown, N., Vanguelova, E., Parnell, S., Broadmeadow, S. & Denman, S. (2018). Predisposition of forests to biotic disturbance: predicting the distribution of Acute Oak Decline using environmental factors. *Forest Ecology and Management* 407, 145–154.

Browne, L., Wright, J.W., Fitz-Gibbon, S., Gugger, P.F. & Sork, V.L. (2019). Adaptational lag to temperature in valley oak (*Quercus lobata*) can be mitigated by genome-informed assisted gene flow. *PNAS* 116, 25179–25185.

Bugalho, M.N., Caldeira, M.C., Pereira, J.S., Aronson, J. & Pausas, J.G. (2011). Mediterranean cork oak savannas require human use to sustain biodiversity and ecosystem services. *Frontiers in Ecology and the Environment* 9, 278–286.

Buggs, R.J.A. (2021). The origin of Darwin's "abominable mystery". *American Journal of Botany* 108, 22–36.

Büntgen, U., Martínez-Peña, F., Aldea, J. *et al.* (2013). Declining pine growth in Central Spain coincides with increasing diurnal temperature range since the 1970s. *Global and Planetary Change* 107, 177–185.

Burger, A.E. (2005). Dispersal and germination of seeds of *Pisonia grandis*, an Indo-Pacific tropical tree associated with insular seabird colonies. *Journal of Tropical Ecology* 21, 263–271.

Burgess, M.D., Smith, K.W., Evans, K.L. *et al.* (2018). Tritrophic phenological match–mismatch in space and time. *Nature Ecology & Evolution* 2, 970–975.

Burgess, S.O., Adams, M.A., Turner, N.C. & Ong, C.K. (1998). The redistribution of soil water by tree root systems. *Oecologia* 115, 306–311.

Burns, R.M. & Honkala, B.H. (1990). *Silvics of North America, Vol. 1, Conifers*. Agriculture Handbook 654, Forest Service, US Department of Agriculture, Washington, DC.

Butler, J., Alexander, K.N.A. & Green, T. (2002). *Decaying Wood: An Overview of its Status and Ecology in the United Kingdom and Continental Europe*. Forest Service, US Department of Agriculture, General Technical Report PSW-GTR-181, Albany, California.

Calladine, A. & Pate, J.S. (2000). Haustorial structure and functioning of the root hemiparasitic tree *Nuytsia floribunda* (Labill.) R.Br. and water relationships with its hosts. *Annals of Botany* 85, 723–731.

Camarero, J.J. & Ortega-Martínez, M. (2019). Sancho, the oldest known Iberian shrub. *Dendrochronologia* 53, 32–36.

Cameron, D.D., Neal, A.L., van Wees, S.C.M. & Ton, J. (2013). Mycorrhiza-induced resistance: more than the sum of its parts? *Trends in Plant Science* 18, 539–545.

Carvalho, M.R., Jaramillo, C., de la Parra, F. *et al.* (2021). Extinction at the end-Cretaceous and the origin of modern Neotropical rainforests. *Science* 372, 63–68.

CAT (2013). *Zero Carbon Britain: Rethinking the Future*. Centre for Alternative Technology, Machynlleth, Powys.

Catney, P. & Henneberry, J. (2022). The political economy of street trees. In J. Woudstra & C.J. Allen (eds.), *The Politics of Street Trees*. Routledge, London, pp. 1–16.

Ceballosa, G., Ehrlich, P.R. & Raven, P.H. (2020). Vertebrates on the brink as indicators of biological annihilation and the sixth mass extinction. *PNAS* 117, 13596–13602.

Chang, C.Y.-Y., Bräutigam, K., Hüner, N.P.A. & Ensminger, E. (2021). Champions of winter survival: cold acclimation and molecular regulation of cold hardiness in evergreen conifers. *New Phytologist* 229, 675–691.

Chano, V., Collada, C. & Soto, A. (2017). Transcriptomic analysis of wound xylem formation in *Pinus canariensis*. *BMC Plant Biology* 17, e234.

Chawla, L. (2020). Childhood nature connection and constructive hope: a review of research on connecting with nature and coping with environmental loss. *People and Nature* 2, 619–642.

Chen, L., Hänninen, H., Rossi, S. *et al* (2020). Leaf senescence exhibits stronger climatic responses during warm than during cold autumns. *Nature Climate Change* 10, 777–780.

Chen, W., Xie, Z. & Zhou, Y. (2019). Proximity to roads reduces acorn dispersal effectiveness by rodents: implication for forest regeneration and management. *Forest Ecology and Management* 433, 625–632.

Chetan, A. & Brueton, D. (1994). *The Sacred Yew*. Penguin, London.

Chmielewski, F.-M. & Rötzer, T. (2001). Response of tree phenology to climate change across Europe. *Agricultural and Forest Meteorology* 108, 101–112.

Choat, B., Cobb, A. & Jansen, S. (2008). Structure and function of bordered pits: new discoveries and impacts on whole plant hydraulic function. *New Phytologist* 177, 608–626.

Choi, W.-G., Hilleary, R., Swanson, S.J., Kim, S.-H. & Gilroy, S. (2016). Rapid, long-distance electrical and calcium signaling in plants. *Annual Review of Plant Biology* 67, 287–307.

Christensen-Dalsgaard, K.K. & Tyree, M.T. (2013). Does freezing and dynamic flexing of frozen branches impact the cavitation resistance of

Malus domestica and the *Populus* clone Walker? *Oecologia* 173, 665–674.

Cirelli, H.D., Jagels, R. & Tyree, M.T. (2008). Toward an improved model of maple sap exudation: the location and role of osmotic barriers in sugar maple, butternut and white birch. *Tree Physiology* 28, 1145–1155.

Clifford, H.T. & Monteith, G.B. (1989). A three phase seed dispersal mechanism in Australian quinine bush (*Petalostigma pubescens* Domin). *Biotropica* 21, 284–286.

Coleman, D., Samuels, A.L., Guy, R.D. & Mansfield, S.D. (2008). Perturbed lignification impacts tree growth in hybrid poplar – a function of sink strength, vascular integrity, and photosynthetic assimilation. *Plant Physiology* 148, 1229–1237.

Coleman, M. (2009). *Wych Elm*. Royal Botanic Garden, Edinburgh.

Collins, M., Crawley, M.J. & McGavin, G.C. (1983). Survivorship of the sexual and agamic generations of *Andricus quercuscalicis* on *Quercus cerris* and *Q. robur*. *Ecological Entomology* 8, 133–138.

Conservation Foundation (2017). *The Great British Elm Experiment*. Retrieved from http://www.conservationfoundation.co.uk/elm

Cornille, A., Antolín, F., Garcia, E. et al. (2019). A multifaceted overview of apple tree domestication. *Trends in Plant Science* 24, 770–782.

Cornille, A., Gladieux, P., Smulders, M.J.M. et al. (2012). New insight into the history of domesticated apple: secondary contribution of the European wild apple to the genome of cultivated varieties. *PLoS Genetics* 8, e1002703.

Costes, E. & García-Villanueva, E. (2007). Clarifying the effects of dwarfing rootstock on vegetative and reproductive growth during tree development: a study on apple trees. *Annals of Botany* 100, 347–357.

Cottrell, J.E., Krystufek, V., Tabbener, H.E. et al (2005). Postglacial migration of *Populus nigra* L.: lessons learnt from chloroplast DNA. *Forest Ecology and Management* 219, 293–312.

Coulter, A., Poulis, B.A.D. & von Aderkas, P. (2012). Pollination drops as dynamic apoplastic secretions. *Flora* 207, 482–490.

Critchfield, W.B. (1971). Shoot growth and heterophylly in *Acer*. *Journal of the Arnold Arboretum* 52, 248–266.

Crowther, T.W., Glick, H.B., Covey, K.R. et al. (2015). Mapping tree density at a global scale. *Nature* 525, 201–205.

Cuny, H.E., Rathgeber, C.B.K., Frank, D., Fonti, P. & Fournier, M. (2014). Kinetics of tracheid development explain conifer tree-ring structure. *New Phytologist* 203, 1231–1241.

Curtis, P.G., Slay, C.M., Harris, N.L., Tyukavina, A. & Hansen, M.C. (2018). Classifying drivers of global forest loss. *Science* 361, 1108–1111.

Cutler, D.F. & Richardson, I.B.K. (1981). *Tree Roots and Buildings*. Construction Press (Longman), London.

D'Andrea, E., Rezaie, N., Battistelli, A. et al. (2019). Winter's bite: beech trees survive complete defoliation due to spring late-frost damage by mobilizing old C reserves. *New Phytologist* 224, 625–631.

Danso, S.K.A., Zapata, F. & Awonaike, K.O. (1995). Measurement of biological N_2 fixation in field-grown *Robinia pseudoacacia* L. *Soil Biology and Biochemistry* 27, 415–419.

Darwin, C.R. (1876). *The Effects of Cross and Self Fertilisation in the Vegetable Kingdom*. John Murray, London.

Dawson, T.E. & Goldsmith, G.R. (2018). The value of wet leaves. *New Phytologist* 219, 1156–1169.

De Candolle, Prof. (1833). On the longevity of trees and the means of ascertaining it. *Edinburgh New Philosophical Journal* 15, 330–348.

Delaux, P.-M. & Schornack, S. (2021). Plant evolution driven by interactions with symbiotic and pathogenic microbes. *Science* 371, eaba6605.

Del Tredici, P. (2000). Aging and rejuvenation in trees. *Arnoldia* 59, 10–16.

De Micco, V., Balzano, A., Wheeler, E.A. & Baas, P. (2016). Tyloses and gums: a review of structure, function, and occurrence of vessel occlusions. *IAWA Journal* 37, 186–205.

De Roo, L., Salomón, R.L., Oleksyn, J. & Steppe, K. (2020). Woody tissue photosynthesis delays drought stress in *Populus tremula* trees and maintains starch reserves in branch xylem tissues. *New Phytologist* 228, 70–81.

Der Weduwen, D. & Ruxton, G.D. (2019). Secondary dispersal mechanisms of winged seeds: a review. *Biological Reviews* 94, 1830–1838.

Di Bitetti, M.S. (2019). Primates bark-stripping trees in forest plantations – a review. *Forest Ecology and Management* 449, e117482.

Di Sacco, A., Hardwick, K.A., Blakesley, D. et al. (2021). Ten golden rules for reforestation to optimize carbon sequestration, biodiversity recovery and livelihood benefits. *Global Change Biology* 27, 1328–1348.

Dixon, H.H. & Jolly, J. (1895). On the ascent of sap. *Philosophical Transactions of the Royal Society of London B* 186, 563–576.

Dominy, N.J., Lucas, P.W., Ramsden, L.W. et al. (2002). Why are young leaves red? *Oikos* 98, 163–176.

Dossa, G.G.O., Schaefer, D., Zhang, J.-L. et al. (2017). The cover uncovered: bark control over wood decomposition. *Journal of Ecology* 106, 2147–2160.

Dox, I., Gričar, J., Marchand, L.J. *et al.* (2020). Timeline of autumn phenology in temperate deciduous trees. *Tree Physiology* 40, 1001–1013.

Duchelle, A.E., Simonet, G., Sunderlin, W.D. & Wunder, S. (2018). What is REDD+ achieving on the ground? *Current Opinion in Environmental Sustainability* 32, 134–140.

Dujesiefken, D., Fay, N., de Groot, J.-W. & de Berker, N. (2016). *Trees – A Lifespan Approach. Contributions to Arboriculture from European Practitioners.* Fundacja EkoRozwoju, Wrocław.

Duursma, E.K. & Boisson, P.R.M. (1993). Global stability of atmospheric oxygen. *European Review* 1, 217–229.

Edwards, P.J., Kollmann, J. & Fleischmann, K. (2003). Life history evolution in *Lodoicea maldivica* (Arecaceae). *Nordic Journal of Botany* 22, 221–237.

Elhacham, E., Ben-Uri, L., Grozovski, J., Bar-On, Y.M. & Milo, R. (2020). Global human-made mass exceeds all living biomass. *Nature* 588, 442–444.

Ellis, E.C., Gauthier, N., Goldewijk, K.K. *et al* (2021). People have shaped most of terrestrial nature for at least 12,000 years. *PNAS* 118, e2023483118.

Ellison, A.M., Orwig, D.A., Fitzpatrick, M.C. & Preisser, E.L. (2018). The past, present, and future of the hemlock woolly adelgid (*Adelges tsugae*) and its ecological interactions with eastern hemlock (*Tsuga canadensis*) forests. *Insects* 9, e172.

Ellmore, G.S., Zanne, A.E. & Orians, C.M. (2006). Comparative sectoriality in temperate hardwoods: hydraulics and xylem anatomy. *Botanical Journal of the Linnean Society* 150, 61–71.

Ennos, R., Cottrell, J., Hall, J. & O'Brien, D. (2019). Is the introduction of novel exotic forest tree species a rational response to rapid environmental change? – a British perspective. *Forest Ecology and Management* 432, 718–728.

Falavigna, V.S., Guitton, B., Costes, E. & Andrés, F. (2019). I want to (bud) break free: the potential role of DAM and SVP-like genes in regulating dormancy cycle in temperate fruit trees. *Frontiers in Plant Science* 9, e1990.

ffrench-Constant, R.H., Somers-Yeates, R., Bennie, J. *et al.* (2016). Light pollution is associated with earlier tree budburst across the United Kingdom. *Proceedings of the Royal Society B* 283, e20160813.

Field, J.P., Breshears, D.D., Bradford, J.B. *et al.* (2020). Forest management under megadrought: urgent needs at finer scale and higher intensity. *Frontiers in Forests and Global Change* 3, article 502669.

Field, T.S. & Brodribb, T.J. (2005). A unique mode of parasitism in the conifer coral tree *Parasitaxus*

ustus (Podocarpaceae). *Plant, Cell and Environment* 28, 1316–1325.

Fine, P.V.A., Miller, Z.J., Mesones, I. *et al.* (2006) The growth-defense trade-off and habitat specialization by plants in Amazonian forests. *Ecology* 87 Suppl., S150–S162.

Fleming, P.A., Hofmeyr, S.D., Nicolson, S.W. & du Toit, J.T. (2006). Are giraffes pollinators or flower predators of *Acacia nigrescens* in Kruger National Park, South Africa? *Journal of Tropical Ecology* 22, 247–253.

Fons, W.L. & Storey, T.G. (1955). Project 3.3. *Blast Effects on Tree Stand.* Forest Service, US Department of Agriculture, Washington, DC.

Forestry Commission (2003). *National Inventory of Woodland and Trees: Great Britain.* Inventory Report. Forestry Commission, Edinburgh.

Fox, J. & Castella, J.C. (2013). Expansion of rubber (*Hevea brasiliensis*) in Mainland Southeast Asia: what are the prospects for smallholders? *Journal of Peasant Studies* 40, 155–170.

Frank, M.H. & Chitwood, D.H. (2016). Plant chimeras: the good, the bad, and the 'Bizzaria'. *Developmental Biology* 419, 41–53.

Friggens, N.L., Hester, A.J., Mitchell, R.J. *et al.* (2020). Tree planting in organic soils does not result in net carbon sequestration on decadal timescales. *Global Change Biology* 26, 5178–5188.

Fu, Z.Q. & Dong, X. (2013). Systemic acquired resistance: turning local infection into global defense. *Annual Review of Plant Biology* 64, 839–863.

Galle, N.J., Nitoslawski, S.A. & Pilla, F. (2019). The internet of nature: How taking nature online can shape urban ecosystems. *The Anthropocene Review* 6, 279–287.

Garbelotto, M., Schmidt, D.J. & Harnik, T.Y. (2007). Phosphite injections and bark application of phosphite + Pentrabark™ control sudden oak death in coast live oak. *Arboriculture & Urban Forestry* 33, 309–317.

García-Suárez, A.M., Butlera, C.J. & Baillie, M.G.L. (2009). Climate signal in tree-ring chronologies in a temperate climate: a multi-species approach. *Dendrochronologia* 27, 183–198.

Garonna, I., De Jong, R., De Wit, A.J.W. *et al.* (2014). Strong contribution of autumn phenology to changes in satellite-derived growing season length estimates across Europe (1982–2011). *Global Change Biology* 20, 3457–3470.

Gartland, J.S., McHugh, A.T., Brasier, C.M. *et al.* (2000). Regeneration of phenotypically normal English elm (*Ulmus procera*) plantlets following transformation with an *Agrobacterium tumefaciens* binary vector. *Tree Physiology* 20, 901–907.

Gebbeken, N., Warnstedt, P. & Rüdiger, L. (2018). Blast protection in urban areas using protective plants. *International Journal of Protective Structures* 9, 226–247.

Genesio, L., Bassi, R. & Miglietta, F. (2021). Plants with less chlorophyll: a global change perspective. *Global Change Biology* 27, 959–967.

Gianoli, E. & Carrasco-Urra, F. (2014). Leaf mimicry in a climbing plant protects against herbivory. *Current Biology* 24, 984–987.

Gibbard, S., Caldeira, K., Bala, G., Phillips, T.J. & Wickett, M. (2005). Climate effects of global land cover change, *Geophysical Research Letters* 32, L23705.

Godfrey, A. (2003). *A Review of the Invertebrate Interest of Coarse Woody Debris in England.* English Nature Research Reports, Number 513, English Nature, Peterborough.

Godoy, J.A. & Jordano, P. (2001). Seed dispersal by animals: exact identification of source trees with endocarp DNA microsatellites. *Molecular Ecology* 10, 2275–2283.

Gömöry, D., Krajmerová, D., Hrivnák, M. & Longauer, R. (2020). Assisted migration vs. close-to-nature forestry: what are the prospects for tree populations under climate change? *Central European Forestry Journal* 66, 63–70.

Goodwin, C.E.D., Hodgson, D.J., Al-Fulaij, N. *et al.* (2017) Voluntary recording scheme reveals ongoing decline in the United Kingdom hazel dormouse *Muscardinus avellanarius* population. *Mammal Review* 47, 183–197.

Gora, E.M. & Esquivel-Muelbert, A. (2021). Implications of size-dependent tree mortality for tropical forest carbon dynamics. *Nature Plants* 7, 384–391.

Gougherty, A.V. & Gougherty, S.W. (2018). Sequence of flower and leaf emergence in deciduous trees is linked to ecological traits, phylogenetics, and climate. *New Phytologist* 220, 121–131.

Grabosky, J. & Bassuk, N. (2016). Seventeen years' growth of street trees in structural soil compared with a tree lawn in New York City. *Urban Forestry and Urban Greening* 16, 103–109.

Grantham, H.S., Duncan, A., Evans, T.D. *et al.* (2020). Anthropogenic modification of forests means only 40% of remaining forests have high ecosystem integrity. *Nature Communications* 11, article 5978.

Greater London Authority (2018). *London Environment Strategy.* Greater London Authority, London.

Greene, D.F. & Johnson, E.A. (1992). Fruit abscission in *Acer saccharinum* with reference to seed dispersal. *Canadian Journal of Botany* 70, 2277–2283.

Groover, A.T. (2005). What genes make a tree a tree? *Trends in Plant Science* 10, 210–214.

Gros-Balthazard, M., Flowers, J.M., Hazzouri, K.M. *et al.* (2021). The genomes of ancient date palms germinated from 2,000 y old seeds. *PNAS* 118, e2025337118.

Grubb, P.J. (1992). Presidential address: a positive distrust in simplicity – lessons from plant defences and from competition among plants and among animals. *Journal of Ecology* 80, 585–610.

Gunton, R.M., van Asperen, E.N., Basden, A. *et al.* (2017) Beyond ecosystem services: valuing the invaluable. *Trends in Ecology & Evolution* 32, 249–257.

Hacket-Pain, A.J., Lageard, J.G.A. & Thomas, P.A. (2017). Drought and reproductive effort interact to control growth of a temperate broadleaved tree species (*Fagus sylvatica*). *Tree Physiology* 37, 744–754.

Hafner, B.D., Hesse, B.D., Bauerle, T.L. & Grams, T.E.E. (2020a). Water potential gradient, root conduit size and root xylem hydraulic conductivity determine the extent of hydraulic redistribution in temperate trees. *Functional Ecology* 34, 561–574.

Hafner, B.D., Hesse, B.D. & Grams, T.E.E. (2020b). Friendly neighbours: hydraulic redistribution accounts for one quarter of water used by neighbouring drought stressed tree saplings. *Plant Cell & Environment* 2020, 1–14.

Hall, C.M., James, M. & Baird, T. (2011). Forests and trees as charismatic megaflora: implications for heritage tourism and conservation. *Journal of Heritage Tourism* 6, 309–323.

Hallmark, A.J., Maurer, G.E., Pangle, R.E. & Litvak, M.E. (2021). Watching plants' dance: movements of live and dead branches linked to atmospheric water demand. *Ecosphere* 12, e03705.

Hamilton, W.D. & Brown, S.P. (2001). Autumn tree colours as a handicap signal. *Proceedings of the Royal Society of London B* 268, 1489–1493.

Hammond, W.M., Yu, K., Wilson, L.A. *et al.* (2019) Dead or dying? Quantifying the point of no return from hydraulic failure in drought-induced tree mortality. *New Phytologist* 223, 1834–1843.

Hansen, M.M., Jones, R. & Tocchini, K. (2017). Shinrin-yoku (forest bathing) and nature therapy: a state-of-the-art review. *International Journal of Environmental Research & Public Health* 14, e851.

Harper, A.L., McKinney, L.V., Nielsen, L.R. *et al.* (2016). Molecular markers for tolerance of European ash (*Fraxinus excelsior*) to dieback disease identified using Associative Transcriptomics. *Scientific Reports* 6, e19335.

Harper, J.L., Lovell, P.H. & Moore, K.G. (1970). The shapes and sizes of seeds. *Annual Review of Ecology and Systematics* 1, 327–356.

Hartmann, H. & Trumbore, S. (2016). Understanding the roles of nonstructural carbohydrates in forest trees – from what we can measure to what we want to know. *New Phytologist* 211, 386–403.

Helmisaari, H.-S., Makkonen, K., Kellomäki, S., Valtonen, E. & Mälkönen, E. (2002). Below- and above-ground biomass, production and nitrogen use in Scots pine stands in eastern Finland. *Forest Ecology and Management* 165, 317–326.

Hertel, D. (2011). Tree roots in canopy soils of old European beech trees – an ecological reassessment of a forgotten phenomenon. *Pedobiologia* 54, 119–125.

Herwitz, S.R. (1991). Aboveground adventitious roots and stemflow chemistry of *Ceratopetalum virchowii* in an Australian montane tropical rain forest. *Biotropica* 23, 210–218.

Hewitt, N. (1998). Seed size and shade-tolerance: a comparative analysis of North American temperate trees. *Oecologia* 114, 432–440.

Hill, J.L., Jr & Hollender, C.A. (2019). Branching out: new insights into the genetic regulation of shoot architecture in trees. *Current Opinion in Plant Biology* 47, 73–80.

Hindson, T.R., Moir, A.K. & Thomas, P.A. (2019). Challenging Constant Annual Increment as a suitable model for estimating the ages of old yews. *Quarterly Journal of Forestry* 113, 184–188.

Hirons, A.D. & Sjöman, H. (2019). *Tree Species Selection for Green Infrastructure*, Issue 1.3. Trees & Design Action Group.

Hirons, A.D. & Thomas, P.A. (2018). *Applied Tree Biology*. Wiley Blackwell, Oxford.

Hirons, A.D., Watkins, J.H.R., Baxter, T.J. *et al.* (2020). Using botanic gardens and arboreta to help identify urban trees for the future. *Plants People Planet* 3, 182–193.

Hoang, N.T. & Kanemoto, K. (2021). Mapping the deforestation footprint of nations reveals growing threat to tropical forests. *Nature Ecology and Evolution* 5, 845–853.

Hoch, G. & Körner, C. (2009). Growth and carbon relations of tree line forming conifers at constant vs. variable low temperatures *Journal of Ecology* 99, 57–66.

Hoch, G., Richter, A. & Körner, C. (2003). Non-structural carbon compounds in temperate forest trees. *Plant Cell Environment* 26, 1067–1081.

Hodge, S.J. (1991). *Improving the Growth of Established Amenity Trees: Fertilizer and Weed Control*. Arboriculture Research Note 103–191, Forestry Commission, Farnham.

Hölttä, T., Dominguez Carrasco, M.D.R., Salmon, Y. *et al.* (2018). Water relations in silver birch during springtime: how is sap pressurised? *Plant Biology* 20, 834–847.

Holyoak, D.T. (1983). The identity and origins of *Picea abies* (L.) Karsten from the Chelford Interstadial (Late Pleistocene) of England. *New Phytologist* 95, 153–157.

Hopper, S.D. (2010). 660. *Nuytsia floribunda*. Loranthaceae. *Curtis's Botanical Magazine* 26, 333–368.

Horn, H.S., Nathan, R. & Kaplan, S.R. (2001). Long-distance dispersal of tree seeds by wind. *Ecological Research* 16, 877–885.

Horton, T.R., Bruns, T.D. & Parker, V. (1999). Ectomycorrhizal fungi associated with *Arctostaphylos* contribute to *Pseudotsuga menziesii* establishment. *Canadian Journal of Botany* 77, 93–102.

Hüppe, B., & Röhrig, E. (1996). Ein Mischbestand mit Bergulmen im Kommunal-Forstamt Haina (Hessen). *Forstarchiv* 67, 207–211.

Ikei, H., Song, C. & Miyazaki, Y. (2016). Physiological effects of wood on humans: a review. *Journal of Wood Science* 63, 1–23.

Illies, I. & Mühlen, W. (2007). The foraging behaviour of honeybees and bumblebees on late blooming lime trees (*Tilia* spec) (Hymenoptera: Apidae). *Entomologia Generalis* 30, 155–165.

IPCC (2018). *Global Warming of 1.5 °C*. Intergovernmental Panel on Climate Change, Geneva..

IPCC (2021). *Climate Change 2021. The Physical Science Basis*. Intergovernmental Panel on Climate Change, Geneva.

Jackson, G. (1974). Cryptogeal germination and other seedling adaptations to the burning of vegetation in savanna regions: the origin of the pyrophytic habit. *New Phytologist* 73, 771–780.

Jackson, J. (2008). *The Thief at the End of the World: Rubber, Power and the Seeds of Empire*. Viking, New York.

Jackson, R.B., Banner, J.L., Jobbágy, E.G., Pockman, W.T. & Wall, D.H. (2002). Ecosystem carbon loss with woody plant invasion of grasslands. *Nature* 418, 623–626.

Jagels, R., Equiza, M.A., Maguire, D.A. & Cirelli, D. (2018). Do tall tree species have higher relative stiffness than shorter species? *American Journal of Botany* 105, 1617–1630.

Jakoby, G., Rog, I., Shtein, I. *et al.* (2020). Tree forensics: Modern DNA barcoding and traditional anatomy identify roots threatening an ancient necropolis. *Plants People Planet* 3, 211–219.

Janzen, D.H. (1969). Seed-eaters versus seed size, number, toxicity and dispersal. *Evolution* 23, 1–27.

Janzen, D.H. (1971). Seed predation by animals. *Annual Review of Ecology and Systematics 2*, 465–492.

Janzen, D.H. & Martin, P.S. (1982). Neotropical anachronisms: the fruits the gomphotheres ate. *Science* 215, 19–27.

Jauhar, P.P. & Joshi, A.B. (1970). The concept of species and "microspecies". *Taxon* 19, 77–79.

Jia, S., Wang, X., Yuan, Z. *et al.* (2020). Tree species traits affect which natural enemies drive the Janzen-Connell effect in a temperate forest. *Nature Communications* 11, e286.

Johns, J.W., Yost, J.M., Nicolle, D., Igic, B. & Ritter, M.K. (2017). Worldwide hemisphere-dependent lean in Cook pines. *Ecology* 98, 2482–2484.

Jones, G.M., Cassidy, N.J., Thomas, P.A., Plante, S. & Pringle, J.K. (2009). Imaging and monitoring tree-induced subsidence using electrical resistivity imaging. *Near Surface Geophysics* 2009, 191–206.

Jordano, P. (2017). What is long-distance dispersal? And a taxonomy of dispersal events. *Journal of Ecology* 105, 75–84.

Joynt, J.L.R. & Kang, J. (2010). The influence of preconceptions on perceived sound reduction by environmental noise barriers. *Science of the Total Environment* 408, 4368–4375.

Kachroo, A. & Kachroo, P. (2020). Mobile signals in systemic acquired resistance. *Current Opinion in Plant Biology* 58, 41–47.

Kainer, K.A., Wadt, L.H.O. & Staudhammer, C.L. (2014). Testing a silvicultural recommendation: Brazil nut responses 10 years after liana cutting. *Journal of Applied Ecology* 51, 655–663.

Kashyap, A., Planas-Marquès, M., Capellades, M., Valls, M. & Coll, N.S. (2021). Blocking intruders: inducible physico-chemical barriers against plant vascular wilt pathogens. *Journal of Experimental Botany* 72, 184–198.

Kay, Q.O.N. (1985). Nectar from willow catkins as a food source for Blue Tits. *Bird Study* 32, 40–45.

Keenan, T.F. & Richardson, A.D. (2015). The timing of autumn senescence is affected by the timing of spring phenology: implications for predictive models. *Global Change Biology* 21, 2634–2641.

Kelly, D. (2020). Mast seeding: the devil (and the delight) is in the detail. *New Phytologist* 229, 1829–1831.

Kelly, P.E., Cook, E.R. & Larson, D.W. (1992). Constrained growth, cambial mortality, and dendrochronology of ancient *Thuja occidentalis* on cliffs of the Niagara escarpment: an eastern version of bristlecone pine? *International Journal of Plant Science* 153, 117–127.

Kembro, J.M., Lihoreau, M., Garriga, J., Raposo, E.P. & Bartumeus, F. (2019). Bumblebees learn foraging routes through exploitation–exploration cycles. *Journal of the Royal Society Interface* 16, e20190103.

Kennedy, C.E.J. & Southwood, T.R.E. (1984). The number of species of insects associated with British trees: A re-analysis. *Journal of Animal Ecology* 53, 455–478.

Keskitalo, J., Bergquist, G., Gardeström, P. & Jansson, S. (2005). A cellular timetable of autumn senescence. *Plant Physiology* 139, 1635–1648.

Kim, H.N., Jin, H.Y., Kwak, M.J. *et al.* (2017). Why does *Quercus suber* species decline in Mediterranean areas? *Journal of Asia-Pacific Biodiversity* 10, 337–341.

Kirk, R.L. & Stenhouse, N.S. (1953). Ability to smell solutions of potassium cyanide. *Nature* 171, 698–699.

Kirk, W.D.J. & Howes, F.N. (2012). *Plants for Bees: A Guide to the Plants that Benefit the Bees of the British Isles*. International Bee Research Association, Cardiff.

Klein, T., Siegwolf, R.T.W. & Körner, C. (2016). Belowground carbon trade among tall trees in a temperate forest. *Science* 352, 342–344.

Klironomos, J.N. & Hart, M.M. (2001). Food-web dynamics: animal nitrogen swap for plant carbon. *Nature* 410, 651–652.

Kobashikawa, S., Trentin, B. & Koike, S. (2019). The benefit of wrapping trees in biodegradable material netting to protect against bark stripping by bears extends to surrounding stands. *Forest Ecology and Management* 437, 134–138.

Koch, H. & Stevenson, P.C. (2017). Do linden trees kill bees? Reviewing the causes of bee deaths on silver linden (*Tilia tomentosa*). *Biology Letters* 13, 20170484.

Koenig, W.D. & Mumme, R.L. (1988). Ecological consequences of acorn storage. In W.D. Koenig & R.L. Mumme (eds), *Population Ecology of the Cooperatively Breeding Acorn Woodpecker*. Monographs in Population Biology 24, Princeton University Press, Princeton, pp. 70–111.

Koike, T. (1990). Autumn coloring, photosynthetic performance and leaf development of deciduous broad-leaved trees in relation to forest succession. *Tree Physiology* 7, 21–32.

Kollas, C., Gutsch, M., Hommel, R., Lasch-Born, P. & Suckow, F. (2017). Mistletoe-induced growth reductions at the forest stand scale. *Tree Physiology* 38, 735–744.

Konter, O., Krusic, P.J., Trouet, V. & Esper, J. (2017). Meet Adonis, Europe's oldest dendrochronologically dated tree. *Dendrochronologia* 42, 12.

Körner, C. (2005). An introduction to the functional diversity of temperate forest trees. In M. Scherer-Lorenzen, C. Körner & E.-D. Schulze (eds), *Forest Diversity and Function*. Ecological Studies 176. Springer, Berlin, pp. 13–37.

Körner, C., Basler, D., Hoch, G. *et al.* (2016). Where, why and how? Explaining the low-temperature range limits of temperate tree species. *Journal of Ecology* 104, 1076–1088.

Kozlowski, T.T. (1968). *Water Deficits and Plant Growth*. Academic Press, New York.

Krisnawati, H., Varis, E., Kallio, M. & Kanninen, M. (2011). Paraserianthes falcataria (*L.) Nielsen: Ecology, Silviculture and Productivity*. Center for International Forestry Research. Retrieved from http://www.jstor.org/stable/resrep02123.5

Lanner, R.M. (1996). *Made for Each Other: A Symbiosis of Birds and Pines*. Oxford University Press, Oxford.

Lanner, R.M. & Connor, K.F. (2001). Does bristlecone pine senesce? *Experimental Gerontology* 36, 675–685.

Larcher, W. (2003). *Physiological Plant Ecology*, 4th edition. Springer, Berlin.

Larson, D.W., Doubt, J. & Matthes-Sears, U. (1994). Radially sectored hydraulic pathways in the xylem of *Thuja occidentalis* as revealed by the use of dyes. *International Journal of Plant Sciences* 155, 569–582.

Lechowicz, M.J. (1984). Why do temperate deciduous trees leaf out at different times? Adaptation and ecology of forest communities. *The American Naturalist* 124, 821–842.

Ledford, H. (2014). Brazil considers transgenic trees. *Nature* 512, 357.

Ledo, A., Paul, K.I., Burslem, D.F.R.P. *et al.* (2018). Tree size and climatic water deficit control root to shoot ratio in individual trees globally. *New Phytologist* 217, 8–11.

Lee, D.W., O'Keefe, J., Holbrook, N.M. & Feild, T.S. (2003). Pigment dynamics and autumn leaf senescence in a New England deciduous forest, eastern USA. *Ecological Research* 18, 677–694.

Lee, J.-E., Oliveira, R.S., Dawson, T.E. & Fung, I. (2005). Root functioning modifies season al climate. *PNAS* 102, 17576–17581.

Leishman, M.R., Wright, I.J., Moles, A.T. & Westoby, M. (2000). The evolutionary ecology of seed size. In M. Fenner (ed.), *Seeds: The Ecology of Regeneration in Plant Communities*, 2nd edition. CABI, Wallingford, pp. 31–57.

Leite, C. & Pereira, H. (2017). Cork-containing barks – a review. *Frontiers in Material* 3, article 63.

Lendvay, B., Hartmann, M., Brodbeck, S. *et al.* (2018). Improved recovery of ancient DNA from subfossil wood – application to the world's oldest Late Glacial pine forest. *New Phytologist* 217, 1737–1748.

Lens, F., Davin, N., Smets, E. & del Arco, M. (2013). Insular woodiness on the Canary Islands: a remarkable case of convergent evolution. *International Journal of Plant Sciences* 174, 992–1013.

Lentink, D., Dickson, W.B., van Leeuwen, L.J. & Dickinson, M.H. (2009). Leading-edge vortices elevate lift of autorotating plant seeds. *Science* 324, 1438–1440.

Leuschner, C., Hertel, D., Coners, H. & Büttner, V. (2001). Root competition between beech and oak: a hypothesis. *Oecologia* 126, 276–284.

Lewis, S.L., Brando, P.M., Phillips, O.L., van der Heijden, G.M.F. & Nepstad, D. (2011). The 2010 Amazon drought. *Science* 331, 554.

Li, L.-F. & Olsen, K.M. (2016). To have and to hold: selection for seed and fruit retention during crop domestication. *Current Topics in Developmental Biology* 119, 63–109.

Li, M., Van Renterghem, T., Kang, J., Verheyen, K. & Botteldooren, D. (2020). Sound absorption by tree bark. *Applied Acoustics* 165, e107328.

Liao, C., Luo, Y., Fang, C. & Li, B. (2010). Ecosystem carbon stock influenced by plantation practice: implications for planting forests as a measure of climate change mitigation. *PLoS ONE* 5, e10867.

Lin, Y.-H., Tsai, C.-C., Sullivan, W.C., Chang, P.-J. & Chang, C.-Y. (2014). Does awareness affect the restorative function and perception of street trees? *Frontiers in Psychology* 5, e906.

Lindenmayer, D.B., Laurance, W.F. & Franklin, J.F. (2012). Global decline in large old trees. *Science* 338, 1305–1306.

Lindsey, P. & Bassuk, N. (1991). Specifying soil volumes to meet the water needs of mature urban street trees and trees in containers. *Journal of Arboriculture* 17, 141–148.

Lipson, D. & Näsholm, T. (2001). The unexpected versatility of plants: organic nitrogen use and availability in terrestrial ecosystems. *Oecologia* 128, 305–316.

Loarie, S.R., Duffy, P.B., Hamilton, H. *et al.* (2009). The velocity of climate change. *Nature* 462, 1052–1055.

Logan, S.A. (2016). *Ancient Relicts in the Limelight: An Evolutionary Study of Diversity and Demographic History in Species of the Broad-leaved Temperate Forest Tree Genus Tilia*. PhD Thesis, Newcastle University, Newcastle.

Logan, S.A., Phuekvilai, P. & Wolff, K. (2015). Ancient woodlands in the limelight: delineation and genetic structure of ancient woodland species *Tilia cordata* and *Tilia platyphyllos* (Tiliaceae) in the UK. *Tree Genetics & Genomes* 11, article 52.

Lonsdale, D. (2013a). *Ancient and Other Veteran Trees: Further Guidance on Management*. The Tree Council, London.

Lonsdale, D. (2013b). The recognition of functional units as an aid to tree management, with particular reference to veteran trees. *Arboricultural Journal* 35, 188–201.

Lopez, D., Franchel, J., Venisse, J.-S. *et al.* (2021). Early transcriptional response to gravistimulation in poplar without phototropic confounding factors. *AoB PLANTS* 13, plaa071.

Lowe, J. (1897). *The Yew Trees of Great Britain and Ireland.* Macmillan, London.

Lusk, C.H., Clearwater, M.J., Laughlin, D.C. *et al.* (2018). Frost and leaf-size gradients in forests: global patterns and experimental evidence. *New Phytologist* 219, 565–573.

Lüttge, U. & Hertel, B. (2009). Diurnal and annual rhythms in trees. *Trees* 23, 683–700.

Lutz, J.A., Furniss, T.J., Johnson, D.J. *et al.* (2018). Global importance of large-diameter trees. *Global Ecology and Biogeography* 27, 849–864.

Lyford, W.H. (1980). *Development of the Root System of Northern Red Oak* (Quercus rubra L.). Harvard Forest Papers, No. 21. Harvard University, Massachusetts.

Mabey, R. (2019). Scribbling in the margins. *British Wildlife* 31, 39.

Mackey, B., Prentice, I.C., Steffen, W. *et al.* (2013). Untangling the confusion around land carbon science and climate change mitigation policy. *Nature Climate Change* 3, 552–557.

Manuel, C. (2020). *The Miyawaki Method – Data and Concepts.* Urban Forests, Belgium. Retrieved from http://urban-forests.com/wp-content/uploads/2020/05/Urban-Forests-report-The-Miyawaki-method---Data-concepts.pdf

Marron, N. & Ceulemans, R. (2006). Genetic variation of leaf traits related to productivity in a *Populus deltoides* × *Populus nigra* family. *Canadian Journal of Forest Research* 36, 390–400.

Marron, N., Dillen, S.Y. & Ceulemans, R. (2007). Evaluation of leaf traits for indirect selection of high yielding poplar hybrids. *Environmental and Experimental Botany* 61, 103–116.

Marschner, P. (2012). *Mineral Nutrition of Higher Plants*, 3rd edition. Academic Press, San Diego.

Marselle, M.R., Bowler, D.E., Watzema, J. *et al.* (2020). Urban street tree biodiversity and antidepressant prescriptions. *Scientific Reports* 10, e22445.

Martínez-Baroja, L., Pérez-Camacho, L., Villar-Salvador, P. *et al.* (2019). Massive and effective acorn dispersal into agroforestry systems by an overlooked vector, the Eurasian magpie (*Pica pica*). *Ecosphere* 10, e02989.

Maskell, L., Henrys, P., Norton, L., Smart, S. & Wood, C. (2013). *Distribution of Ash Trees* (Fraxinus excelsior) *in Countryside Survey Data.* Centre for Ecology and Hydrology, Wallingford.

Mason, B., Hampson, A. & Edwards, C. (2004). *Managing the Pinewoods of Scotland.* Forestry Commission, Edinburgh.

Mattheck, C. & Breloer, H. (1994). *The Body Language of Trees: A Handbook for Failure Analysis.* Research for Amenity Trees, No. 4, HMSO, London.

Maurer, K.D., Bohrer, G., Medvigy, D. & Wright, S.J. (2013). The timing of abscission affects dispersal distance in a wind-dispersed tropical tree. *Functional Ecology* 27, 208–218.

McCormack, L.M., Adams, T.S., Smithwick, E.A. & Eissenstat, D.M. (2012). Predicting fine root lifespan from plant functional traits in temperate trees. *New Phytologist* 195, 823–831.

McCormack, M.L., Adams, T.S., Smithwick, E.A. & Eissenstat, D.M. (2014). Variability in root production, phenology, and turnover rate among 12 temperate tree species. *Ecology* 95, 2224–2235.

McCulloh, K., Sperr, J.S., Lachenbruch, B. *et al.* (2010). Moving water well: comparing hydraulic efficiency in twigs and trunks of coniferous, ring-porous, and diffuse-porous saplings from temperate and tropical forests. *New Phytologist* 186, 439–450.

McDonald, A.G., Bealey, W.J., Fowler, D. et al. (2007). Quantifying the effect of urban tree planting on concentrations and depositions of PM10 in two UK conurbations. *Atmospheric Environment* 41, 8455–8467.

McElrone, A.J., Choat, B., Gambetta, G.A. & Brodersen, C.R. (2013). Water uptake and transport in vascular plants. *Nature Education Knowledge* 4, article 6.

McKendrick, S.L., Leake, J.R. & Read, D.J. (2000). Symbiotic germination and development of myco-heterotrophic plants in nature: transfer of carbon from ectomycorrhizal *Salix repens* and *Betula pendula* to the orchid *Corallorhiza trifida* through shared hyphal connections. *New Phytologist* 145, 539–548.

Mellanby, K. (1968). The effects of some mammals and birds on regeneration of oak. *Journal of Applied Ecology* 5, 59–366.

Menezes-Silva, P.E., Loram-Lourenço, L., Ferreira, R.D. *et al.* (2019). Different ways to die in a changing world: consequences of climate change for tree species performance and survival through an ecophysiological perspective. *Ecology and Evolution* 9, 11979–11999.

Midgley, G.F., Aranibar, J.N., Mantlana, K.B. & Macko, S. (2004). Photosynthetic and gas exchange characteristics of dominant woody plants on a moisture gradient in an African savanna. *Global Change Biology* 10, 309–317.

Mills, D.J., Bohlman, S.A., Putz, F.E. & Andreu, M.G. (2019). Liberation of future crop trees from lianas in Belize: completeness, costs, and timber-yield benefits. *Forest Ecology and Management* 439, 97–104.

Ministry of Housing, Communities & Local Government (2014). *Tree Preservation Orders and Trees in Conservation Areas*, Guidance. Retrieved from https://www.gov.uk/guidance/tree-preservation-orders-and-trees-in-conservation-areas.

Mitchell, A.F. (1978). *Trees of Britain and Northern Ireland*, 2nd edition. Collins Field Guide, Collins, London.

Mitchell, R.J., Bailey, S., Beaton, J.K. *et al.* (2014a). *The Potential Ecological Impact of Ash Dieback in the UK*. JNCC Report No. 483, Joint Nature Conservation Committee, Peterborough.

Mitchell, R.J., Beaton, J.K., Bellamy, P.E. *et al.* (2014b). Ash dieback in the UK: a review of the ecological and conservation implications and potential management options. *Biological Conservation* 175, 95–109.

Moir, A.K. (1999). The dendrochronological potential of modern yew *(Taxus baccata)* with special reference to yew from Hampton Court Palace, UK. *New Phytologist* 144, 479–488.

Moomaw, W.R., Masino, S.A. & Faison, E.K. (2019). Intact forests in the United States: proforestation mitigates climate change and serves the greatest good. *Frontiers in Forests and Global Change* 2, article 27.

Moore, J. (2011). *Wood Properties and Uses of Sitka Spruce in Britain*. Forestry Commission Research Report. Forestry Commission, Edinburgh.

Morand, S. & Lajaunie, C. (2021). Outbreaks of vector-borne and zoonotic diseases are associated with changes in forest cover and oil palm expansion at global scale. *Frontiers in Veterinary Science* 8, article 661063.

Moreira, X., Nell, C.S., Katsanis, A., Rasmann, S. & Mooney, K.A. (2018). Herbivore specificity and the chemical basis of plant–plant communication in *Baccharis salicifolia* (Asteraceae). *New Phytologist* 220, 70–713.

Morris, H., Hietala, A.M., Jansen, S. *et al.* (2020). Using the CODIT model to explain secondary metabolites of xylem in defence systems of temperate trees against decay fungi. *Annals of Botany* 125, 701–720.

Mosedale, J.R. (1995). Effects of oak wood on the maturation of alcoholic beverages with particular reference to whisky. *Forestry* 68, 203–230.

Motley, K., Havill, N.P., Arsenault-Benoit, A.L. *et al.* (2017). Feeding by *Leucopis argenticollis* and *Leucopis piniperda* (Diptera: Chamaemyiidae) from the western USA on *Adelges tsugae* (Hemiptera: Adelgidae) in the eastern USA. *Bulletin of Entomological Research* 107, 699–704.

Mullaney, J., Lucke, T. & Trueman, S.J. (2015). A review of benefits and challenges in growing street trees in paved urban environments. *Landscape and Urban Planning* 134, 157–166.

Muñoz-Villers, L.E., Geris, J., Alvarado-Barrientos, M.S., Holwerda, F. & Dawson, T. (2020). Coffee and shade trees show complementary use of soil water in a traditional agroforestry ecosystem. *Hydrology and Earth System Sciences* 24, 1649–1668.

Newbould, P.J. (1960). *The Age and Structure of the Yew Wood at Kingley Vale*. Report, Nature Conservancy Council, Wye.

Newhouse, A.E., Schrodt, F., Liang, H., Maynard, C.A. & Powell, W.A. (2007). Transgenic American elm shows reduced Dutch elm disease symptoms and normal mycorrhizal colonization. *Plant Cell Reports* 26, 977–987.

Newton, A.C., Allnutt, T.R., Gillies, A.C.M., Lowe, A.J. & Ennos, R.A. (1999). Molecular phylogeography, intraspecific variation and the conservation of tree species. *Trends in Ecology and Evolution* 14, 140–145.

NFI. (2016). *NFI Preliminary Estimates of the Presence and Extent of Rhododendron in British Woodlands*. National Forestry Inventory, Forestry Commission, Edinburgh.

Nitoslawski, S.A., Galle, N.J., Konijnendijk Van Den Boscha, C. & Steenberg, J.W.M. (2019). Smarter ecosystems for smarter cities? A review of trends, technologies, and turning points for smart urban forestry. *Sustainable Cities and Society* 51, e101770.

Nowak, D.J., Hirabayashi, S., Bodine, A. & Greenfield, E. (2014). Tree and forest effects on air quality and human health in the United States. *Environmental Pollution* 193, 119–129.

Ochiai, H., Ikei, H., Song, C. *et al.* (2015). Physiological and psychological effects of a forest therapy program on middle-aged females. *International Journal of Environmental Research and Public Health* 12, 15222–15232.

Offenthaler, I., Hietz, P. & Richter, H. (2001). Wood diameter indicates diurnal and long-term patterns of xylem water potential in Norway spruce. *Trees* 15, 215–221.

Ohashi, K. & Thomson, J.D. (2007). Trapline foraging by pollinators: its ontogeny, economics and possible consequences for plants. *Annals of Botany* 103, 1365–1378.

Ohya, I., Nanami, S. & Itoh, A. (2017). Dioecious plants are more precocious than cosexual plants: a comparative study of relative sizes at the onset of sexual reproduction in woody species. *Ecology and Evolution* 7, 5660–5668.

Oldlist (2021). *Oldlist, A Database of Old Trees*. Rocky Mountain Tree-Ring Research. Retrieved from http://www.rmtrr.org/oldlist.htm.

Ollerton, J., Erenler, H., Edwards, M. & Crockett, R. (2014). Extinctions of aculeate pollinators in Britain and the role of large-scale agricultural changes. *Science* 346, 1360–1362.

Ollerton, J., Winfree, R. & Tarrant, S. (2011). How many flowering plants are pollinated by animals? *Oikos* 120, 321–326.

Otto, C. & Nilsson, L.M. (1981). Why do beech and oak trees retain leaves until spring? *Oikos* 37, 387–390.

Palacio, S., Hoch, G., Sala, A., Körner, C. & Millard, P. (2014). Does carbon storage limit tree growth? *New Phytologist* 201, 1096–1100.

Palmer, T.M., Stanton, M.L., Young, T.P. *et al.* (2008). Breakdown of an ant-plant mutualism follows the loss of large herbivores from an African savanna. *Science* 319, 192–195.

Pan, R. & Tyree, M.T. (2019). How does water flow from vessel to vessel? Further investigation of the tracheid bridge concept. *Tree Physiology* 39, 1019–1031.

Pan, Y., Birdsey, R.A., Fang, J. *et al.* (2011). A large and persistent carbon sink in the world's forests. *Science* 333, 988–993.

Partap, U.M.A., Partap, T.E.J. & Yonghua, H.E. (2001). Pollination failure in apple crop and farmers' management strategies in Hengduan Mountains, China. *Acta Horticulturae* 561, 225–230.

Patrut, A., Woodborne, S., Patrut, R.T. *et al.* (2018). The demise of the largest and oldest African baobabs. *Nature Plants* 4, 423–426.

Pau, S., Detto, M., Kim, Y. & Still, C.J. (2018). Tropical forest temperature thresholds for gross primary productivity. *Ecosphere* 9, e02311.

Pearse, I.S., LaMontagne, J.M. & Koenig, W.D. (2017). Inter-annual variation in seed production has increased over time (1900–2014). *Proceedings of the Royal Society B* 284, 20171666.

Pearson, C., Manning, S.W., Coleman, M. & Jarvis, K. (2005). Can tree-ring chemistry reveal absolute dates for past volcanic eruptions? *Journal of Archaeological Science* 32, 1265–1274.

Pena-Novas, I. & Archetti, M. (2020). A comparative analysis of the photoprotection hypothesis for the evolution of autumn colours. *Journal of Evolutionary Biology* 33, 1669–1676.

Pennant, T. (1771). *A Tour in Scotland; 1769.* Benjamin White, London.

Peñuelas, J. (2005). A big issue for trees. *Nature* 437, 965–966.

Percival, G.C. (2013). Influence of pure mulches on suppressing *Phytophthora* root rot pathogens. *Journal of Environmental Horticulture* 31, 221–226.

Percival, G.C., Barrow, I., Noviss, K., Keary, I. & Pennington, P. (2011). The impact of horse-chestnut leaf miner (*Cameraria ohridella* Deschka and Dimic; HCLM) on vitality, growth and reproduction of *Aesculus hippocastanum* L. *Urban Forestry & Urban Greening* 10, 11–17.

Perea, R., San Miguel, A. & Gil, L. (2011). Leftovers in seed dispersal: ecological implications of partial seed consumption for oak regeneration. *Journal of Ecology* 99, 194–201.

Pérez-Méndez, P., Andersson, G.K.S., Requier, F. *et al.* (2020). The economic cost of losing native pollinator species for orchard production. *Journal of Applied Ecology* 57, 599–608.

Peter, G.F. (2018). Breeding and engineering trees to accumulate high levels of terpene metabolites for plant defense and renewable chemicals. *Frontiers in Plant Science* 9, article 1672.

Peterken, G.F. (1996). *Natural Woodland: Ecology and Conservation in Northern Temperate Regions.* Cambridge University Press, Cambridge.

Peterken, G.F., Craven, T. & Payne, C. (2020). *Art Meets Ecology: The Arborealists in Lady Park Wood.* Samson & Co, Clifton, Bristol.

Petit, G., Anfodillo, T., Carraro, V., Grani, F. & Carrer, M. (2010). Hydraulic constraints limit height growth in trees at high altitude. *New Phytologist* 189, 241–252.

Phillips, B., Bullock, J.M., Osborne, J.L. & Gaston, K.J. (2021). Spatial extent of road pollution: a national analysis. *Science of the Total Environment* 773, e145589.

Pichersky, E. & Raguso, R.A. (2018). Why do plants produce so many terpenoid compounds? *New Phytologist* 220, 692–702.

Pichot, C., El Maâtaoui, M., Raddi, S. & Raddi, P. (2001). Surrogate mother for endangered *Cupressus. Nature* 412, 39.

Piermattei, A., Crivellaro, A., Carrer, M. & Urbinati, C. (2015). The "blue ring": anatomy and formation hypothesis of a new tree-ring anomaly in conifers. *Trees* 29, 613–620.

Pigott, C.D. (1995). The radial growth-rate of yews (*Taxus baccata*) at Hampton Court, Middlesex. *Garden History* 23, 249–252.

Pigott, C.D. (1989). Factors controlling the distribution of *Tilia cordata* Mill at the northern limits of its geographical range. *New Phytologist* 112, 117–121.

Piovesan, G. & Biondi, F. (2021). On tree longevity. *New Phytologist* 231, 1318–1337.

Pirone, P.P. (1970). *Diseases and Pests of Ornamental Plants,* 4th edition. Ronald Press, New York.

Pither, J., Pickles, B.J., Simard, S.W., Ordonez, A. & Williams, J.W. (2018). Below-ground biotic interactions moderated the postglacial range dynamics of trees. *New Phytologist* 220, 1148–1160.

Pons, J. & Pausas, J.G. (2007a). Not only size matters: acorn selection by the European jay (*Garrulus glandarius*). *Acta Oecologica* 31, 353–360.

Pons, J. & Pausas, J.G. (2007b). Acorn dispersal estimated by radio-tracking. *Oecologia* 153, 903–911.

Poorter, H., Bühler, J., van Dusschoten, D., Climent, J. & Postma, J.A. (2012). Pot size matters: a meta-analysis of the effects of rooting volume on plant growth. *Functional Plant Biology* 39, 839–850.

Powell, W.A., Newhouse, A.E. & Coffey, V. (2019). Developing blight-tolerant American chestnut trees. *Cold Springs Harbor Perspectives in Biology* 11, a034587.

Prevedello, J.A., Almeida-Gomes, M. & Lindenmayer, D.B. (2018). The importance of scattered trees for biodiversity conservation: a global meta-analysis. *Journal of Applied Ecology* 55, 205–214.

Prieto, I. & Querejeta, J.I. (2019). Simulated climate change decreases nutrient resorption from senescing leaves. *Global Change Biology* 26, 1795–1807.

Proctor, M.C.F. & Yeo, P. (1973). *The Pollination of Flowers*. Collins, London.

Puchałka, R., Koprowski, M., Gričar, J. & Przybylak, R. (2017). Does tree-ring formation follow leaf phenology in pedunculate oak (*Quercus robur* L.)? *European Journal of Forest Research* 136, 259–268.

Putkinen, A., Siljanen, H.M.P., Laihonen, A. *et al.* (2021). New insight to the role of microbes in the methane exchange in trees: evidence from metagenomic sequencing. *New Phytologist* 231, 524–536.

Putz, F.E. (1984). How trees avoid and shed lianas. *Biotropica* 16, 19–23.

Qu, X.-J., Fan, S.-J., Wicke, S. & Yi, T.-S. (2019). Plastome reduction in the only parasitic gymnosperm *Parasitaxus* is due to losses of photosynthesis but not housekeeping genes and apparently involves the secondary gain of a large inverted repeat. *Genome Biology and Evolution* 11, 2789–2796.

Queenborough, S.A., Mazer, S.J., Vamosi, S.M. *et al.* (2009). Seed mass, abundance and breeding system among tropical forest species: do dioecious species exhibit compensatory reproduction or abundances? *Journal of Ecology* 97, 555–566.

Rackham, O. (1986). *The History of the Countryside*. Dent, London.

Ramage, M.H., Burridge, H., Busse-Wicher, M. *et al.* (2017). The wood from the trees: The use of timber in construction. *Renewable and Sustainable Energy Reviews* 68, 333–359.

Ramírez-Valiente, J.A., Valladares, F., Gil, L. & Aranda, I. (2009). Population differences in juvenile survival under increasing drought are mediated by seed size in cork oak (*Quercus suber* L.). *Forest Ecology and Management* 257, 1676–1683.

Rauf, A., Imran, M., Abu-Izneid, T. *et al.* (2019). Proanthocyanidins: a comprehensive review. *Biomedicine & Pharmacotherapy* 116, 108999.

Read, D.J., Freer-Smith, P.H., Morison, J.I.L. *et al.* (2009). *Combating Climate Change – A Role for UK Forests. An Assessment of the Potential of the UK's Trees and Woodlands to Mitigate and Adapt to Climate Change. The Synthesis Report.* The Stationery Office, Edinburgh.

Reid, C., Hornigold, K., McHenry, E. *et al.* (2021). *State of the UK's Woods and Trees 2021.* Woodland Trust, Grantham.

Renner, S.S. & Zohner, C.M. (2019). The occurrence of red and yellow autumn leaves explained by regional differences in insolation and temperature. *New Phytologist* 224, 1464–1471.

Reynolds, E.R.C. (1975). Tree rootlets and their distribution. In J.G. Torrey & D.T. Clarkson (eds), *The Development and Function of Roots*. Academic Press, London, pp. 163–177.

Rich, T.C.G., Houston, L., Robertson, A. & Proctor, M.C.F. (2010). *Whitebeams, Rowans and Service Trees of Britain and Ireland. A Monograph of British and Irish Sorbus L.* BSBI Handbook No. 14, Botanical Society of the British Isles, London.

Richardson, A.D., Anderson, R.S., Arain, M.A. *et al.* (2012). Terrestrial biosphere models need better representation of vegetation phenology: results from the North American Carbon Program Site Synthesis. *Global Change Biology* 18, 566–584.

Richardson, A.D., Carbone, M.S., Hugget, B.A. *et al.* (2015). Distribution and mixing of old and new nonstructural carbon in two temperate trees. *New Phytologist* 206, 590–597.

Rivas-San Vicente, M. & Plasencia, J. (2011). Salicylic acid beyond defence: its role in plant growth and development. *Journal of Experimental Botany* 62, 3321–3338.

Rivers, M., Beech, E., Bazos, I., *et al.* (2019). *European Red List of Trees.* International Union for Conservation of Nature and Natural Resources, Brussels.

Robbins, W.J. (1964). Topophysis, a problem in somatic inheritance. *Proceedings of the American Philosophical Society* 108, 395–403.

Rodell, M., Beaudoing, H.K., L'ecuyer, T.S. *et al.* (2015). The observed state of the water cycle in the early Twenty-First Century. *Journal of Climate* 28, 8289–8318.

Roetzer, T., Wittenzeller, M., Haeckel, H. & Nekovar, J. (2000). Phenology in Central Europe – differences and trends of spring phenophases in urban and rural areas. *International Journal of Biometeorology* 44, 60–66.

Rollinson, C.R. (2020). Surplus and stress control autumn timing. Climate change might cause an early shedding of leaves if trees have stored enough carbon. *Science* 370, 1030–1031.

Rose, D. & Webber, J. (2011). *De-icing Salt Damage to Trees.* Forest Research, Pathology Advisory Note No. 11, Forestry Commission, Edinburgh.

Rossi, S., Deslauriers, A., Anfodillo, T. et al. (2006). Conifers in cold environments synchronize maximum growth rate of tree-ring formation with day length. *New Phytologist* 170, 301–310.

Roy, S., Byrne, J. & Pickering, C. (2012). A systematic quantitative review of urban tree benefits, costs, and assessment methods across cities in different climatic zones. *Urban Forestry and Urban Greening* 11, 351–363.

Sakai, A. (1965). Survival of plant tissue at super-low temperatures III. Relation between effective prefreezing temperatures and the degree of frost hardiness. *Plant Physiology* 40, 882–887.

Salzer, M.W., Pearson, C.L. & Baisan, C.H. (2019). Dating the Methuselah Walk bristlecone pine floating chronologies. *Tree-Ring Research* 75, 61–66.

Sánchez-Romero, C. (2019). Somatic embryogenesis in *Olea* spp. *Plant Cell, Tissue and Organ Culture* 138, 403–426.

Säumel, I. & Kowarik, I. (2010). Urban rivers as dispersal corridors for primarily wind-dispersed invasive tree species. *Landscape and Urban Planning* 94, 244–249.

Säumel, I. & Kowarik, I. (2013). Propagule morphology and river characteristics shape secondary water dispersal in tree species. *Plant Ecology* 214, 1257–1272.

Saveyn, A., Steppe, K., Ubierna, N. & Dawson, T.E. (2010). Woody tissue photosynthesis and its contribution to trunk growth and bud development in young plants. *Plant, Cell and Environment* 33, 1949–1958.

Schaberg, P.G., Murakami, P.F., Turner, M.R., Heitz, H.K. & Hawley, G.J. (2008). Association of red coloration with senescence of sugar maple leaves in autumn. *Trees* 22, 573–578.

Schenk, H.J., Jansen, S. & Hölttä, T. (2020). Positive pressure in xylem and its role in hydraulic function. *New Phytologist* 230, 27–45.

Schepaschenko, D., Shvidenko, A., Usoltsev, V. et al. (2017). A dataset of forest biomass structure for Eurasia. *Scientific Data* 4, 1–11.

Schmitt, C., Parola, P. & de Haro, L. (2013). Painful sting after exposure to *Dendrocnide* sp: two case reports. *Wilderness & Environmental Medicine* 24, 471–473.

Schulman, E. (1954). Longevity under adversity in conifers. *Science* 119, 396–399.

Schweingruber, F.H. (1996). *Tree Rings and Environment: Dendroecology.* Haupt, Berne.

Scofield, D.G., Sork, V.L. & Smouse, P.E. (2010). Influence of acorn woodpecker social behaviour on transport of coast live oak (*Quercus agrifolia*) acorns in a southern California oak savanna. *Journal of Ecology* 98, 561–572.

Scott, A.C. & Chaloner, W.G. (1983). The earliest fossil conifer from the Westphalian B of Yorkshire. *Proceedings of the Royal Society B* 220, 163–182.

Sellin, A. (1994). Sapwood-heartwood proportion related to tree diameter, age, and growth rate in *Picea abies*. *Canadian Journal of Forest Research* 24, 1022–1028.

Shenkin, A., Chandler, C.J., Boyd, D.S. et al. (2019). The world's tallest tropical tree in three dimensions. *Frontiers in Forests and Global Change* 2, article 32.

Shestakov, A.L., Filippov, B.Y., Zubrii, N.A. et al. (2020). Doubling of biomass production in European boreal forest trees by a four-year suppression of background insect herbivory. *Forest Ecology and Management* 462, e117992.

Silva, R.M.G., Brigatti, J.G.F., Santos, V.H.M., Mecina, G.F. & Silva L.P. (2013). Allelopathic effect of the peel of coffee fruit. *Scientia Horticulturae* 158, 39–44.

Silvertown, J. (2015). Have ecosystem services been oversold? *Trends in Ecology & Evolution* 30, 641–648.

Simard, S.W., Perry, D.A., Jones, M.D. et al. (1997). Net transfer of carbon between ectomycorrhizal tree species in the field. *Nature* 388, 579–582.

Sinclair, W.T., Morman, J.D. & Ennos, R.A. (1999). The postglacial history of Scots pine (*Pinus sylvestris* L.) in western Europe: evidence from mitochondrial DNA variation, *Molecular Ecology* 8, 83–88.

Sivarajah, S., Thomas, S.C. & Smith, S.M. (2020). Evaluating the ultraviolet protection factors of urban broadleaf and conifer trees in public spaces. *Urban Forestry & Urban Greening* 51, e126679.

Sjöman, H. & Östberg, J. (2019). Vulnerability of ten major Nordic cities to potential tree losses caused by longhorned beetles. *Urban Ecosystems* 22, 385–395.

Sjöman, H. & Watkins, J.H.R. (2020). What do we know about the origin of our urban trees? – A north European perspective. *Urban Forestry & Urban Greening* 56, e126879.

Slater, D. (2016). *Assessment of Tree Forks: Assessment of Junctions for Risk Management.* Arboricultural Association, Stroud.

Smiley, E.T., Calfee, L., Fraedrich, B.R. & Smiley, E.J. (2006). Comparison of structural and noncompacted soils for trees surrounded by pavement. *Arboriculture & Urban Forestry* 32, 164–169.

Snowdonia Rhododendron Partnership (2015). *The Ecosystem Benefits of Managing the Invasive Non-native Plant* Rhododendron ponticum *in Snowdonia*. Snowdonia Rhododendron Partnership, Snowdonia.

Sobel, D. (1996). *Longitude*. Fourth Estate, London.

Solheim, H., Eriksen, R. & Hietala, A.M. (2011). Dutch elm disease has currently a low incidence on Wych elm in Norway. *Forest Pathology* 41, 182–188.

Sonesson, L.K. (1994). Growth and survival after cotyledon removal in *Quercus robur* seedlings, grown in different natural soil types. *Oikos* 69, 65–70.

Song, C., Ikei, H. & Miyazaki, Y. (2016). Physiological effects of nature therapy: a review of the research in Japan. *International Journal of Environmental Research and Public Health* 13, e781.

Song, G.-q., Walworth, A.E. & Loescher, W.H. (2015). Grafting of genetically engineered plants. *Journal of the American Society for Horticultural Science* 140, 203–213.

Southwood, T.R.E. (1961). The number of species of insect associated with various trees. *Journal of Animal Ecology* 30, 1–8.

Spengler, R.N. (2019). Origins of the apple: the role of megafaunal mutualism in the domestication of *Malus* and Rosaceous trees. *Frontiers in Plant Science* 10, e617.

Springmann, S., Rogers, R. & Spiecker, H. (2011). Impact of artificial pruning on growth and secondary shoot development of wild cherry (*Prunus avium* L.). *Forest Ecology and Management* 261, 764–769.

Staal, A., Tuinenburg, O.A., Bosmans, J.H.C. et al. (2018). Forest-rainfall cascades buffer against drought across the Amazon. *Nature Climate Change* 8, 539–543.

Stace, C. (2019). *New Flora of the British Isles*, 4th edition. C & M Floristics, Suffolk.

Steiger, S.S., Fidler, A.E., Valcu, M. & Kempenaers, B. (2008). Avian olfactory receptor gene repertoires: evidence for a well-developed sense of smell in birds? *Proceedings of the Royal Society B* 275, 2309–2317.

Stevenson, P.C., Bidartondo, M.I., Blackhall-Miles, R. et al. (2020). The state of the world's urban ecosystems: what can we learn from trees, fungi, and bees? *Plants, People, Planet* 2, 482–498.

St. George, S. (2014). An overview of tree-ring width records across the Northern Hemisphere. *Quaternary Science Reviews* 95, 132–150.

Straw, N.A. & Williams, D.T. (2013). Impact of the leaf miner *Cameraria ohridella* (Lepidoptera: Gracillariidae) and bleeding canker disease on horse-chestnut: direct effects and interaction. *Agricultural and Forest Entomology* 15, 321–333.

Strutt, J.G. (1822). *Sylva Britannica; or, Portraits of Forest Trees, Distinguished for their Antiquity, Magnitude, or Beauty*. Privately published, London.

Sun, L., Bukovac, M.J., Forsline, P.L. & van Nocker, S. (2009). Natural variation in fruit abscission-related traits in apple (*Malus*). *Euphytica* 165, 55–67.

Svendsen, C.R. (2001). Effects of marcescent leaves on winter browsing by large herbivores in northern temperate deciduous forests. *Alces* 37, 475–482.

Tajuddin, Ahmad, S., Latif, A. & Qasmi, I.A. (2003). Aphrodisiac activity of 50% ethanolic extracts of *Myristica fragrans* Houtt. (nutmeg) and *Syzygium aromaticum* (L) Merr. & Perry. (clove) in male mice: a comparative study. *BMC Complementary and Alternative Medicine* 3, article 6.

Takahashi, F. & Shinozaki, K. (2019). Long-distance signaling in plant stress response. *Current Opinion in Plant Biology* 47, 106–111.

Takos, I., Varsamis, G., Avtzis, D. et al. (2008). The effect of defoliation by *Cameraria ohridella* Deschka and Dimic (Lepidoptera: Gracillariidae) on seed germination and seedling vitality in *Aesculus hippocastanum* L. *Forest Ecology and Management* 255, 830–835.

Taylor, J.D., Kline, K.N. & Morzillo, A.T. (2019). Estimating economic impact of black bear damage to western conifers at a landscape scale. *Forest Ecology and Management* 432, 599–606.

TDAG (2012). *Trees in the Townscape. A Guide for Decision Makers*. Trees & Design Action Group. Retrieved from http://www.tdag.org.uk/uploads/4/2/8/0/4280686/tdag_treesinthetownscape.pdf.

Teskey, R.O., Saveyn, A., Steppe, K. & McGuire, M.A. (2008). Origin, fate and significance of CO_2 in tree stems. *New Phytologist* 177, 17–32.

Thomas, B.A. & Cleal, C.J. (2018). Arborescent lycophyte growth in the late Carboniferous coal swamps. *New Phytologist* 218, 885–890.

Thomas, P.A. (2014). *Trees: Their Natural History*, 2nd edition. Cambridge University Press, Cambridge.

Thomas, P.A. (2016). Biological Flora of the British Isles: *Fraxinus excelsior*. *Journal of Ecology* 104, 1158–1209.

Thomas, P.A. (2017). Biological Flora of the British Isles: *Sorbus torminalis*. *Journal of Ecology* 105, 1806–1831.

Thomas, P.A., Alhamd, O., Iszkuło, G., Dering, M. & Mukassabi, T.A. (2019). Biological Flora of the British Isles: *Aesculus hippocastanum*. *Journal of Ecology* 107, 992–1030.

Thomas, P.A., El-Barghathi, M. & Polwart, A. (2007). Biological Flora of the British Isles: *Juniperus communis* L. *Journal of Ecology* 95, 1404–1440.

Thomas, P.A., El-Barghathi, M. & Polwart, A. (2011). Biological Flora of the British Isles: *Euonymus europaeus* L. *Journal of Ecology* 99, 345–365.

Thomas, P.A., Leski, T., La Porta, N., Dering, M. & Iszkuło, G. (2021). Biological Flora of the British Isles: *Crataegus laevigata*. *Journal of Ecology* 109, 572–596.

Thomas, P.A. & Mukassabi, T.A. (2014). Biological Flora of the British Isles: *Ruscus aculeatus*. *Journal of Ecology* 102, 1083–1100.

Thomas, P.A. & Packham, J.R. (2007). *Ecology of Woodlands and Forests*. Cambridge University Press, Cambridge.

Thomas, P.A. & Polwart, A. (2003). Biological Flora of the British Isles: *Taxus baccata*. *Journal of Ecology* 91, 489–524.

Thomas, P.A., Stone, D. & La Porta, N. (2018). Biological Flora of the British Isles: *Ulmus glabra*. *Journal of Ecology* 106, 1724–1766.

Thompson, K. (2018). *Darwin's Most Wonderful Plants*. Profile Books, London.

Thompson, S.E. & Katul, G.G. (2013). Implications of nonrandom seed abscission and global stilling for migration of wind-dispersed plant species. *Global Change Biology* 19, 1720–1735.

Thorpe, T.A. (2007). History of plant tissue culture. *Molecular Biotechnology* 37, 169–180.

Tiberi, R., Branco, M., Bracalini, M., Croci, F. & Panzavolta, T. (2016). Cork oak pests: a review of insect damage and management. *Annals of Forest Science* 73, 219–232.

Tilman, D. (1978). Cherries, ants and tent caterpillars: timing of nectar production in relation to susceptibility of caterpillars to ant predation. *Ecology* 59, 686–692.

Tisné, S., Denis, M., Domonhédo, H. *et al.* (2020). Environmental and trophic determinism of fruit abscission and outlook with climate change in tropical regions. *Plant-Environment Interactions* 1, 17–28.

Tng, D.Y.P., Williamson, G.J., Jordan, G.J. & Bowman, D.M.J.S. (2012). Giant eucalypts: globally unique fire-adapted rain-forest trees? *New Phytologist* 196, 1001–1014.

Treseder, K.K., Allen, E.B., Egerton-Warburton, L.M. *et al.* (2018). Arbuscular mycorrhizal fungi as mediators of ecosystem responses to nitrogen deposition: a trait-based predictive framework. *Journal of Ecology* 106, 480–489.

Trewavas, A. (2014). *Plant Behaviour and Intelligence*. Oxford University Press, Oxford.

Tsen, E.W.J., Sitzia, T. & Webber, B.L. (2016). To core, or not to core: the impact of coring on tree health and a best-practice framework for collecting dendrochronological information from living trees. *Biological Reviews* 91, 899–924.

Turgut-Kara, N., Arikan, B. & Celik, H. (2020). Epigenetic memory and priming in plants. *Genetica* 148, 47–54.

Tuskan, A., Difazio, S., Jansson, S. *et al.* (2006). The genome of black cottonwood, *Populus trichocarpa* (Torr. & Gray). *Science* 313, 1596–1604.

Tweddle, J.C., Dickie, J.B., Baskin, C.C. & Baskin, J.M. (2003). Ecological aspects of seed desiccation sensitivity. *Journal of Ecology* 91, 294–304.

Tyree, M.T., Cochard, H., Cruiziat, P., Sinclair, B. & Ameglio, T. (1993). Drought-induced shedding in walnut: evidence for vulnerability segmentation. *Plant, Cell and Environment* 16, 879–882.

UK National Ecosystem Assessment (2011). *The UK National Ecosystem Assessment: Synthesis of the Key Findings*. UNEP-WCMC, Cambridge.

Ulmer, M., Wolf, K.L., Backman, D.R. *et al.* (2016). Multiple health benefits of urban tree canopy: the mounting evidence for a green prescription. *Health & Place* 42, 54–62.

Ulrich, R.S. (1984). View through a window may influence recovery from surgery. *Science* 224, 420–421.

UNEP (2009). *2008 Annual Report*. United Nations Environment Programme.

United Nations (2019). *World Urbanization Prospects. The 2018 Revision (ST/ESA/SER.A/420)*. Department of Economic and Social Affairs, Population Division, United Nations, New York.

Urli, M., Porté, A.J., Cochard, H. *et al.* (2013). Xylem embolism threshold for catastrophic hydraulic failure in angiosperm trees. *Tree Physiology* 33, 672–683.

Vahdati, K., McKenna, J.R., Dandekar, A.M. *et al.* (2002). Rooting and other characteristics of a transgenic walnut hybrid (*Juglans hindsii × J. regia*) rootstock expressing rolABC. *Journal of the American Society for Horticultural Science* 127, 724–728.

Vallejo-Marín, M. (2019). Buzz pollination: studying bee vibrations on flowers. *New Phytologist* 224, 1068–1074.

van der Kooi, C.J. & Ollerton, J. (2020). The origins of flowering plants and pollinators. *Science* 368, 1306–1308.

Vander Wall, S.B. (1994). Removal of wind-dispersed pine seeds by ground-foraging vertebrates. *Oikos* 69, 125–132.

Vander Wall, S.B. & Balda, R.P. (1977). Coadaptations of the Clark's nutcracker and the piñon pine for efficient seed harvest and dispersal. *Ecological Monographs* 47, 89–111.

Van Renterghem, T., Botteldooren, D. & Verheyen, K. (2012). Road traffic noise shielding by vegetation belts of limited depth. *Journal of Sound and Vibration* 331, 2404–2425.

Vellinga, E.C., Wolfe, B.E. & Pringle, A. (2009). Global patterns of ectomycorrhizal introductions. *New Phytologist* 181, 960–973.

Vergutz, L., Manzoni, S., Porporato, A., Novais, R.F. & Jackson, R.B. (2012). Global resorption efficiencies and concentrations of carbon and nutrients in leaves of terrestrial plants. *Ecological Monographs* 82, 205–220.

Visseren-Hamakers, I.J., Gupta, A., Herold, M., Peña-Claros, M. & Vijge, M.J. (2012). Will REDD+ work? The need for interdisciplinary research to address key challenges. *Current Opinion in Environmental Sustainability* 4, 590–596.

Vitasse, Y. & Basler, D. (2013). What role for photoperiod in the bud burst phenology of European beech? *European Journal of Forest Research* 132, 1–8.

Vogel, S. (2009). Leaves in the lowest and highest winds: temperature, force, shape. *New Phytologist* 183, 13–26.

Vogel, S. (2012). *The Life of a Leaf.* University of Chicago Press, Chicago.

Walas, Ł., Mandryk, W., Thomas, P.A., Tyrała-Wierucka, Ż. & Iszkuło, G. (2018). Sexual systems in gymnosperms: a review. *Basic and Applied Ecology* 31, 1–9.

Walker, A.P., De Kauwe, M.G., Bastos, A. *et al.* (2020). Integrating the evidence for a terrestrial carbon sink caused by increasing atmospheric CO_2. *New Phytologist* 229, 2413–2445.

Wallace, A.R. (1869). *The Malay Archipelago.* Macmillan, London.

Wang, L., Cui, J., Jin, B. *et al.* (2020). Multifeature analyses of vascular cambial cells reveal longevity mechanisms in old *Ginkgo biloba* trees. *PNAS* 117, 2201–2210.

Ward, L.K. & King, M. (2006). Decline of juniper in Sussex. *Quarterly Journal of Forestry* 100, 263–272.

Waring, B., Neumann, M., Prentice, I.C. *et al.* (2020). Forests and decarbonization – roles of natural and planted forests. *Frontiers in Forests and Global Change* 3, article 58.

Watkins, J.H.R., Cameron, R.W.F., Sjöman, H. & Hitchmough, J.D. (2020). Using big data to improve ecotype matching for Magnolias in urban forestry. *Urban Forestry & Urban Greening* 48, e126580.

Watson, M., White, B., Lanigan, J., Slatter, T. & Lewis, R. (2020). The composition and friction-reducing properties of leaf layers. *Proceedings of the Royal Society A* 476, e20200057.

Wearn, J.A. (2016). Seeds of change – polemobotany in the study of war and culture. *Journal of War & Culture Studies* 9, 271–284.

Wearn, J.A., Budden, A.P., Veniard, S.C. & Richardson, D. (2017). The flora of the Somme battlefield: a botanical perspective on a post-conflict landscape. *First World War Studies* 8, 63–77.

Weber, K. & Mattheck, C. (2006). The effects of excessive drilling diagnosis on decay propagation in trees. *Trees* 20, 224–228.

Weber, R., Gessler, A. & Hoch, G. (2019). High carbon storage in carbon-limited trees. *New Phytologist* 222, 171–182.

Weber, R., Schwendener, A., Schmid, S. *et al.* (2018). Living on next to nothing: tree seedlings can survive weeks with very low carbohydrate concentrations. *New Phytologist* 218, 107–118.

Wegner, L.H. (2014). Root pressure and beyond: energetically uphill water transport into xylem vessels? *Journal of Experimental Botany* 65, 381–393.

Welp, L.R., Keeling, R.F., Meijer, H.A.J. *et al.* (2011). Interannual variability in the oxygen isotopes of atmospheric CO_2 driven by El Niño. *Nature* 477, 579–582.

Wheeler, E.A., Gasson, P.E. & Baas, P. (2020). Using the InsideWood web site: potentials and pitfalls. *IAWA Journal* 41, 412–462.

Wheeler, T.D. & Stroock, A.D. (2008). The transpiration of water at negative pressures in a synthetic tree. *Nature* 455, 208–212.

White, J. (1998). *Estimating the Age of Large and Veteran Trees in Britain.* Information Note, Forestry Commission, Edinburgh.

White, T.C.R. (2009). Catching a red herring: autumn colours and aphids. *Oikos* 118, 1610–1612.

Wieder, W.R., Cleveland, C.C., Kolby Smith, W. & Todd-Brown, K. (2015). Future productivity and carbon storage limited by terrestrial nutrient availability. *Nature Geoscience* 8, 441–444.

Williams, G.W. (2008). The Civilian Conservation Corps and The National Forests. *FS Today.* Retrieved from https://www.fs.USDA.gov/Internet/FSE_DOCUMENTS/fsbdev3_004791.pdf.

Williamson, R. (1978). *The Great Yew Forest – The Natural History of Kingley Vale.* Macmillan, London.

Willis, K.G., Garrod, G., Scarpa, R. *et al.* (2003). *The Social and Environmental Benefits of Forests in Great Britain.* Forestry Commission, Edinburgh.

Wisniewski, M., Nassuth, A. & Arora, R. (2018). Cold hardiness in trees: a mini-review. *Frontiers in Plant Science* 9, e1394.

Wolfe, B.T., Sperry, J.S. & Kursar, T.A. (2016). Does leaf shedding protect stems from cavitation during seasonal droughts? A test of the hydraulic fuse hypothesis. *New Phytologist* 212, 1007–1018.

Woods, R.G. & Coppins, B.J. (2012). *A Conservation Evaluation of British Lichens and Lichenicolous Fungi*. Joint Nature Conservation Committee, Peterborough.

Wright, I.J., Reich, P.B., Westoby, M. *et al*. (2004). The worldwide leaf economics spectrum. *Nature* 428, 821–827.

Wurdack, K.J. & Farfan-Rios, W. (2017). *Incadendron*: a new genus of Euphorbiaceae tribe Hippomaneae from the sub-Andean cordilleras of Ecuador and Peru. *PhytoKeys* 85, 69–86.

Wyse, S.V. & Dickie, J.B. (2017). Predicting the global incidence of seed desiccation sensitivity. *Journal of Ecology* 105, 1082–1093.

Wyse, S.V., Dickie, J.B. & Willis, K.J. (2018). Seed banking not an option for many threatened plants. *Nature Plants* 4, 848–850.

Xia, J., Guo, Z., Yang, Z. *et al*. (2021). Whitefly hijacks a plant detoxification gene that neutralizes plant toxins. *Cell* 184, 1–13.

Xie, R.J., Deng, L., Jing, L. *et al*. (2013). Recent advances in molecular events of fruit abscission. *Biologia Plantarum* 57, 201–209.

Yang, B., Meng, X., Singh, A.K. *et al*. (2020). Intercrops improve surface water availability in rubber-based agroforestry systems. *Agriculture, Ecosystems & Environment* 298, e106937.

Yang, D., Zhang, Y.-J., Song, J., Niu, C.-Y. & Hao, G.-Y. (2019). Compound leaves are associated with high hydraulic conductance and photosynthetic capacity: evidence from trees in Northeast China. *Tree Physiology* 39, 729–739.

Yang, F., Liu, F.-h. & Rowland, G. (2013). Effects of diurnal temperature range and seasonal temperature pattern on the agronomic traits of fibre flax (*Linum usitatissimum* L.). *Canadian Journal of Plant Science* 93, 1249–1255.

Yi, X., Bartlow, A.W., Curtis, R., Agosta, S.J. & Steele, M.A. (2019). Responses of seedling growth and survival to post-germination cotyledon removal: an investigation among seven oak species. *Journal of Ecology* 107, 1817–1827.

Yi, X.F. & Yang, Y.Q. (2010). Large acorns benefit seedling recruitment by satiating weevil larvae in *Quercus aliena*. *Plant Ecology* 209, 291–300.

Yih, D. (2012). Land bridge travelers of the Tertiary: the Eastern Asian–Eastern North American floristic disjunction. *Arnoldia* 69, 14–23.

Yin, X.-Y., Sterck, F. & Hao, G.-Y. (2018). Divergent hydraulic strategies to cope with freezing in co-occurring temperate tree species with special reference to root and stem pressure generation. *New Phytologist* 219, 530–541.

Young, S.N.R., Sack, L., Sporck-Koehler, M.J. & Lundgren, M.R. (2020). Why is C_4 photosynthesis so rare in trees? *Journal of Experimental Botany* 71, 4629–4638.

Yuan, Z.Y. & Chen, H.Y.H. (2015). Negative effects of fertilization on plant nutrient resorption. *Ecology* 96, 373–380.

Zani, D., Crowther, T.W., Mo, L., Renner, S.S. & Zohner, C.M. (2020). Increased growing-season productivity drives earlier autumn leaf senescence in temperate trees. *Science* 370, 1066–1071.

Zanne, A.E., Sweeney, K., Sharma, M. & Orians, C.M. (2006). Patterns and consequences of differential vascular sectoriality in 18 temperate tree and shrub species. *Functional Ecology* 20, 200–206.

Zeppel, M., Macinnis-Ng, C.M.O., Ford, C.R. & Eamus, D. (2008). The response of sap flow to pulses of rain in a temperate Australian woodland. *Plant and Soil* 305, 121–130.

Zhang, Y., Commane, R., Zhou, S., Williams, A.P. & Gentine, P. (2020). Light limitation regulates the response of autumn terrestrial carbon uptake to warming. *Nature Climate Change* 10, 739–743.

Zhu, K., Zhang, J., Niu, S., Chu, C. & Luo, Y. (2018). Limits to growth of forest biomass carbon sink under climate change. *Nature Communications* 9, e2709.

Zlinszky, A., Molnár, B. & Barfod, A.S. (2017). Not all trees sleep the same—high temporal resolution terrestrial laser scanning shows differences in nocturnal plant movement. *Frontiers in Plant Science* 8, article 1814.

Zohner, C.M. & Renner, S.S. (2017). Innately shorter vegetation periods in North American species explain native–non-native phenological asymmetries. *Nature Ecology & Evolution* 1, 1655–1660.

Zotz, G., Wilhelm, K. & Becker, A. (2011). Heteroblasty – a review. *The Botanical Review* 77, 109–151.

Zwart, D.C. & Kim, S.-H. (2012). Biochar amendment increases resistance to stem lesions caused by *Phytophthora* spp. in tree seedlings. *HortScience* 47, 1736–1740.

Index

Page numbers in **bold** refer to information contained in captions and map locations. Page numbers in *italics* refer to information contained in tables.

Species listed in tables have not been indexed.

SPECIES INDEX

GENERAL INDEX